AF380649

Nanoenergetic Materials

Nanoenergetic Materials: Preparation, Properties, and Applications

Editors

Djalal Trache
Luigi T. DeLuca

MDPI • Basel • Beijing • Wuhan • Barcelona • Belgrade • Manchester • Tokyo • Cluj • Tianjin

Editors
Djalal Trache
Ecole Militaire Polytechnique
Algeria

Luigi T. Deluca
Politecnico di Milano
Italy

Editorial Office
MDPI
St. Alban-Anlage 66
4052 Basel, Switzerland

This is a reprint of articles from the Special Issue published online in the open access journal *Nanomaterials* (ISSN 2079-4991) (available at: https://www.mdpi.com/journal/nanomaterials/special_issues/nanoenergetic).

For citation purposes, cite each article independently as indicated on the article page online and as indicated below:

LastName, A.A.; LastName, B.B.; LastName, C.C. Article Title. *Journal Name* **Year**, *Volume Number*, Page Range.

ISBN 978-3-0365-0010-2 (Hbk)
ISBN 978-3-0365-0011-9 (PDF)

Contents

About the Editors

Djalal Trache has been working as an Associate Professor at Ecole Militaire Polytechnique (EMP), Algeria, since 2016. He received his Engineer degree in chemical engineering, Magister in applied chemistry and Doctor of Sciences in chemistry at EMP. He has made several presentations at national and international conferences, published over 90 scientific papers in international peer reviewed journals in the field of chemical sciences/materials science, eight book chapters, and two books. He is a reviewer of more than 60 international reputed journals. Prof. Trache has particular expertise in energetic materials, bio-based materials, polymer composites and their characterization. He also has interests in nanomaterials and their applications, phase equilibria and kinetics. Besides, he has successfully supervised many engineer, MSc and doctoral students. He sits on the editorial board of several journals (e.g. Arabian Journal of Chemistry, Carbohydrate Polymer Technologies and Applications).

Luigi T. DeLuca (Professor, Ret) got his Ph.D. in Aerospace and Mechanical Sciences under the supervision of Prof. M. Summerfield at Princeton University, Princeton, NJ, USA. In 1975 he founded the Space Propulsion Laboratory (SPLab) at Politecnico di Milano, Milan, MI, Italy. During most of his career, Dr. L.T. DeLuca was a Full Professor of Aerospace Propulsion at Politecnico di Milano. After his retirement in 2014, he was a Visiting Professor of Rocket Propulsion at Nanjing University of Science & Technology (NUST), Nanjing, China, and a Visiting Professor of Propulsion and Combustion at Konkuk University, Seoul, Korea. Overall, Dr. L.T. DeLuca edited 18 books and 6 special issues of journals for international publishers. In addition, he has authored hundreds of research and review papers. The scientific activity was mostly devoted to fundamental combustion problems of solid-phase energetic materials, from both experimental and theoretical viewpoints. Recently, his interests moved to nanoenergetics for propulsion, performance of metallized formulations, agglomeration and aggregation, solid and hybrid rocket motors, space launchers, and in-space propulsion.

Editorial

Nanoenergetic Materials: Preparation, Properties, and Applications

Djalal Trache [1,*] and Luigi T. DeLuca [2,†]

[1] Teaching and Research Unit of Energetic Processes, Energetic Materials Laboratory, Ecole Militaire Polytechnique, BP 17, Bordj El-Bahri, Algiers 16046, Algeria
[2] Department of Aerospace Science and Technology, Politecnico di Milano, 20, 156 Milan, Italy; luigi.t.deluca@gmail.com
* Correspondence: djalaltrache@gmail.com
† Retired Professor.

Received: 22 November 2020; Accepted: 23 November 2020; Published: 26 November 2020

Energetic materials (EMs) are considered to be pure components or mixtures of chemical substances, which consist of both fuel and oxidizer that could release a large amount of energy or gas upon ignition. EMs can broadly be classified into propellants, explosives, and pyrotechnics with a wide range of applications in ordnance, rockets, missiles, space technology, fireworks, gas generators, automobile airbags, deconstruction, welding, and mining, to cite a few [1]. Explosives generate supersonic detonation velocity but low energy density, whereas propellants and pyrotechnics provide high energy densities by subsonic deflagration process. Typically, EMs can be produced as either monomolecular materials or as composites. The first class, which contains reactants within the same molecule, exhibits a fast reaction process but presents low performance, whereas the second class, which displays better performance, suffers from the slow reaction process mainly due to the limited mass transport rate between species. Several additives such as catalysts, coolants, stabilizers, and plasticizers, in few percent ratios, could be added to the EM formulations to improve their peculiar features and tailor their performance [2].

The advancement in the synthesis approaches and the advent of material characterization tools at multiple length scales have pushed the energetic materials community to explore new opportunities. During the past two decades, several significant achievements in research on nanoenergetic materials (nEMs) have been realized, thanks to the technological novelties in the field of nanoscience and nanotechnology. The principle of nanoenergetics is the enhancement of the specific surface area and intimacy with chemical components to improve the reaction rate while reducing the ignition delay at an acceptable level of safety. Nanoenergetics started with the manufacturing of nano-sized metal particles, mainly aluminum, which was mainly used for rocket propulsion, since the second half of the 20th century. During the last two decades, the physical mixing of oxidizers and fuels is considered as the second stage of the development of nanoenergetics at the nanoscale for which the diffusion distances between the chemical species is improved and the surface-over-volume ratio is enhanced, currently reaching the advanced third stage, where modern technologies, which allowed producing novel types of reactive nanocomposites structures and morphology with tunable features, are applied [3,4].

nEMs, which are composed of nano-sized fuel and oxidizer with or without additives, have been found to be potential sources of extremely high heat release rates and tailored burning rates, reliability, and extraordinary combustion efficiency. Nowadays, they play a vital role in widespread applications such as miniaturized electro-explosive devices, the attitude control of micro/nano satellites, and actuation in lab-on-a-chip devices, to name a few. The improvement of properties and the discovery of new functionalities and methodologies are key goals that cannot be reached without a better understanding of the preparation, characterization, manufacturing, and properties that constitute the starting points of the design of specific and adequate systems. The investigation of nanoenergetic materials has demonstrated both academic as well as technological importance and offered great research opportunities within cross-disciplinary areas.

In this framework, the present Special Issue in *Nanomaterials* aims to further contribute to the momentum of research and development in nanoenergetic materials, by featuring eleven (11) original research articles, two (2) review articles as well as one (1) short communication, authored and reviewed by experts in the field. This targets a broad readership of materials scientists, chemists, physicists, and nanotechnologists, among others. The potential topics address issues of the synthesis, characterization, properties, modifications, and applications of nanoenergetic materials as well as the incorporation of nano-additives (e.g., catalysts) for energetic material formulations. Most of the research papers highlight theoretical concepts and practical approaches of interest for real-world applications related to nanoenergetic materials.

Four interesting research papers dedicated to the synthesis, modification and characterization of nanoscale ingredients for new energetic formulations are published in this Special Issue. Luo et al. prepared an energetic composite fiber, in which 2,6-diamino-3,5-dinitropyrazine-1-oxide (LLM-105) nanoparticles are intimately incorporated with a nitrocellulose/glycidyl azide polymer (NC/GAP) fiber through the electrospinning method [5]. Compared to pure NC/GAP and LLM-105 nanoparticles, the obtained insensitive three-dimensional nanofibers with a large specific surface area exhibited a lower decomposition temperature but a higher decomposition rate. Such nanofibers displayed outstanding performance with a higher combustion chamber temperature and improved specific impulse as well. The authors claimed that such nanofibers might find potential application in the field of solid rocket propellant systems. One paper by Wang et al. explored the ultra-low temperature spray method to produce 2D network structures of nanoscale ammonium perchlorate (AP) and ammonium nitrate (AN) [6], which are widely used as oxidizer of solid rocket propellants [7]. The authors demonstrated that the obtained nano-AP was more sensitive than the raw micro-AP, whereas either nano- or micro-sized particles of AN were both insensitive to friction and impact stimuli. They reported that the thermal decomposition of nanometric AP and AN occurred at lower temperatures with respect to their raw materials, respectively. The decomposition products of nano-AP corresponded to NO_2, N_2O, HCl, H_2O with a small amount if $NOCl$, whereas those of nano-AN were mainly N_2O and H_2O with a small amount of NH_3. The study by Dobrynin et al. assessed the potential of supercritical antisolvent (SAS) processing to produce nano-sized nitrocellulose (NC) [8], which is a workhorse of numerous energetic formulations [9]. The authors prepared nano-NC with an average particles size of 190 nm from raw NC of 20 µm in diameter. The obtained nano-NC displayed low friction sensitivity and lower decomposition temperature compared to the raw NC. The authors prepared nano-NC/carbon nanotubes/nano-Fe_2O_3 nanocomposites for which the combustion process exhibited an increase in the burning rate of 20% at 12 MPa compared to the pure nano-NC. In addition, the authors revealed that the employment of SAS processing to produce the nanocomposite offered higher performance, lower sensitivity and better stability compared to nanocomposites prepared by conventional dry mixing. In another research work, the Dreizin research group prepared fuel-rich composite powders combining elemental Si with metal fluoride oxidizers BiF_3 and CoF_2, using the common approach of arrested reactive milling [10]. The reactivity assessment of the obtained powders was carried out by means of thermo-analytical techniques in both inert argon (Ar) and oxidizing (Ar/O_2) atmospheres. In addition, the obtained powders were ignited by an electrically heated filament using CO_2 laser as an ignition source under ambient atmosphere. They mentioned that both composites lead to fast ignition and effective oxidation under oxidizing atmosphere, while elemental Si could not ignite using either laser or heated filament under the same conditions. It was also demonstrated that the combustion process, which occurred at lower temperature, is more effective on BiF_3 for which both oxidation and fluorination occur nearly simultaneously. However, during the use of CoF_2, it was found that the oxidation of Si occurs before its fluorination.

Four papers are dedicated to the evaluation of the reactivity and performance of some nanoenergetic materials through controllable manufacturing and/or modifications approaches. A paper by the Rossi research group introduced one area of interest by describing the technology of the deposition of Al/CuO multilayers through sputter-deposition [11]. Such multilayers are expected to be employed as tunable igniters and actuators for defense, space and infrastructure safety applications. The same research group described in another paper a joint experimental/theoretical investigation dealing with the aging of reactive Al/CuO nanolaminates through the assessment of the structural changes as well as the combustion features [12]. The group reported that the nanolaminates remained stable even after decades of storage at ambient temperature. The authors revealed that the aging transition occurred at 200 °C for 14 days for which the interfacial modification is attributed to the stack dimensional characteristics. They also demonstrated that the burn rate of Al/CuO nanolaminates with bilayer thickness greater than 500 nm was not affected, whereas it decreased by ~25% for a thickness of 300 nm. In another study, Guo et al. designed prominent nano-Al/MoO$_3$ metastable intermolecular composite (MIC) chips with the homogeneous distribution of particles through a suitable and high-effective electrophoretic deposition (EPD) method at room temperature and under ambient pressure conditions [13]. The obtained nano-MIC chips exhibited interesting heat-release performance. It was shown that the MIC ship initiator could be successfully ignited with a typical capacitor charge/discharge ignition device, displaying excellent detonation performance. The authors claim that the developed fabrication method is fully compatible with micro-electromechanical systems, which can be employed for micro-ignition/propulsion applications. The report by Vorozhtsov et al. proposed an interesting approach, which is based on the coating process of nano-aluminum using hydroxy terminated polybutadiene (HTPB), to enhance the performance of the nano-fuel within a propellant formulation [14]. The authors demonstrated that the coating process, which maintains a high reactivity of nano-Al, has been successfully performed. The authors report that the prepared composite solid propellant based on nano-Al coated with HTPB provided a higher burning rate with an increase in the burning stability at low pressure compared to the propellant supplemented with nano-Al without HTPB coating.

Another increasing area of interest in nanoenergetic materials addressed in this issue concerns the incorporation of various additives such as catalysts to tailor the properties and improve the performance of energetic formulations. A paper by the Yan research group reported on the analysis of the gaseous products during the decompositi0on of ammonium perchlorate (AP) supplemented with graphene oxide (GO)-based additives through the employment of a tandem analytic tool, which consists of thermogravimetry coupled with mass spectrometry [15]. They proved that the GO-based catalysts improved the catalytic decomposition efficiency due to the capability in increasing the conversion rate of NH$_3$ and H$_2$O through the excessive O elements being transferred to react with NH$_4{}^+$, which enhanced the decomposition heat. The same research group investigated the morphology, particle size, and composition of condensed combustion products (CCP) during the combustion of modified double-base solid propellants, which contain nano-catalysts, under various pressures [16]. The authors report that the average particle size of CCPs with various morphologies decreased with the increase in pressure. The surface of CCP particles contained various chemical elements such as C, N, Al, Cu, Pb and Si. In another research work, Yao et al. prepared bamboo leaf-like CuO and flaky-shaped CuO by the hydrothermal method, which are combined with Al nanoparticles through ultrasonic dispersion methods [17]. The obtained composites have been assessed as catalysts of nitrocellulose. It was demonstrated that the decomposition process of NC with or without catalysts follows the same kinetic mechanism as the Avrami–Erofeev equation. The authors proved that the microstructure of CuO affected the thermolysis process, where the presence of flaky-shaped CuO/Al lead to an easier ignition of NC. Within the same subject, the Trache research group produced hematite nanoparticles decorated on carbon mesospheres and assessed their effect on the thermal decomposition of nitrocellulose [18]. It was revealed that the obtained nanoparticles had a minor effect on the decomposition temperature of nitrocellulose, whereas an obvious decrease in the activation energy was acquired. Moreover, the decomposition reaction mechanism of NC is influenced by the incorporation of the nano-catalyst.

Metal nanoparticles have receive tremendous attention from the scientific community [19,20]. They display interesting features in various applications in the combustion and explosion processes of energetic materials. For such a type of nanomaterials, two interesting review papers have been published in the current issue. The first review article by Pang et al. describes the recent advance in the preparation, characterization and performance of Al-based nano-sized composite energetic materials (CEMs) [21]. The authors emphasize the most important approaches that are currently applied worldwide to improve the performance of Al-based nano CEMs, which are revealed to differ from those based on micro CEMs. The remaining issues and challenges for the future research directions have been deeply discussed. The last paper in the issue by Zarko and Glazunov is dedicated to an overview of experimental methods used to measure the ignition and combustion characteristics of metal nanoparticles [22]. The authors focus on the methods employed to determine the nanoparticle size, their heat-exchange parameters as well as the ignition delay and combustion time. They reveal that despite the advance of the analytic method to investigate the metal nanoparticles features, other issues exist concerning the comparison of the data of ignition and burning time with respect to particle size dependencies. They recommend continuing the development of effective approaches in the future to correctly describe the particles' dynamic behavior.

In summary, the present Special Issue advances not only our understanding of the emerging and significant role of nanoenergetic materials for the future of pyrotechnical science, but also that of challenges and the future research directions to fully explore their potential features according to their practical utilization. Despite the fact that we are still struggling to translate the fundamental advances reported in the scientific literature into tangible applications, it is expected that this Special Issue will encourage multidisciplinary research activities on nanoenergetic materials to overcome the current technological limitations. Thus, further experimental, simulation and calculation works should be carried out in the future through the focus on the scaling-up of the systems, the economic viability, stability and aging, coating and dispersion, material safety, hazard level during manufacturing, recycling needs, and the control of ignition processes. Finally, it is anticipated that nanoenergetic materials as the next generation of materials will be fruitfully integrated into defense, special and other civilian fields.

Author Contributions: D.T. wrote this Editorial Letter. L.T.D. provided his feedback, which was assimilated into the Letter. All authors have read and agreed to the published version of the manuscript.

Acknowledgments: We want to thank all authors for their outstanding contributions. The time and effort devoted by reviewers of the articles and their constructive comments are also highly appreciated. Furthermore, we would like to acknowledge the members of the editorial office of *Nanomaterials* for their help, promptness, administrative and editorial support during all of this long period from the point of designing the issue and throughout its implementation and completion.

Conflicts of Interest: The authors declare no conflict of interest.

References

1. Benhammada, A.; Trache, D. Thermal decomposition of energetic materials using tg-FTIR and TG-MS: A state-of-the-art review. *Appl. Spectrosc. Rev.* **2020**, *55*, 724–777. [CrossRef]
2. Pang, W.; DeLuca, L.T.; Gromov, A.; Cumming, A.S. *Innovative Energetic Materials: Properties, Combustion Performance and Application*; Springer: Singapore, 2020.
3. DeLuca, L.T. A survey of nanotechnology for rocket propulsion: Promises and challenges. *ELSI Handb. Nanotechnol. Risk Saf. ELSI Commer.* **2020**, 277–332. [CrossRef]
4. Yan, Q.-L.; He, G.-Q.; Liu, P.-J.; Gozin, M. *Nanomaterials in Rocket Propulsion Systems*; Elsevier: Amsterdam, The Netherlands, 2019.
5. Luo, T.; Wang, Y.; Huang, H.; Shang, F.; Song, X. An electrospun preparation of the NC/GAP/nano-LLM-105 nanofiber and its properties. *Nanomaterials* **2019**, *9*, 854. [CrossRef] [PubMed]
6. Wang, Y.; Song, X.; Li, F. Nanometer ammonium perchlorate and ammonium nitrate prepared with 2D network structure via rapid freezing technology. *Nanomaterials* **2019**, *9*, 1605. [CrossRef] [PubMed]

7.   Trache, D.; Klapötke, T.M.; Maiz, L.; Abd-Elghany, M.; DeLuca, L.T. Recent advances in new oxidizers for solid rocket propulsion. *Green Chem.* **2017**, *19*, 4711–4736. [CrossRef]

8.   Dobrynin, O.S.; Zharkov, M.N.; Kuchurov, I.V.; Fomenkov, I.V.; Zlotin, S.G.; Monogarov, K.A.; Meerov, D.B.; Pivkina, A.N.; Muravyev, N.V. Supercritical antisolvent processing of nitrocellulose: Downscaling to nanosize, reducing friction sensitivity and introducing burning rate catalyst. *Nanomaterials* **2019**, *9*, 1386. [CrossRef] [PubMed]

9.   Trache, D.; Tarchoun, A.F. Differentiation of stabilized nitrocellulose during artificial aging: Spectroscopy methods coupled with principal component analysis. *J. Chemom.* **2019**, *33*, e3163. [CrossRef]

10.   Valluri, S.K.; Schoenitz, M.; Dreizin, E. Preparation and characterization of silicon-metal fluoride reactive composites. *Nanomaterials* **2020**. Accepted for publication.

11.   Salvagnac, L.; Assié-Souleille, S.; Rossi, C. Layered Al/CuO thin films for tunable ignition and actuations. *Nanomaterials* **2020**, *10*, 2009. [CrossRef] [PubMed]

12.   Estève, A.; Lahiner, G.; Julien, B.; Vivies, S.; Richard, N.; Rossi, C. How thermal aging effects ignition and combustion properties of reactive Al/CuO nanolaminates: A joint theoretical/experimental study. *Nanomaterials* **2020**, *10*, 2087. [CrossRef] [PubMed]

13.   Guo, X.; Sun, Q.; Liang, T.; Giwa, A. Controllable electrically guided nano-Al/MoO$_3$ energetic-film formation on a semiconductor bridge with high reactivity and combustion performance. *Nanomaterials* **2020**, *10*, 955. [CrossRef] [PubMed]

14.   Vorozhtsov, A.; Lerner, M.; Rodkevich, N.; Sokolov, S.; Perchatkina, E.; Paravan, C. Preparation and characterization of Al/HTPB composite for high energetic materials. *Nanomaterials* **2020**, *10*, 2222. [CrossRef] [PubMed]

15.   Chen, S.; An, T.; Gao, Y.; Lyu, J.-Y.; Tang, D.-Y.; Zhang, X.-X.; Zhao, F.; Yan, Q.-L. Gaseous products evolution analyses for catalytic decomposition of ap by graphene-based additives. *Nanomaterials* **2019**, *9*, 801. [CrossRef] [PubMed]

16.   Li, J.-Q.; Liu, L.; Fu, X.; Tang, D.; Wang, Y.; Hu, S.; Yan, Q.-L. Transformation of combustion nanocatalysts inside solid rocket motor under various pressures. *Nanomaterials* **2019**, *9*, 381. [CrossRef] [PubMed]

17.   Yao, E.; Zhao, N.; Qin, Z.; Ma, H.; Li, H.; Xu, S.; An, T.; Yi, J.; Zhao, F. Thermal decomposition behavior and thermal safety of nitrocellulose with different shape CuO and Al/CuO nanothermites. *Nanomaterials* **2020**, *10*, 725. [CrossRef]

18.   Benhammada, A.; Trache, D.; Kesraoui, M.; Chelouche, S. Hydrothermal synthesis of hematite nanoparticles decorated on carbon mesospheres and their synergetic action on the thermal decomposition of nitrocellulose. *Nanomaterials* **2020**, *10*, 968. [CrossRef] [PubMed]

19.   Zachariah, M.R. Nanoenergetics: Hype, reality and future. *Propellants Explos. Pyrotech.* **2013**, *38*, 7.

20.   Kline, D.J.; Rehwoldt, M.C.; Turner, C.J.; Biswas, P.; Mulholland, G.W.; McDonnell, S.M.; Zachariah, M.R. Spatially focused microwave ignition of metallized energetic materials. *J. Appl. Phys.* **2020**, *127*, 055901. [CrossRef]

21.   Pang, W.; Fan, X.; Wang, K.; Chao, Y.; Xu, H.; Qin, Z.; Zhao, F. Al-based nano-sized composite energetic materials (nano-CEMs): Preparation, characterization, and performance. *Nanomaterials* **2020**, *10*, 1039. [CrossRef] [PubMed]

22.   Zarko, V.; Glazunov, A. Review of experimental methods for measuring the ignition and combustion characteristics of metal nanoparticles. *Nanomaterials* **2020**, *10*, 2008. [CrossRef]

 *nanomaterials* 

Article

An Electrospun Preparation of the NC/GAP/Nano-LLM-105 Nanofiber and Its Properties

Tingting Luo [1], Yi Wang [1,*], Hao Huang [2], Feifei Shang [3] and Xiaolan Song [4]

[1] School of Materials Science and Engineering, North University of China, Taiyuan 030051, China; luotingting1002@163.com
[2] China North Industries Group Corporation Limited, Beijing 100821, China; huanghao-xiang@163.com
[3] Teaching and Research Support Center, Army Academy of Armored Forces, Beijing 100072, China; feiniaozi@126.com
[4] School of Environment and Safety Engineering, North University of China, Taiyuan 030051, China; songxiaolan00@126.com
* Correspondence: wangyi528528@aliyun.com; Tel.: +86-134-5345-8592

Received: 28 April 2019; Accepted: 31 May 2019; Published: 4 June 2019

Abstract: In this work, an energetic composite fiber, in which 2,6-diamino-3,5-dinitropyrazine-1-oxide (LLM-105) nanoparticles intimately incorporated with a nitrocellulose/glycidyl azide polymer (NC/GAP) fiber, was prepared by the electrospinning method. The morphology and structure of the nanofiber was characterized by scanning electron microscopy (SEM), energy dispersive X-Ray (EDX), fourier transform infrared spectroscopy (IR), X-ray diffraction (XRD), X-ray photoelectron spectroscopy (XPS), and Brunauer–Emmett–Teller (BET). The nanofibers possessed a three-dimensional (3D) net structure and a large specific surface area. Thermal analysis, energetic performance, and sensitivities were investigated, and they were compared with NC/GAP and LLM-105 nanoparticles. The NC/GAP/nano-LLM-105 nanofibers show higher decomposition rates and lower decomposition temperatures. The NC/GAP/nano-LLM-105 decomposed to CO_2, CO, H_2O, N_2O, and a few NO, $-CH_2O-$, and $-CH-$ fragments, in the thermal-infrared spectrometry online (TG-IR) measurement. The NC/GAP/nano-LLM-105 nanofibers demonstrated a higher standard specific impulse (I_{sp}), a higher combustion chamber temperature (T_c), and a higher specialty height (H_{50}). The introduction of nano-LLM-105 in the NC/GAP matrix results in an improvement in energetic performance and safety.

Keywords: electrospinning; NC/GAP/nano-LLM-105; thermolysis; energetic performance; sensitivity

---

1. Introduction

2,6-Diamino-3,5-dinitropyrazine-1-oxide (LLM-105) is an essential ingredient in many propellant and explosive formulas for its low sensitivity, high energy, high density, and excellent thermal stability [1,2]. The properties of low sensitivity and excellent thermal stability are attributed to the π-conjugated system. Due to the intense intramolecular hydrogen bonding, the LLM-105 possesses good compatibility with the common components of propellant and explosives, such as hexahydro-1,3,5-trinitro-1,3,5-triazine (RDX), nitrocellulose (NC), etc. [3–6]. Nanoscale LLM-105 has a superior energy release rate and a higher reaction rate, compared with conventional LLM-105 [7–10]. However, the agglomeration of nano-LLM-105 causes a decrease in performance and limits its application [11,12]. It is feasible that energetic matrix is utilized to support LLM-105 nanoparticles, which can effectively avoid agglomerating.

Electrospinning is a universal technology that is used to obtain multifarious nanocomposite [13–16]. The as-spun 3D nanofibers, with high specific surface areas and porosities are desired carrier for supporting nanoparticles [17–19]. However, the application of electrospinning technology in composite energetic materials is rarely performed [20]. For instance, nitrocellulose/aluminum-cupric oxide

(NC/Al-CuO) and nitrocellulose/2,4,6,8,10,12-hexanitro-2,4,6,8,10,12-hexaazaisowurtzitane (NC/CL-20) nanofibers with high burning rates were obtained via electrospinning [21,22]. Li also fabricated nanoboron/nitrocellulose (B/NC) electrospun nanofibers with excellent thermostability [23]. Similarly, researchers selected single NC as electrospinning matrix. However, simplex NC has a relatively large viscosity, which affects the morphology of the nanofiber, and results in low loading of the nano-explosive. In addition, a high spinning voltage could generate electric sparks and bring great safety risks. The composite energetic matrix can compensate these deficiencies. GAP is a high-energy prepolymer with low viscosity and high density. Moreover, it has more flexible segments, a lower glass transition temperature (T_g), and higher mechanical properties than NC. Currently, there is no report on using GAP as a matrix to load explosives with electrospinning [24–26]. In this work, ball milling nano-LLM-105 is assembled onto a NC/GAP composite matrix by electrospun technology, to create a new type of energetic 3D nanocomposite. In fact, it is not dangerous to prepare explosive materials by electrospinning and ball milling. This is because that the energetic material is very stable at a normal temperature and pressure, and under the protection of solvent. At this time, they are no different from inert materials. Further experiments suggest that the NC/GAP/nano-LLM-105 nanofibers possess lower sensitivity and remarkable thermal decomposition and energy performance, which makes the nanofibers have application potentials in the field of solid propellants.

2. Materials and Methods

2.1. Materials

2,6-Diamino-3,5-dinitropyrazine-1-oxide (LLM-105) was provided by Gansu Yinguang Chemical Co., Ltd. (Baiyin city, Gansu province, P.R. China). Glycidyl azide polymer (GAP, Mn = 4000, hydroxyl value of 0.49 mmol·g^{-1}) was obtained from the 42nd Institute of the Fourth Academy of China Aerosce Science and Technology Corporation. Nitrocellulose (NC, 12.6% N, industrial grade) was provided by Foshan Junyuan Chemical Co., Ltd. (Foshan city, Guangdong province, P.R. China). Ethanol (EtOH) and acetone were purchased from Tianjin Guangfu Chemical Co., Ltd. (Tianjin city, China).

2.2. Fabrication of Nanofibers

Firstly, nano-LLM-105 was prepared by the high-energy ball milling method. The ingredients, including 200 g balls, 6 g LLM-105, 60 mL deionized water, and 60 mL ethanol, are added into a mill jar. The four jars are sealed and immobilized on the ball mill. The mill rotates at 300 rpm for 6 hr. Then 0.3 g nano-LLM-105 is dissolved in 4.4 g acetone to get nano-LLM-105 suspension. Exactly 0.45 g NC and 0.45 g GAP were added into 4.4 g acetone to obtain a NC/GAP solution. The above-prepared nano-LLM-105 suspension and NC/GAP solutions were blended to obtain a NC/GAP/nano-LLM-105 precursor (12 wt %). The mass ratio of NC, GAP, and nano-LLM-105, respectively was set to 3:3:2. As a contrast, the NC/GAP precursor solution (12 wt %) was obtained by dissolving 0.6 g NC and 0.6 g GAP into 8.8 g acetone. The mass ratio of NC and GAP is set to 1:1. For the electrospinning process of these two nanofibers, the inner diameter of the stainless steel needle is 0.8 mm. The ambient humidity was controlled at 40–50%. The applied voltage was maintained at 12–18 kV. Additionally, the flow rate was fixed at 3–5 mL·hr^{-1}. Aluminum foil was used to collect the fibers, which were placed 12 cm away from the needle. The preparation scheme is described in Figure 1.

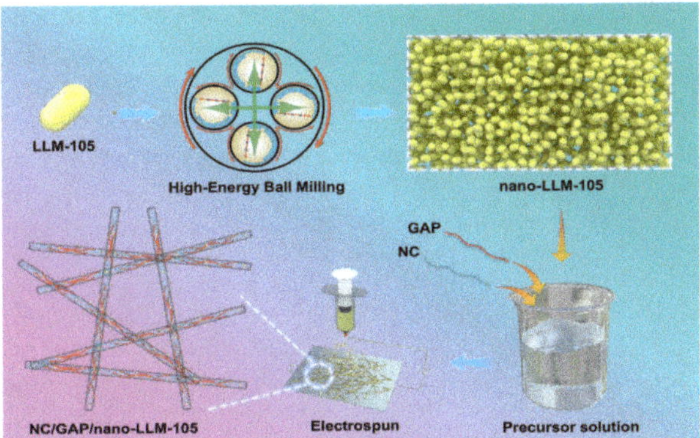

Figure 1. Sketch for the synthesis of the nitrocellulose/glycidyl azide polymer/nano 2,6-diamino-3,5-dinitropyrazine-1-oxide (NC/GAP/nano-LLM-105) composite nanofiber.

2.3. Characterization

The analyses of SEM, EDS, IR, XRD, and XPS were performed in order to investigate the morphology and structure of NC/GAP, nano-LLM-105, and NC/GAP/nano-LLM-105. Scanning electron microscopy (SEM) was performed on a Hitachi SU8010. The diameters of particles and fibers were measured by Nano Measurer 1.2 software. X-ray diffraction (XRD) analysis was performed on a DX-2700 X-ray diffractometer (Hao yuan) with Cu Kα radiation. The IR spectrum was obtained on an infrared spectrometer (American Thermo Fisher Scientific Nicolet 6700). XPS was conducted with X-ray photoelectron spectroscopy (XPS) and a PHI5000 Versa-Probe (ULVAC-PHI). The BET measurements of NC/GAP and NC/GAP/nano-LLM-105 were performed, utilizing nitrogen adsorption with a Micromeritics ASAP 2010 instrument. Thermal analyses for NC/GAP, nano-LLM-105, and NC/GAP/nano-LLM-105 were conducted on a differential scanning calorimeter (DSC, TA Model Q600) at heating rates of 5, 10, 15, and 20 °C/min. thermal-infrared spectrometry online (TG-IR) analyses of NC/GAP and NC/GAP/nano-LLM-105 were performed on a thermal analyzer system (TG/DSC, Mettler Toledo) coupled with a Fourier transform infrared spectrometer in a nitrogen atmosphere. The temperature range that we considered was 50 °C to 400 °C. The impact sensitivity was tested by using HGZ-1 impact equipment. In each test, 25 drop tests were carried out to calculate the H_{50}, and each portion was performed three times to obtain a mean value and a standard deviation.

3. Results and Discussion

3.1. Morphology and Structure

Figure 2a,b reveals that there are some weaker agglomerates rather than hard agglomerates for LLM-105 nanoparticles, and there are no bridge between the particles. The particle diameter distribution is obtained by measuring a diameter of ~100 particles, and the results are displayed in Figure 2c–d. We acquired the volume curve by integrating the frequency curve. The mean diameter calculated from the frequency curve is 152 nm, which is same as the median diameter (d_{50} = 152 nm) calculated from the volume curve. For the SEM images of NC/GAP and NC/GAP/nano-LLM-105 (Figure 3a,b), it is clearly observed that both of the two nanofibers reveal 3D reticulate structures. The surface of the NC/GAP nanofiber is smooth and uniform. On the contrary, the surface of the NC/GAP/nano-LLM-105 nanofiber is rough and uneven. The difference in morphology is primarily caused by two factors. The addition of LLM-105 nanoparticles results in an inhomogeneity of precursor solution. Furthermore, partial LLM-105 nanoparticles are agglomerated during the electrospinning process. From Figure 3c–f, for NC/GAP nanofibers, the mean diameter and median diameter are 469 nm and 478 nm, respectively. By comparison, the mean diameter and median diameter of NC/GAP/nano-LLM-105 are 758 nm and

764 nm, respectively. It is apparent that the diameter of NC/GAP/nano-LLM-105 is larger than that of the NC/GAP nanofibers. The difference in mean diameter of these two nanofibers is due to the fact that the LLM-105 nanoparticles are loaded onto the surface of NC/GAP.

Figure 2. (**a**,**b**) SEM image of LLM-105 nanoparticles; (**c**,**d**) diameter distribution of the LLM-105 nanoparticles.

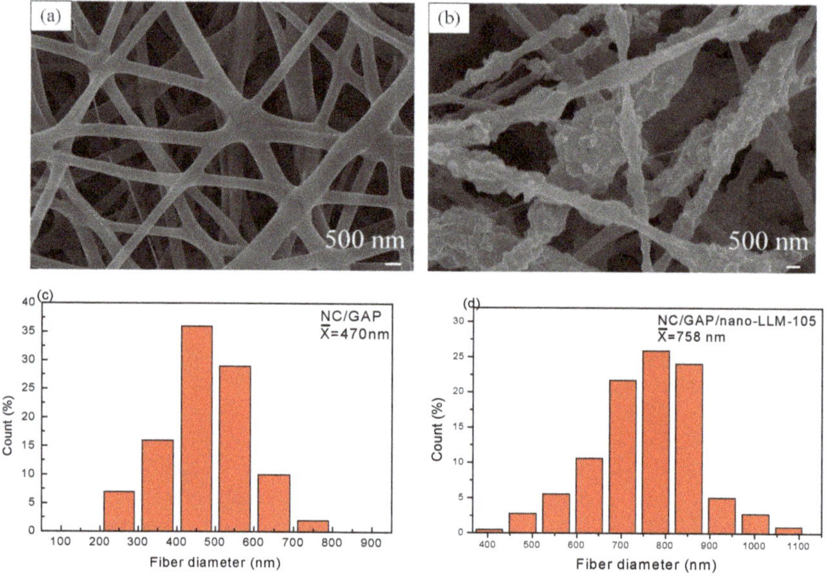

Figure 3. *Cont.*

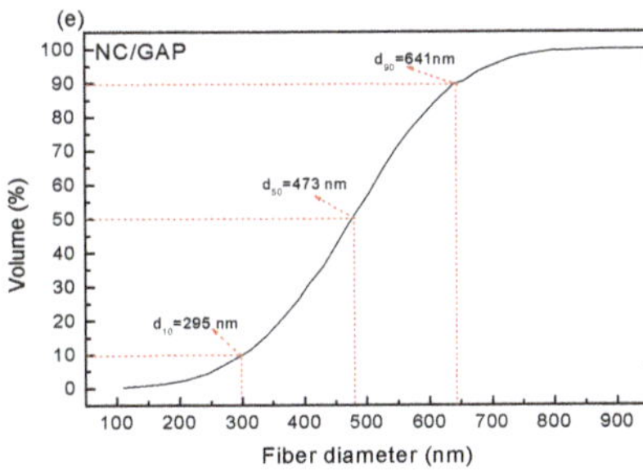
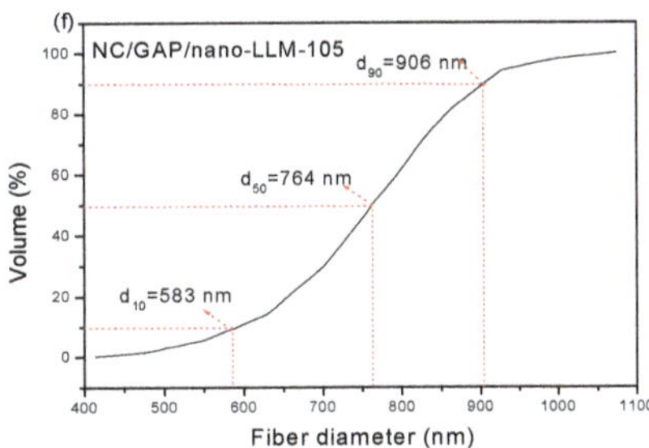

Figure 3. Images and diameter distributions of samples: (**a**) for NC/GAP; (**b**) for NC/GAP/nano-LLM-105; (**c–f**) the diameter distribution.

EDS analyses were performed to probe the surface elements of the nanofibers; the results are exhibited in Figure 4. From Figure 4a,b, the peaks at about 2 Kv belonged to the gold element sprayed during the test. There were only O, C, and N elements that were presented on the surfaces of those two fibers, illustrating that impurities were not introduced in the process of ball milling and electrospinning. The theoretical elemental contents are tabulated in Table 1. After the addition of the LLM-105 nanoparticles, the O content hardly varied, the N content appreciably increased, and the C content declined.

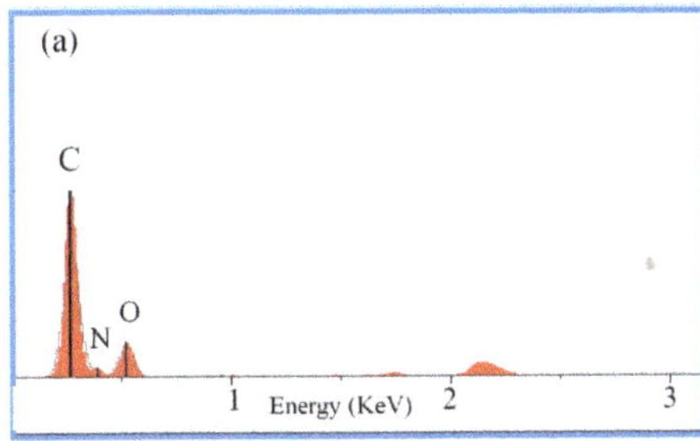
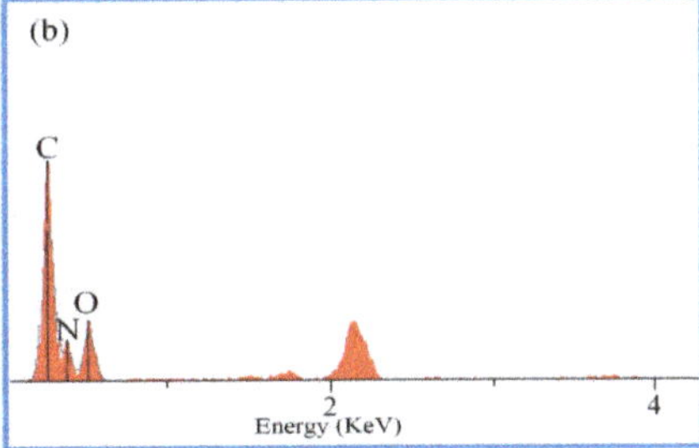

Figure 4. EDS spectra of samples: (**a**) for NC/GAP; (**b**) for NC/GAP/nano-LLM-105.

Table 1. Theoretical elemental contents from EDS analyses.

Sample	C Content (%)	N Content (%)	O Content (%)
NC/GAP	31.45	27.5	37.2
NC/GAP/nano-LLM-105	29.14	30.20	37.16

The IR spectra of NC/GAP, NC/GAP/nano-LLM-105, and nano-LLM-105 are contrasted in Figure 5a. For NC/GAP/nano-LLM-105, the peaks at 3437, 3405, 3285 3233, and 1648 cm^{-1} respectively indicated symmetric, anti-symmetric stretching vibrations and deformation vibrations of –NH$_2$ in nano-LLM-105; two strong absorption peaks located at 1480 and 1448 cm^{-1} corresponded to the stretching vibrations of the C=C skeleton in the ring of nano-LLM-105; the peak at 1577 cm^{-1} indicated anti-symmetric stretching vibrations of –NO$_2$ in nano-LLM-105; the peaks at 1351 cm^{-1} and 1383 cm^{-1} corresponded to symmetric stretching vibrations of –NO$_2$ in nano-LLM-105 [27]; the peak at 2101 cm^{-1} was ascribed to the stretching vibrations of –N$_3$ from the GAP that was present [28]; the peaks at 1280 and 1648 cm^{-1} reflected the symmetric and anti-symmetric stretching vibrations of -ONO$_2$ in NC, respectively [29]; the peak at 1075 cm^{-1} corresponded to the out-of-plane bending vibrations of C–H. Hence, the functional groups for NC/GAP/nano-LLM-105 were in accord with nano-LLM-105 and NC/GAP, indicating that the LLM-105 nanoparticles were well-combined with NC/GAP. These peaks were the same as the published report. Overall, the molecular structures of the nano-LLM-105 and NC/GAP do not alter in the process of electrospinning. There were no new groups generated, indicating that NC, GAP, and nano-LLM-105 do not react with each other. Figure 5b shows the XRD patterns of samples. There were two main peaks at 28.4 and 33.2° in pattern of nano-LLM-105 [10]. Also, there were no diffraction peaks in the pattern of NC/GAP. This is because LLM-105 is a type of crystal, and NC/GAP is a type of polymer. The peak positions of the LLM-105 nanoparticles were in line with those of the NC/GAP/nano-LLM-105 nanofibers, which means that the crystal phase of the LLM-105 nanoparticles does not transform by electrospinning. This is a superior feature of the electrospinning compared with recrystallization to prepare energetic materials. For the recrystallization method, if the solvent is not properly selected, the crystal phase is liable to transform. Song prepared 1,3,5,7-Tetranittro-1,3,5,7-tetrazocane (HMX) by solvent/non-solvent method, and the crystal phase of HMX changed from β-HMX to γ-HMX [30]. In this work, acetone is chosen as the solvent, and the LLM-105 nanoparticles are suspended in it, which avoids the recrystallization of the LLM-105.

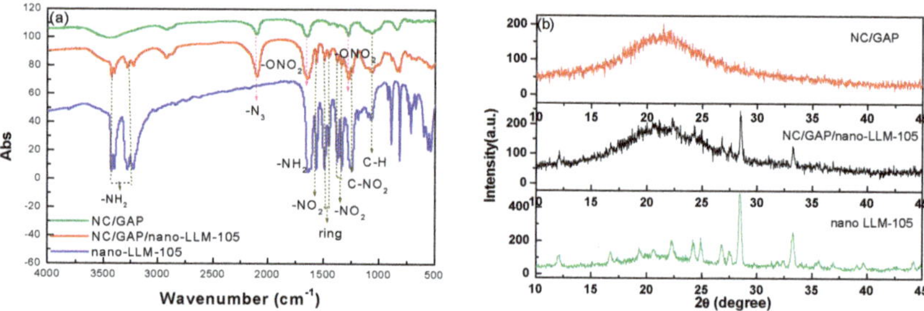

Figure 5. IR spectra (**a**) and XRD patterns (**b**) of NC/GAP, nano-LLM-105 and NC/GAP/nano-LLM-105.

The XPS spectra of NC/GAP, nano-LLM-105, and NC/GAP/nano-LLM-105 are displayed in Figure 6. From Figure 6a–c, typical signals of C, N, and O were clearly detected. For NC/GAP/nano-LLM-105, the O1s spectra presented five features with binding energies of 531.3 eV, 532.3 eV, 533.1 eV, 534.2 eV, and 534.8 eV; the peaks at 532.3 eV and 534.8 eV were ascribed to -NO$_2$ and N-O in the ring of LLM-105; the peaks at 533.1 eV and 534.2 eV belonged to -O*-NO$_2$ and -O-NO*$_2$ in NC; the peak at 531.3 eV was related to the -C-O-C group of NC and GAP [31]. For the C1s spectrum of NC/GAP/nano-LLM-105, the peak was fitted to six peaks. The peaks located at 284.5 eV, 286.3 eV, and 288.2 eV were assigned to -C-C, C-N$_3$, and -C-ONO$_2$ of GAP and NC. The peaks at 284.7 eV, 286.8 eV, and 287.6 eV corresponded to -C-C, -C-NH$_2$, and -C-NO$_2$ in nano-LLM-105 [32]. The XPS spectrum of N1s consisted of seven peaks at 400.4 eV, 401.1 eV, 404.1 eV, 404.2 eV, 406.9 eV, 407.7 eV, and 408.2 eV, which corresponded respectively to –$\underline{N}$=N=N, –NH$_2$, –N=$\underline{N}$=N, C-N in ring, -NO$_2$, –ONO$_2$, and N-O in the ring [28]. The -N$_3$ and –ONO$_2$ groups were ascribed to NC and GAP, respectively. Finally, the groups of C-N in ring, N-O in ring, –NH$_2$ and -NO$_2$ belong to LLM-105. Hence, we infer the existence of LLM-105 nanoparticles on the surface of NC/GAP, and there are no new chemical bonds being produced on the surface of the nanofibers.

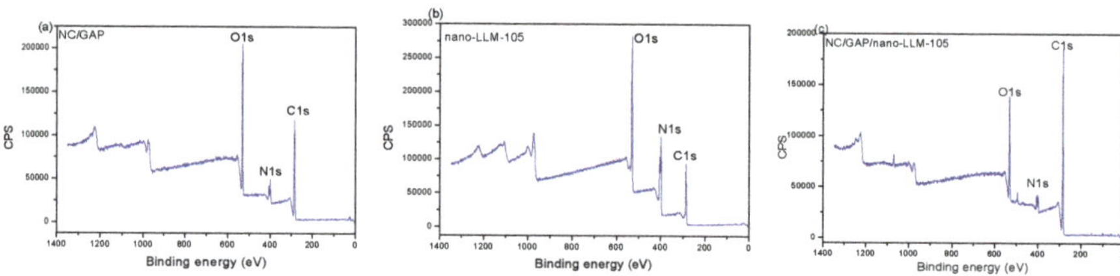

Figure 6. *Cont.*

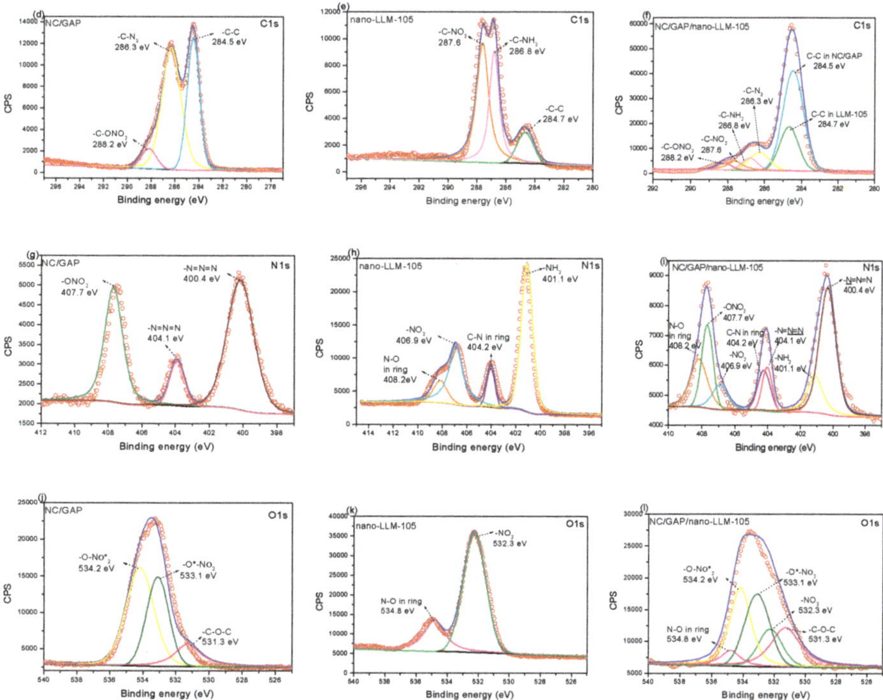

Figure 6. (**a–c**) XPS spectra of NC/GAP, nano-LLM-105 and NC/GAP/nano-LLM-105; high resolution XPS spectra (**d–f**) for C 1s of samples; for (**g–i**) N 1s of samples and for (**j–l**) O 1s of samples.

The nitrogen adsorption–desorption isotherms of NC/GAP and NC/GAP/nano-LLM-105 are displayed in Figure 7. The specific surface areas, pore volumes, and pore sizes of the samples are shown in Table 2. The isotherms are considered as class IV (H3-type hysteresis loop), indicating that the prepared nanofibers were mesoporous materials. At a low p/p^o, there is the first steep portion of the isotherm, as the p/p^o increases, adsorption of multiple layers begins. In the multi-layer adsorption process, capillary condensation a common accompaniment (IV isotherms). Capillary condensation and capillary evaporation generally do not occur at identical p/p^o, resulting in the generation of hysteresis loops. The specific surface areas of the NC/GAP/nano-LLM-105 and NC/GAP nanofibers were 6.0545 and 4.3573, respectively. The higher surface area value of NC/GAP/nano-LLM-105 is ascribed it having a rough surface. Compared to the energetic composite prepared by other methods, energetic nanofibers have a larger specific surface area. For example, the specific surface area of NC/GAP/CL-20 when prepared by a sol-gel-supercritical method, is 2.7 [33].

Table 2. BET surface area and the pore structure parameters of the nanofibers.

Samples	BET Surface Area (m^2·g^{-1})	Pore Volume (cm^3·g^{-1})	Pore Size (nm)
NC/GAP	4.3573	0.004422	4.05911
NC/GAP/nano-LLM-105	6.0545	0.007664	5.06352

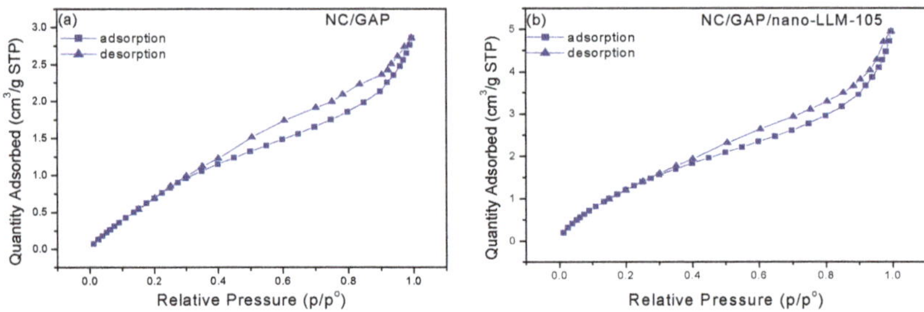

Figure 7. The BET data of samples: (**a**) NC/GAP and (**b**) NC/GAP/nano-LLM-105 nanofiber.

3.2. Thermal Analysis

The DSC thermograms of samples collected at different heating rates are displayed in Figure 8a–c. The kinetic and thermodynamic parameters for thermal decomposition are calculated with the DSC data; the results are displayed in Table 3. For all samples, the exothermic peak temperature increases with the increase of heating rate. For NC/GAP/nano-LLM-105, there was only one exothermic peak, demonstrating that NC, GAP, and nano-LLM-105 decompose synchronously. In addition, the exothermic peak temperature of NC/GAP/nano-LLM-105 is slightly lower than that of NC/GAP, and it is markedly lower than those of the LLM-105 nanoparticles. This manifests that the thermolysis of NC, LLM-105 nanoparticles, and GAP cooperate with each other.

The activation energy (E_K), pre-exponential factor (A_K), and rate constant (k), are calculated by the Kissinger equation (Equation (1)) [29] and the Arrhenius equation (Equation (2)) [34]. The E_K of NC/GAP/nano-LLM-105 (185.139 kJ·mol^{-1}) is lower than that of the LLM-105 nanoparticles and is higher than that of NC/GAP. The k of NC/GAP/nano-LLM-105 is higher than the k of the NC/GAP and LLM-105 nanoparticles, implying that NC/GAP/nano-LLM-105 has a higher decomposition rate.

$$\ln \frac{\beta}{T_p^2} = \ln \frac{R \cdot A_K}{E_K} - \frac{E_K}{R} \cdot \frac{1}{T_p} \tag{1}$$

$$k = A_K \cdot Exp\left(-\frac{E_K}{T_p \cdot R}\right) \tag{2}$$

$$A_K \exp\left(-\frac{E_K}{RT_P}\right) = \frac{K_B T_P}{h} \exp\left(-\frac{\Delta G^{\neq}}{RT_P}\right) \tag{3}$$

$$\Delta H^{\neq} = E_K - RT_P \tag{4}$$

$$\Delta G^{\neq} = \Delta H^{\neq} - T_P \Delta S^{\neq} \tag{5}$$

where T_p is the peak temperature in the DSC trace, with a heating rate of 15 °C·min^{-1}; K_B and h are the Boltzmann ($K_B = 1.381 \times 10^{-23}$ J/K) and Planck constants (h = 6.626×10^{-34} J/s), respectively; β is the heating rate.

The thermal decomposition of the energetic molecules originates from the activation and rupture of the weakest bond, which is quite significant for the decomposition process. As the temperature of the explosive increases, the molecular thermal motion is enhanced. When the temperature attains a critical point, the weakest bond will be stretched. Subsequently, a rupture occurs. This activation process could be described by the parameters of activation enthalpy ($\Delta H^{\neq}$), activation free energy ($\Delta G^{\neq}$), and activation entropy ($\Delta S^{\neq}$), as calculated by Equations (3)–(5) [27]. $\Delta H^{\neq}$ is the energy that the molecules absorb to transform from a common state to an activated state. Compared with nano-LLM-105, NC/GAP/nano-LLM-105 needs a lower level of energy to be activated. $\Delta G^{\neq}$ is the chemical potential of the activation course. For all of these samples, the values of $\Delta G^{\neq}$ are positive numbers, indicating that none of the activation courses proceed spontaneously [28]. Figure 8e shows a kinetic compensation effect during the thermolysis of NC/GAP, nano-LLM-105, and NC/GAP/nano-LLM-105. The three points do not present a linear relationship, which means that the three samples have disparate kinetic mechanisms of decomposition. In addition, Wang prepared 1,3,5,7-tetranittro-1,3,5,7-tetrazocane/nitrocellulose (HMX/NC) and NC/GAP/CL-20 by the sol-gel method. Similarly, the samples do not have just one exothermic peak, indicating that thermal decomposition for them is carried out in multiple steps. Moreover, E_K of HMX/NC iii and NC/GAP/CL-20 i are 277.68 kJ·mol^{-1} and 296 kJ·mol^{-1}, respectively, which are significantly higher than the E_K of NC/GAP/nano-LLM-105 [29,33]. This is because, compared with the nanofibers that are prepared by electrospinning, the components of an energetic composite prepared by the sol-gel method cannot be tightly combined with each other, and there is a lower mutual promotion of the components for thermal decomposition.

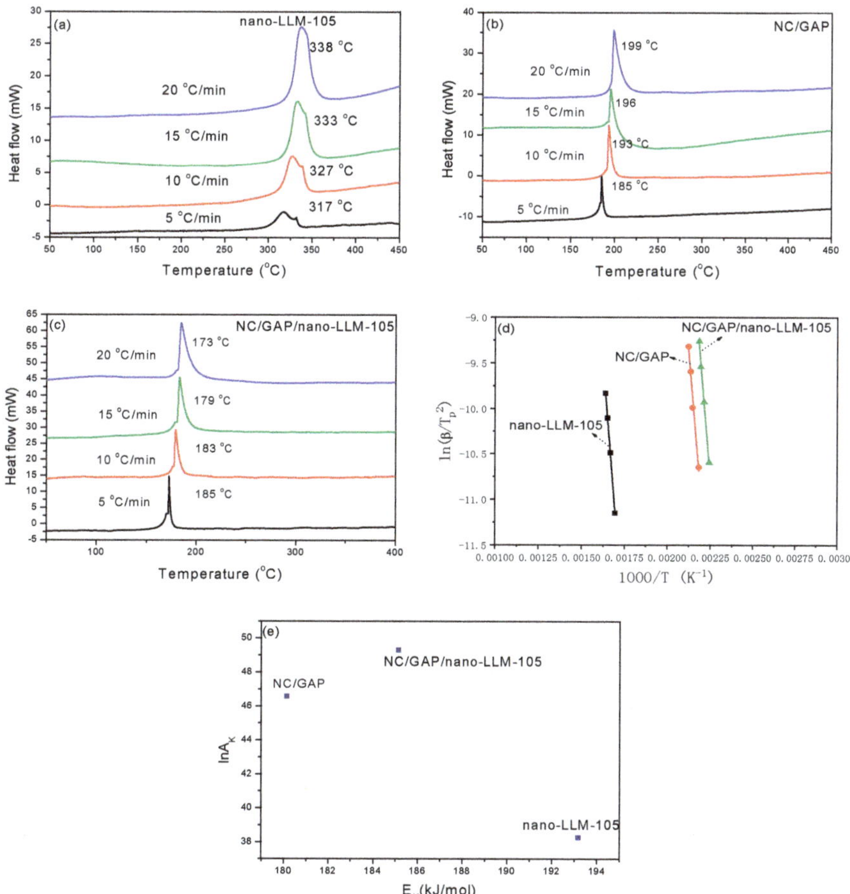

Figure 8. Thermal analyses of the samples. (**a–c**) DSC thermograms of samples collected at different heating rates. (**d**) Kissinger plots of ln(β/Tp^2) to 1000/Tp; (**e**) for the kinetic compensation effect.

Table 3. Thermodynamics and kinetics deduced from DSC traces.

Samples	Tp (K)	Thermodynamics			Kinetics		
		$\Delta H^{\neq}$ (kJ·mol^{-1})	$\Delta G^{\neq}$ (kJ·mol^{-1})	$\Delta S^{\neq}$ (J·mol^{-1}·K^{-1})	E_K (kJ·mol^{-1})	lnA_K	k (s^{-1})
Nano-LLM-105	606	188	152	59	193	38	1.0
NC/GAP	468	176	115	130	180	47	1.4
NC/GAP/Nano-LLM-105	456	181	111	153	185	49	1.7

The products for the thermal decomposition of NC/GAP/nano-LLM-105 and NC/GAP were investigated by TG-IR. The TG and DTG curves are displayed in Figure 9a,b, and the IR spectra at different temperatures are shown in Figure 9c,d. For NC/GAP, the decomposition began at 177 °C, and the decomposition rate reached its maximum at 193.11 °C. The decomposition almost finished at 196.6 °C. For NC/GAP/nano-LLM-105, the initial decomposition temperature rose to 188.65 °C. Composite nanofibers decomposed at the fastest rate at 195.07 °C. Also, the decomposition was generally accomplished at 201.04 °C. This means that the decomposition of NC/GAP/nano-LLM-105 is more concentrated. The IR spectrum, indicated that the main products were diverse. The main peaks and attributions are listed in Table 4. For clarity, the curves are vertically offset in Figure 9c,d. The strong peaks at 2309–2360 cm^{-1} indicated the presence of much CO_2 gas. The peaks in 2113–2199 cm^{-1} represented the appearance of CO. The weak peak located at 2239 cm^{-1} manifested the existence of a very low amount of N_2O gas. The existence of NO explicates the peaks that were located at 1901–1924 cm^{-1}. Moreover, the peaks in 3271–3379 cm^{-1} and 1691–1788 cm^{-1} corresponded to the fragments of -CH and -CH$_2$O, respectively. The –N$_3$ is the energetic group of GAP, which is decomposed to N_2. In reality, N_2 is a nonpolar molecule, and it cannot be probed by IR. For those two nanofibers, the positions of the main peaks are mainly identical. The only difference is the generation of -C-O-C- fragments for NC/GAP nanofibers. However, there are no -C-O-C- fragments for NC/GAP/nano-LLM-105. In addition, compared with NC/GAP, the peak intensities of the -CH$_2$O fragments decreased significantly. This indicates that for NC/GAP/nano-LLM-105, the fragments of -CH$_2$O and -C-O-C- further reacted to generate CO_2, CO, and H_2O. This is because the interposition of the LLM-105 nanoparticles in NC/GAP improved the oxygen balance.

Table 4. Peaks and attributions.

Samples	Peaks and Attribution (cm^{-1})							
	3739–3560	2309–2360	2339	2113–2199	1901–1924	3271–3379	1691–1788	1077–1130
NC/GAP	H_2O	CO_2	N_2O	CO	NO	-C-H	-CH$_2$O	C-O-C
NC/GAP/nano-LLM-105	H_2O	CO_2	N_2O	CO	NO	-C-H	-CH$_2$O	no

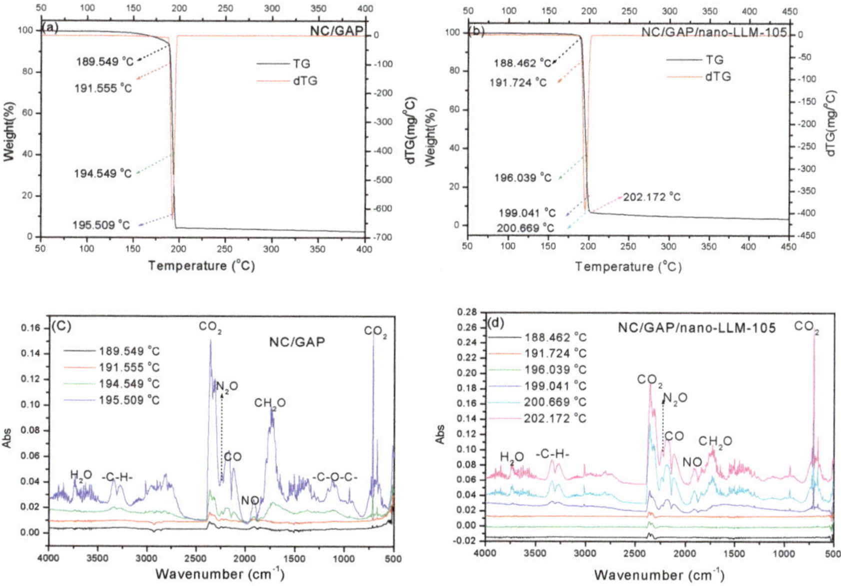

Figure 9. TG-IR analysis of samples: (**a**,**b**) TG and DTG curves; (**c**,**d**) IR spectra of the decomposition products at different temperatures.

3.3. Energetic Performance and Sensitivities

To further explore the energy properties of NC/GAP/nano-LLM-105, their energy performances and impact sensitivities were evaluated. The standard specific impulse (I_{sp}), characteristic velocity (C^*), combustion chamber temperature (T_c), and average molecular weight (M_c) were calculated, the results are listed in Table 5. The functional relationships between the energy performances of NC/GAP/LLM-105 nanofibers and the weight percentage of LLM-105 are shown in Figure 10. For nano-LLM-105, NC/GAP, and NC/GAP/nano-LLM-105, the standard deviations of H_{50} were 4.5, 4.0 and 4.4, respectively. The impact sensitivities of the samples were tested, and the results are displayed in Table 4. Furthermore, the combustion products and their mass molar ratios were calculated, and the results are shown in Figure 11. The feature height (H_{50}) for NC/GAP/nano-LLM-105 was significantly higher than H_{50} for NC/GAP, and it was a little lower than nano-LLM-105. This indicates that the impact sensitivity of NC/GAP/nano-LLM-105 was distinctly lower than that of NC/GAP, and slightly higher than that of nano-LLM-105. For energy performance, the standard specific impulse (I_{sp}) of NC/GAP was 2013.8 N·s·kg^{-1}. For NC/GAP/nano-LLM-105, the I_{sp} value increased to 2032.4 N·s·kg^{-1}. This was not attributed to the high energy of LLM-105, but it was rather due to the higher formation enthalpy (ΔH_f) and C/H values of LLM-105, in contrast to those of NC/GAP. The formation enthalpy of LLM-105 (-13 kJ·mol^{-1}) was observably larger than that of NC/GAP (-294.6 kJ·mol^{-1}). Therefore, after LLM-105 was introduced into NC/GAP, the energy performance was dramatically improved. The value of oxygen balance (OB_{CO2}) was a crucial element for assessing the energy performance. As the OB_{CO2} increased, the energy performance is enhanced. The OB_{CO2} of LLM-105 (-37.03) is higher than that of NC/GAP (-76.1). Hence, the introduction of LLM-105 is in favor of improvement for the energy performance. Additionally, the C/H mass ratios of nano-LLM-105 and NC/GAP are 12 and 8.24, respectively. We infer that the higher OB_{CO2} and C/H are beneficial to enhancement under combustion temperature (T_c) [33,35]. T_c represents the chemical energy storage of an energetic formulation, which is proportional to the explosive heat of propellants. Figure 10c shows that the value of T_c increases as the weight percentage of LLM-105 increases. The energy performance is mainly determined by the heat released from combustion, and the energy conversion efficiency. The latter is related to the hydrogen content in the molecules. The decrease of hydrogen content leads to a decrease of H_2 content. Therefore, the average molecular weight (M_c) increases, which is disadvantageous to energy conversion efficiency. From Figure 11, for NC/GAP, H_2 accounts for 28% in combustion products, and the H_2 proportion of LLM-105 is 19%. Hence, the M_c of LLM-105 is higher than that of NC/GAP in Table 4. Figure 10d shows that the value of M_c increases as the weight percent of LLM-105 increases. In this case, although M_c of LLM-105 is higher than that of NC/GAP, the I_{sp} of LLM-105 is still significantly higher than I_{sp} of NC/GAP. This is because the negative effect of low energy conversion efficiency is offset by the high storage of chemical energy. The higher chemical energy storage of LLM-105 is attributed to its higher C/H mass ratio, OB_{CO2}, and the formation enthalpy. Therefore, the interposition of LLM-105 improves the energy performance of NC/GAP.

Table 5. Impact sensitivity and energy performance of the samples.

Samples	Impact Sensitivity	Energy Performance			
	H_{50} (cm)	I_{sp} (N·s·kg^{-1})	C* (m·s^{-1})	T_c (K)	M_c (g·mol^{-1})
NC(50%)/GAP(50%)	60	2014	1240	1556	22
LLM-105(100%)	113	2171	1393	2453	24
NC(37.5%)/GAP(37.5%)/LLM-105(25%)	78	2027	1253	1640	23

I_{sp} is the standard specific impulse; C* is the characteristic speed; T_c is the combustion chamber temperature; M_c is the average molecular weight of the combustion products. All of the parameters were calculated by the means of software ProPep 3.0 at condition of $P_c/P_e = 70/1$ ($P_e = 0.1$ MPa) and $T_0 = 298$ K.

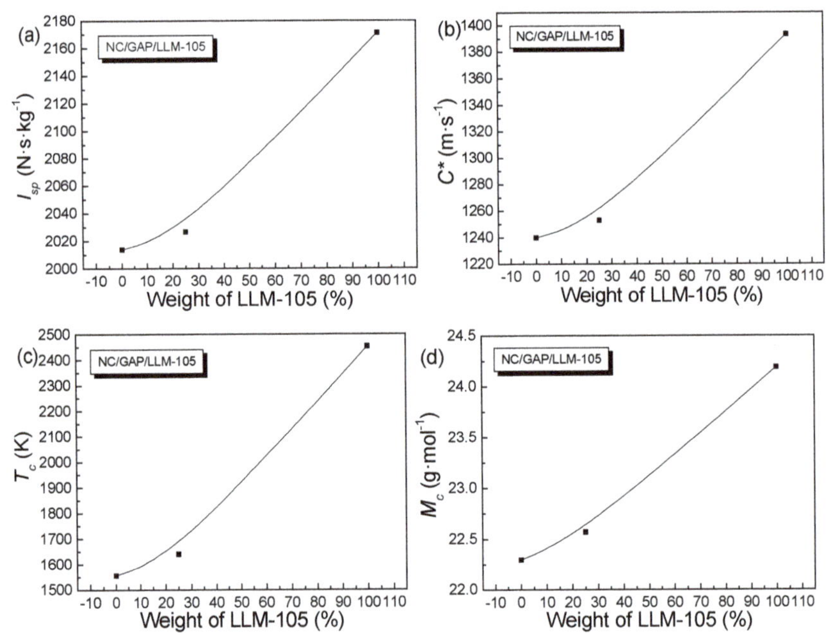

Figure 10. Energy performances of NC/GAP/LLM-105 nanofibers as a function of the weight percentage of LLM-105: (**a**) for I_{sp}; (**b**) for C*; (**c**) for T_c and (**d**) for M_c.

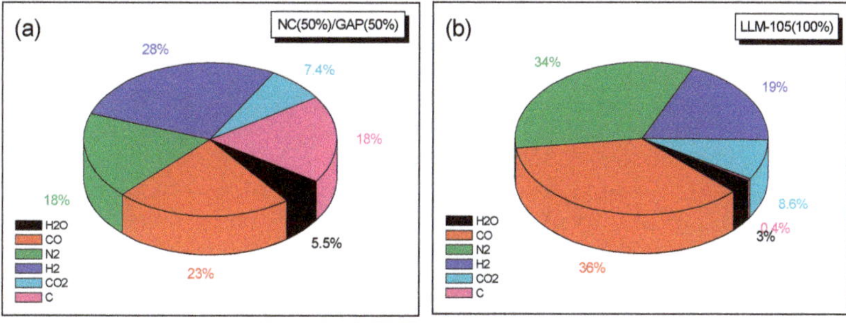
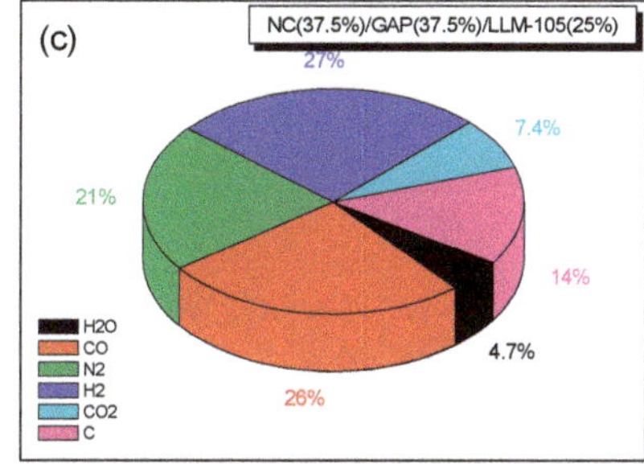

Figure 11. (**a–c**) Combustion products and their molar ratios for NC/GAP/LLM-105 nanofibers. The results in Figure 11 were calculated by the means of the ProPep 3.0 software under conditions of $P_c/P_e = 70/1$ ($P_e = 0.1$ MPa) and $T_0 = 298$ K.

4. Conclusions

The NC/GAP/nano-LLM-105 composite nanofiber with a large specific surface area was prepared by an electrospinning technique. Compared with NC/GAP and LLM-105 nanoparticles, NC/GAP/nano-LLM-105 nanofibers have lower decomposition temperature and distinctly higher decomposition rate. The activation energy of NC/GAP/nano-LLM-105 for thermolysis is lower than that of LLM-105 nanoparticles. These indicate that NC/GAP/nano-LLM-105 decompose relatively easily and violently. If it is used in solid rocket propellant systems, it will decompose first, and then induce the decomposition of other components.

The I_{sp} and T_c of NC/GAP/nano-LLM-105 are higher than those of NC/GAP, which means that NC/GAP/nano-LLM-105 possesses a distinguished energy performance. In addition to the energy performance, safety is another rather crucial factor for energetic materials. The impact sensitivity of NC/GAP/nano-LLM-105 is signally lower than that of NC/GAP. Hence, it has a good safety performance rating. NC/GAP/nano-LLM-105 possess both energy performance and low impact sensitivity. Therefore, this composite nanofiber has enormous potential in the field of solid rocket-propellant systems. This versatile preparation method may provide a concept for synthesizing energetic nanocomposites.

Author Contributions: Y.W. and T.L. conceived and designed the experiments; X.S. and H.H. conducted the experiments; T.L., Y.W., X.S., H.H. and F.S. analyzed the data; F.S. contributed reagents/materials/analysis tools; Y.W. and T.L. wrote the paper.

Funding: This research was supported by the Weapon Equipment Pre-Research Fund of China (Grant No.: 6140656020201).

References

1. Xu, W.Z.; An, C.W.; Wang, J.Y.; Dong, J.; Geng, X.H. Preparation and Properties of An Insensitive Booster Explosive Based on LLM-105. *Propell. Explos. Pyrot.* **2013**, *38*, 136–141. [CrossRef]
2. Ma, H.X.; Song, J.R.; Zhao, F.Q.; Gao, H.X.; Hu, R.Z. Crystal Structure, Safety Performance and Density-Functional Theoretical Investigation of 2,6-Diamino-3, 5-dinitropyrazine-1-oxide (LLM-105). *Chin. J. Chem.* **2008**, *26*, 1997–2002. [CrossRef]
3. Pagoria, P.; Zhang, M.X.; Zuckerman, N.; Lee, G.; Mitchell, A.; DeHope, A.; Clifford, G.; Patrick, C.; Gallagher, P. Synthetic Studies of 2,6-Diamino-3, 5-Dinitropyrazine-1-Oxide (LLM-105) from Discovery to Multi-Kilogram Scale. *Propell. Explos. Pyrot.* **2017**, *42*, 1–14. [CrossRef]
4. Tarver, C.M.; Urtiew, P.A.; Tran, T.D. Sensitivity of 2,6-Diamino-3, 5- Dinitropyrazine-1-Oxide. *J. Energ. Mater.* **2005**, *23*, 183–203. [CrossRef]
5. Manaa, M.R.; Kuo, I.F.W.; Fried, L.E. First-principles high-pressure unreacted equation of state and heat of formation of crystal 2,6-diamino-3, 5-dinitropyrazine-1-oxide (LLM-105). *J. Chem. Phys.* **2014**, *141*, 064702. [CrossRef] [PubMed]
6. Williamson, D.W.; Sue, G.; Taylor, N.E.; Walley, S.M.; Jardine, A.P.; Glauser, A.; French, S.; Wortley, S. Characterisation of the impact response of energetic materials: observation of a lowlevel reaction in 2,6-diamino-3,5-dinitropyrazine-1-oxide (LLM-105). *RSC Adv.* **2016**, *10*, 27896–27900. [CrossRef]
7. Huang, C.; Liu, J.; Ding, L.; Wang, D.J.; Yang, Z.J.; Nie, F. Facile Fabrication of Nanoparticles Stacked 2,6-diamino-3,5-dinitropyrazine-1-oxide(LLM-105) Sub-microspheres via Electrospray Deposition. *Propell. Explos. Pyrot.* **2017**, *42*, 1–7. [CrossRef]
8. Zhang, J.; Wu, P.; Yang, Z.J.; Gao, B.; Zhang, J.H.; Wang, P.; Nie, F.; Longyu Liao, L.H. Preparation and Properties of Submicrometer-Sized LLM-105 via Spray-Crystallization Method. *Propell. Explos. Pyrot.* **2014**, *39*, 653–657. [CrossRef]
9. Deng, P.; Liu, Y.; Luo, P.; Wang, J.X.; Liu, Y.; Wang, D.J.; He, Y. Two-steps synthesis of sandwich-like graphene oxide/LLM-105 nanoenergetic composites using functionalized graphene. *Mater. Lett.* **2017**, *194*, 156–159. [CrossRef]
10. Zhuang, X.B.; Huang, B.; Gao, B.; Qiao, Z.Q. Preparation of rectangular micro-rods by nano-LLM-105 self-assembly. *Chin. J. Energ. Mater.* **2016**, *24*, 433–438. [CrossRef]
11. Kucheyev, S.O.; Gash, A.E.; Lorenz, T. Deformation and fracture of LLM-105 molecular crystals studied by nanoindentation. *Mater. Res. Express.* **2014**, *1*, 025036. [CrossRef]
12. Zuckerman, N.B.; Shusteff, M.; Pagoria, P.F.; Gash, A.E. Microreactor flow synthesis of the secondary high explosive 2, 6-diamino-3, 5-dinitropyrazine-1- oxide (LLM-105). *J. Flow Chem.* **2015**, *5*, 178–182. [CrossRef]
13. Babu, V.J.; Nair, A.S.; Zhu, P.N.; Ramakrishna, S. Synthesis and characterization of rice grains like Nitrogen-doped TiO_2 nanostructures by electrospinning photocatalysis. *Mater. Lett.* **2011**, *65*, 3064–3068. [CrossRef]
14. Kim, C.H.; Kim, B.H.; Yang, K.S. TiO_2 nanoparticles loaded on graphene/carbon composite nanofibers by electrospinning for increased photocatalysis. *Carbon* **2012**, *50*, 2472–2481. [CrossRef]
15. He, T.S.; Zhou, Z.F.; Xu, W.B.; Ren, F.M.; Ma, H.H.; Wang, J. Preparation and photocatalysis of TiO_2–fluoropolymer electrospun fiber nanocomposites. *Polymer* **2009**, *50*, 3031–3036. [CrossRef]
16. An, S.; Joshi, B.N.; Lee, M.W.; Kim, N.Y.; Yoon, S.S. Electrospun graphene-ZnO nanofiber mats for photocatalysis applications. *Appl. Surf. Sci.* **2014**, *294*, 24–28. [CrossRef]
17. Sorayani Bafqi, M.S.; Bagherzadeh, R.; Latifi, M. Fabrication of composite PVDF-ZnO nanofiber mats by electrospinning for energy scavenging application with enhanced efficiency. *J. Polym. Res.* **2015**, *22*, 130. [CrossRef]
18. Bhardwaj, N.; Kundu, S.C. Electrospinning: A fascinating fiber fabrication technique. *Biotech. Adv.* **2010**, *28*, 325–347. [CrossRef]
19. Doshi, J.; Reneker, D.H. Electrospinning process and applications of electrospun fibers. *J. Electrost.* **1995**, *35*, 151–160. [CrossRef]
20. Pourmortazavi, S.M.; Kohsari, I.; Zandavar, H.; Koudehi, M.F.; Mirsadeghi, S. Electrospinning and thermal characterization of nitrocellulose nanofibers containing a composite of diaminofurazan, aluminum nano-powder and iron oxide nanoparticles. *Cellulose* **2019**, *26*, 4405–4415. [CrossRef]

21. Yan, S.; Jian, G.; Zachariah, M.R. Electrospun Nanofiber-Based Thermite Textiles and their Reactive Properties. *ACS Appl. Mater. Inter.* **2012**, *4*, 6432–6435. [CrossRef] [PubMed]

22. Li, M.Y.; Huang, R.H.; Yan, S. Preparation of NC/CL-20 composite fibers by electrospinning. In Proceedings of the 2016 International Workshop on Material Science and Environmental Engineering (IWMSEE2016), Wuhan, China, 22–24 January 2016; pp. 165–172. [CrossRef]

23. Li, Y.C.; Yang, H.T.; Hong, Y.; Yang, Y.; Cheng, Y.; Chen, H.H. Electrospun nanofiber-based nanoboron/nitrocellulose composite and their reactive properties. *J. Therm. Anal. Calorim.* **2017**, *130*, 1063–1068. [CrossRef]

24. Selim, K.; Özkar, S.; Yilmaz, L. Thermal characterization of glycidyl azide polymer (GAP) and GAP-based binders for composite propellants. *J. Appl. Polym. Sci.* **2000**, *77*, 538–546. [CrossRef]

25. You, J.S.; Kweon, J.O.; Kang, S.C.; Noh, S.T. A kinetic study of thermal decomposition of glycidyl azide polymer (GAP)-based energetic thermoplastic polyurethanes. *Macromol. Res.* **2010**, *18*, 1226–1232. [CrossRef]

26. Nazare, A.N.; Asthana, S.N.; Singh, H. Glycidyl Azide Polymer (GAP)-An Energetic Component of Advanced Solid Rocket Propellants-A Review. *J. Energ. Mater.* **1992**, *10*, 43–63. [CrossRef]

27. Li, J.Y.; Zhang, H.B.; Xu, J.J.; Sun, J. IR Absorption Peaks Assignments of LLM-105 by Temperature Dependent FT-IR Spectroscopy. *Chin. J. Energ. Mater.* **2015**, *23*, 507–510. [CrossRef]

28. Shin, W.G.; Han, D.; Park, Y.; Hyun, H.S.; Sung, H.G.; Sohn, Y.K. Combustion of boron particles coated with an energetic polymer material. *Korean J. Chem. Eng.* **2016**, *33*, 3016–3020. [CrossRef]

29. Wang, Y.; Song, X.L.; Song, D.; Liang, L.; An, C.W.; Wang, J.Y. Synthesis, thermolysis, and sensitivities of HMX/NC energetic nanocomposites. *J. Hazard. Mater.* **2016**, *312*, 73–83. [CrossRef]

30. Song, X.L.; Wang, Y.; An, C.W.; Guo, X.D.; Li, F.S. Dependence of particle morphology and size on the mechanical sensitivity and thermal stability of octahydro-1,3,5,7-tetranitro-1,3,5,7-tetrazocine. *J. Hazard. Mater.* **2008**, *159*, 222–229. [CrossRef]

31. Lin, C.M.; Gong, F.Y.; Yang, Z.J.; Pan, L.P.; Liu, S.J.; Li, J.; Guo, S.Y. Bio-inspired fabrication of core@shell structured TATB/polydopamine microparticles via in situ polymerization with tunable mechanical properties. *Polym. Test.* **2018**, *68*, 126–134. [CrossRef]

32. Lin, Z.Y.; Waller, G.H.; Liu, Y.; Liu, M.L.; Wong, C.P. 3D Nitrogen-doped graphene prepared by pyrolysis of graphene oxide with polypyrrole for electrocatalysis of oxygen reduction reaction. *Nano Energy* **2013**, *2*, 241–248. [CrossRef]

33. Wang, Y.; Zhang, M.; Song, X.L.; Huang, H.; Li, F.S. Characteristics and properties of nitrocellulose/glycidyl azide polymer/2, 4, 6, 8, 10, 12-hexanitro-2, 4, 6, 8, 10, 12-hexaazaisowurtzitane nanocomposites synthesized using a sol–gel supercritical method. *Nanomater. Nanotechno.* **2019**, *9*, 1–12. [CrossRef]

34. Ye, B.Y.; An, C.W.; Zhang, Y.R.; Song, C.K.; Geng, X.H.; Wang, J.Y. One-Step Ball Milling Preparation of Nanoscale CL-20/Graphene Oxide for Significantly Reduced Particle Size and Sensitivity. *Nanoscale Res. Lett.* **2018**, *13*, 42. [CrossRef] [PubMed]

35. Wang, Y.; Song, X.L.; Li, F. Thermal Behavior and Decomposition Mechanism of Ammonium Perchlorate and Ammonium Nitrate in the Presence of Nanometer Triaminoguanidine Nitrate. *ACS Omega* **2019**, *4*, 214–225. [CrossRef]

 *nanomaterials*

Brief Report

Nanometer Ammonium Perchlorate and Ammonium Nitrate Prepared with 2D Network Structure via Rapid Freezing Technology

Yi Wang [1,*,†], Xiaolan Song [2,*,†] and Fengsheng Li [3]

[1]   School of Materials Science and Engineering, North University of China, Taiyuan 030051, China
[2]   School of Environment and Safety Engineering, North University of China, Taiyuan 030051, China
[3]   School of Chemical Engineering, Nanjing University of Science and Technology, Nanjing 210094, China; lifengsheng424@163.com
*   Correspondence: wangyi528528@aliyun.com (Y.W.); songxiaolan00@126.com (X.S.)
†   These authors contributed equally to this work.

Received: 30 October 2019; Accepted: 7 November 2019; Published: 12 November 2019

Abstract: Nanometer (nano) ammonium perchlorate (AP) and ammonium nitrate (AN) were prepared with 2D network structures by the ultra-low temperature spray method. Scanning electron microscopy (SEM), X-ray diffractometry (XRD), differential scanning calorimetry (DSC) and thermogravimetric analysis/infrared spectrometry (TG-IR) were employed to probe the micron structure, crystal phase, and thermal decomposition of nano AP and nano AN. SEM images revealed that the sizes of nano AP and AN were in the nanometer scale (<100 nm) in one dimension. XRD patterns showed that the crystal phases of nano AP and AN were in accordance with those of raw AP and raw AN, respectively. DSC traces indicated that the thermal decomposition process of AP depended on its particle size, while the thermolysis of AN was independent of the particle size of AN. TG-IR analyses illustrated that the decomposition products of nano AP were NO_2, N_2O, HCl and H_2O, with a small amount of NOCl, and the main decomposition products of nano AN were N_2O and H_2O, with a small amount of NH_3. The results of mechanical sensitivity tests indicated that nano AP was more sensitive than raw AP and both nano AN and raw AN were very insensitive to impact and friction stimuli.

Keywords: nano AP; nano AN; liquid nitrogen; freeze drying; thermolysis

---

1. Introduction

Ammonium perchlorate (AP) and ammonium nitrate (AN) are the most commonly used oxidizers applied in solid propellants [1–3]. In addition to their use in propellants, AP and AN are also employed as oxygen-enriched ingredients in explosives and pyrotechnics [4–6]. Researchers have found that the performance of AP and AN is a function of their particle sizes. For example, the burning rate of a propellant using superfine AP as an oxidizer is obviously higher than that of the propellant using coarse AP as an oxidizer [7]. However, in contrast to AP, the burning rate of propellants that use AN as an oxidizer is independent of the particle size of AN [8]. Interestingly, a decrease in particle size of AN is beneficial as it decreases the combustion pressure exponent of AN-based propellants [9]. The mechanism concerning this is still obscure, of course, and is also not the theme of this work.

Kumari et al. prepared nanometer (nano) AP by a precipitation method, and the nano AP was precipitated out in the form of nanoparticles in an HTPB (hydroxy terminated polybutadiene) matrix [10]. The HTPB matrix limited the growth of the particles, resulting in a mean size of 33 nm for the nano AP. This method was very efficient because the yield of nano AP was up to 85% (8.5 g per batch). Frolov et al. prepared nano-sized powders of ammonium nitrate through vacuum deposition on a cooled quartz substrate [11]. The AN crystallites were fairly uniform in size and shape and less than 50 nm in diameter. Aside from these two studies, the literature provides a number of other examples on how to prepare superfine AP and AN, which are also of great reference significance. For instance, Ma et al. used acetone as a solvent and ethyl acetate as a non-solvent to prepare superfine AP particles with sizes of 5~10 μm [12]. Wu et al. used an air jet pulverization method to fabricate superfine AP particles with very narrow particle size distribution [13]. Song et al. fabricated spherical superfine AP particles ($d_{50} \approx 5$ μm) via two methods [14]. Specifically, coarse AP particles were first comminuted to fine particles ($d_{50} \approx 20$ μm) by an air jet pulverization method, and then the fine particles were further pulverized by a mechanical milling method to obtain superfine particles. The high spheroidization degree of such particles resulted in a high density propellant. Compared with coarse AP, the impact sensitivity and friction sensitivity of the superfine AP deceased by 32% and 22%, respectively. Furthermore, differential scanning calorimetry (DSC) indicated that decomposition of the superfine AP occurred earlier than it did for coarse AP, and that the decomposition activation energy of the superfine AP was lower than that of coarse AP. There are few reports on superfine AN. Chai et al. fabricated superfine AN with a particle size of 2~10 μm by a mechanical milling method [15]. In this case, acetone was used as a grinding medium and, after milling, nitrocellulose was coated on the surface of the superfine particle to reduce the hygroscopicity of AN.

The abovementioned researches suggest that mechanical milling may be the best method for fabricating nano AP and nano AN. However, so far, this method has been incapable of pulverizing coarse AP and AN into nanoparticles because the inherent texture of AP and AN crystals determine that they cannot be pulverized to nanoparticles by a mechanical method. In fact, many nano explosives have been fabricated using a mechanical milling method, such as nano RDX [14], nano HMX [16,17], nano CL-20 [18,19], nano PETN [19], nano HNS [20,21], nano TATB [20,22], and nano TAGN [23], with the exception of nano AP and nano AN. Thus, the current study aimed to fabricate AP and AN with nanoscale structures via a liquid nitrogen-assisted and vacuum freeze-drying method.

2. Experimental Method

2.1. Materials

For this study, ammonium perchlorate (AP) and ammonium nitrate (AN) were purchased from Tianjin Guangfu Chemical Co., Ltd. (Tianjin, China) and liquid nitrogen was purchased from Taiyuan Taineng Gas Co., Ltd. (Taiyuan, China). A mini high pressure atomization pump (ZEKUN, ZK-PW-XT, Shanghai, China), used in the preparation process, was manufactured by Shanghai Zekun Environmental Protection Technology Co., Ltd. (Shanghai, China).

2.2. Fabrication of Nano AP and Nano AN

We dissolved 10 g of AP (or 20 g AN) in 190 g (or 180 g for AN) deionized water. The resulting aqueous solution was loaded into the mini high pressure atomization pump. An open container was used to hold about 200 mL of liquid nitrogen. The nozzle of the atomization pump was then aligned with the liquid nitrogen and the pump was started. After all aqueous solutions were sprayed into the liquid nitrogen, the pumped was turned off. As the temperature of the liquid nitrogen was −196 °C and the water and liquid nitrogen were not miscible, all the droplets were frozen to their freezing points instantly. In the preparation process, the liquid nitrogen boiled violently. When all the liquid nitrogen evaporated off, the open container was placed in a freeze dryer and drying commenced. One week later, when all the ice had sublimated, nano AP (or nano AN) was obtained.

2.3. Characterization and Tests

Surface morphologies of nano AP and nano AN were investigated using scanning electron microscopy (SEM; JEOL JSM-7500, Tokyo, Japan), while the crystal phases were studied with X-ray diffractometry (XRD; Bruker Advance D8, Karlsruhe, Germany). Further, differential scanning calorimetry (DSC; DSC-100, Nanjing DAZHAN company, Nanjing, China) was employed to investigate the thermal decomposition of samples and thermogravimetric analysis/infrared spectrometry (TG-IR; Mettler Toledo, Zurich, Switzerland) was used to investigate the decomposition products of samples.

The impact sensitivity of the samples was tested with an HGZ-1 impact instrument (North University of China, Taiyuan, China). The special height (H_{50}) represents the height at which a 5 kg drop-hammer will cause an explosive event in 50% of the trials. In each determination, 25 drop tests were carried out to calculate the H_{50}. The friction sensitivity of the samples was tested with a WM-1 friction instrument. In each determination, 50 samples were tested, and the explosion probability (P, %) was obtained.

3. Results and Discussion

SEM images of nano AP and nano AN are shown in Figure 1. In Figure 1a it is clear that there are many spherical-like "particles" with sizes of 5~10 μm. After increasing the magnification of the image, as shown in Figure 1b, it can be seen that the "particles" are hollow with rough surfaces. After further increasing the magnification of the image, as shown in Figure 1c, it can be observed that the surfaces of the "particles" present a 2D network structure with one-dimensional nanometer size (<100 nm). Figure 1d–f reveals that the micron morphology and structure of nano AN is similar to that of nano AP.

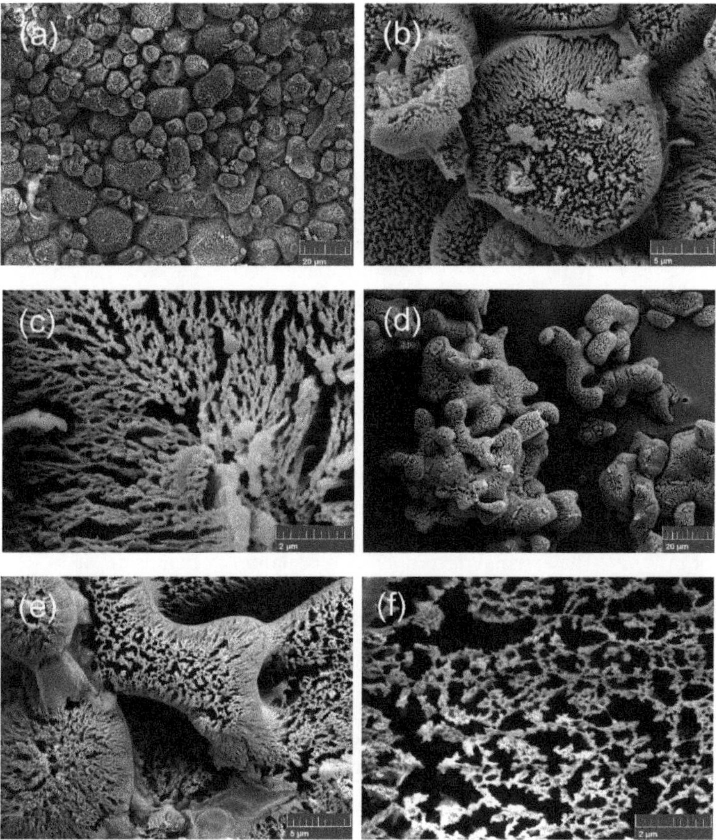

Figure 1. Scanning electron microscopy (SEM) images of samples: (**a–c**) nanometer (nano) ammonium perchlorate (AP) and (**d–f**) nano ammonium nitrate (AN).

In the fabrication process, the only raw materials were the oxidizer, water, and liquid nitrogen. The water and liquid nitrogen were subsequently evaporated. Thus, it is suggested that no impurity was introduced. However, it was not clear whether the crystal phases of AP and AN transformed in the process. In terms of their practical application, phase transformation is undesired. Consequently, we performed XRD on raw AP, nano AP, raw AN, and nano AN, with the patterns from these analyses illustrated in Figure 2. It is clear that the crystal phases of nano AP and nano AN are consistent with the phases of raw AP and raw AN, respectively. Therefore, we were able to conclude that no phase transformation occurred in the fabrication process.

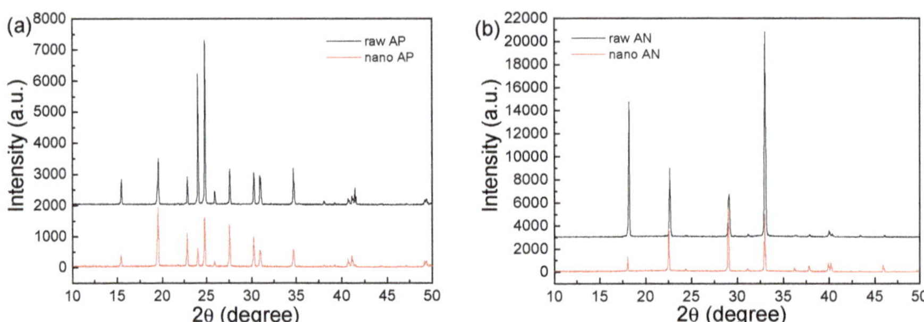

Figure 2. X-ray diffractometry (XRD) patterns of samples: (**a**) raw and nano AP and (**b**) raw and nano AN.

The DSC traces of raw AP, nano AP, raw AN, and nano AN are shown in Figure 3, produced using a heating rate of 20 °C/min. In Figure 3a, each curve has an endothermic peak at 245 °C, corresponding to the phase transformation of AP. For raw AP, low temperature decomposition occurs in the range of 310–360 °C. However, for nano AP, low temperature decomposition does not occur. Further, the high temperature decomposition of raw AP begins at 446.7 °C and the peak point appears at 451 °C, but the high temperature decomposition of nano AP originates at 387.8 °C and the peak point appears at 439.2 °C. This means that the thermolysis occurred earlier for nano AP than raw AP. It can be seen from Figure 3b that nano AN is quite similar to raw AN in terms of DSC trace. For raw AN, the peaks at 60.4 and 135.1 °C correspond to a phase transformation of AN. The peak at 170.1 °C is related to the melting point of AN. The large endothermic peak at 301.7 °C may be ascribed to the thermal decomposition of raw AN. Upon careful observation, it may be observed that there are four small endothermic peaks in the DSC curve of nano AN. For nano AN, an extra phase transformation occurs at 96.7 °C, which does not exist in the DSC curve of raw AN. The melting point of nano AN is 169.9 °C, which is roughly the same as that of raw AN. The thermal decomposition of nano AN begins at 245.9 °C, slightly lower than raw AN's 254.2 °C. The peak temperature for thermolysis of nano AN (300.2 °C) is quite close to that of raw AN (301.7 °C), indicating that the decomposition process does not change when the particle size of AN decreases from the micron scale to the nano scale.

With reference to our results, we noted that the thermal decomposition of AN was independent of its particle size, while the thermolysis of AP depended on its particle size. This may be attributed to differences in the decomposition mechanisms of AP and AN. In the first step of their decomposition, AP and AN dissociate to NH_3 and $HClO_4$ (for AP) and NH_3 and HNO_3 (for AN) when the temperature reaches a critical point. NH_3 cannot directly react with $HClO_4$ or HNO_3 at low temperatures. So, further decomposition of $HClO_4$ or HNO_3 is essential for sustained thermolysis of AP or AN. However, the decomposition of $HClO_4$ is much easier than that of HNO_3, as the decomposition process of $HClO_4$ is very exothermic and the decomposition process of HNO_3 is very endothermic [24,25]. Therefore, the rate limiting step of the decomposition of AP lies in the reaction between NH_3 and the pyrolysis products of $HClO_4$, while the controlling step of AN's decomposition is the pyrolysis of HNO_3 [26]. In fact, the inferior decomposition mechanism of AN accounts for its remarkably lower burning rate compared to AP. The problem for the decomposition of AP lies in that the NH_3 and pyrolysis products of $HClO_4$ react with each other only on the surface of the decomposing AP particles. The excessive accumulation of NH_3 gas on the surfaces of AP particles forms a thick ammonia cage, thereby hindering further decomposition reactions [24]. Thus, increasing the specific surface area of AP particles is conducive to alleviating the negative effects of the ammonia cage, therefore promoting the reaction between NH_3 and the pyrolysis products of $HClO_4$. Aside from the case of AP, the reaction between NH_3 and the pyrolysis products of HNO_3 is instantaneous, which is partly responsible for the detonable property of AN [27]. Hence, thermolysis of AN does not have problems like the ammonia cage. Please note that AP is not detonable partly because the reaction between NH_3 and the pyrolysis products of $HClO_4$ is not instantaneous. These mechanisms illustrate the discrepancy observed in the DSC results in Figure 3a,b.

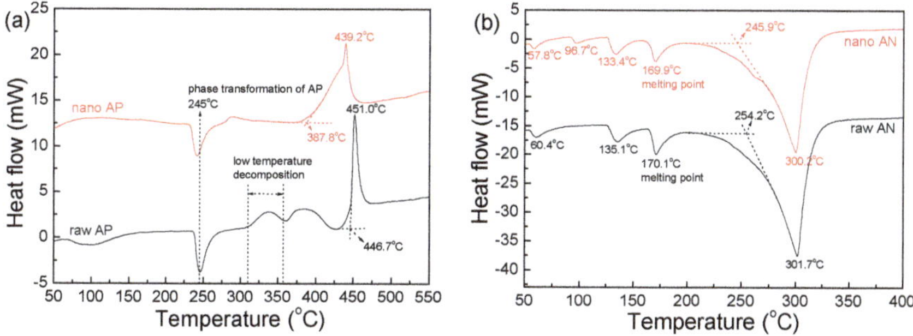

Figure 3. Differential scanning calorimetry (DSC) traces of samples: (**a**) raw and nano AP and (**b**) raw and nano AN.

To further investigate the thermal decomposition of nano AP and nano AN, TG-IR analysis was conducted at a heating rate of 10 °C/min, the results of which are shown in Figure 4. As observable in Figure 4a, which shows the TG curve of nano AP, decomposition begins at 360 °C and ends at 434 °C. This result is slightly different from the DSC trace (Figure 3a) due to the different heating rates employed in the DSC and TG-IR analyses. The IR spectra at 400, 414, 436, and 450 °C were extracted and are shown in Figure 4b. From the figure, it is obvious that the main gas products of nano AP are NO_2, N_2O, HCl, and H_2O, while some NOCl is also detected. For nano AN, its decomposition products are N_2O and a massive amount of H_2O (steam). It should be pointed out that some NH_3 is produced in the case of nano AN, but it is not present in the decomposition products of nano AP. NH_3 is produced as a product of the dissociation of AN [28]. In fact, most ammonia is instantaneously oxidized by the NO_2 produced from the pyrolysis of HNO_3, which gives rise to the formation of large amounts of N_2O and H_2O [29]. Nevertheless, it is hard for NH_3 to react with N_2O at low temperatures. So, when the concentration of N_2O is low, a small amount of NH_3 remains. In fact, TG-IR analysis of the decomposition of micron AP and AN was conducted in our other papers, and the papers of other authors reported similar results. For example, in Ref. [30], Liu et al. found that micron AP decomposed to NO_2, N_2O, HCl, H_2O, and NOCl, which is in accordance with the results of this work. In Ref. [26], the micron AN decomposed to N_2O and H_2O, which is also consistent with the results of this paper. Therefore, on the aspect of final decomposition products, it is believed that the decomposition of AP or AN happens at the molecular level, independent of the particle sizes of AP and AN.

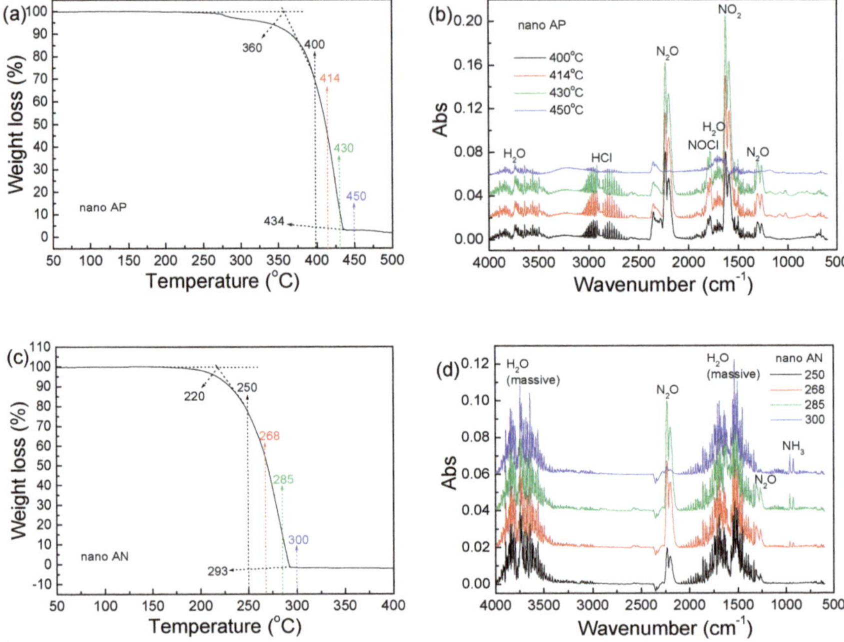

Figure 4. Thermogravimetric analysis/infrared spectrometry (TG-IR) spectra of samples: (**a,c**) TG curves and (**b,d**) IR spectra for decomposition products of nano AP and nano AN.

Impact and friction sensitivities of raw AP, nano AP, raw AN, and nano AN were also tested and the results are presented in Table 1. The results show that when the particle size of AP or AN was reduced to the nanometer scale, their sensitivities change significantly. The impact sensitivity of nano AP was slightly higher than that of raw AP, and the friction sensitivity of nano AP was noticeable higher than that of raw AP. However, for AN, the increase in sensitivity was not obvious. The impact sensitivity of nano AN was similar to that of raw AN, and the friction sensitivity of nano AN was somewhat higher than that of raw AN. This is because AN itself is a very insensitive, energetic material. Of course, these results are not consistent with the results reported by Luo or Dobrynin [31,32]. In Luo's work, the addition of nano LLM-105 was beneficial as it decreased the impact sensitivity of NC/GAP (nitrocellulose/glycidyl azide polymer) fibers, and in Dobrynin's work nano NC presented lower friction sensitivity than micron NC. Now, it cannot yet be elucidated why nano AP is more sensitive than raw AP, but nano NC is less sensitive than raw NC. However, at this time, this discrepancy may simply be attributed to the different decomposition mechanisms of AP and NC (NC is nitrocellulose as a nitrate ester) [33]. The decomposition of AP begins when it dissociates to NH_3 and $HClO_4$, while decomposition of NC originates from the rupture of $O–NO_2$ bonds. Further research on this topic will be continued in our next work.

Table 1. Impact and friction sensitivities of samples.

Samples	Impact Sensitivity (H_{50}, cm)	Friction Sensitivity (P, %)
raw AP	95	26
nano AP	83	94
raw AN	>120	0
nano AN	>120	8

4. Conclusions

Nano AP and nano AN, with structures that were in the nanometer scale in one dimension, were prepared with an ultra-low temperature spray method. The micron morphologies of nano AP and nano AN were 2D network structures. XRD analysis confirmed that crystal phase transformation did not occur in the fabrication process. For DSC curves, compared with raw AP, the peak temperature of nano AP was larger by 12 °C. The thermal decomposition of AN was found to be independent of its particle size; that is, the DSC peak point of nano AN was roughly the same as that of raw AN. The decomposition products of nano AP were NO_2, N_2O, HCl, and H_2O, as well as a small amount of NOCl, while the main decomposition products of nano AN were N_2O and H_2O, with a small amount of NH_3. The impact sensitivity of nano AP was somewhat higher than that of raw AP and the friction sensitivity of nano AP was noticeably higher than that of raw AP. But, in the case of AN, the increase of sensitivity was not obvious.

Author Contributions: Conceptualization, Y.W. and X.S.; methodology, Y.W.; validation, Y.W., X.S. and F.L.; formal analysis, Y.W. and X.S.; investigation, Y.W. and X.S.; resources, Y.W.; data curation, Y.W.; writing—original draft preparation, Y.W.; writing—review and editing, X.S.; F.L. analyzed the sensitivity experiments.

Funding: This research received no external funding.

Conflicts of Interest: The authors declare no conflict of interest.

References

1. Dennis, C.; Bojko, B. On the combustion of heterogeneous AP/HTPB composite propellants: A review. *Fuel* **2019**, *254*, 115646. [CrossRef]
2. Kohga, M.; Togo, S. Catalytic Effect of added Fe_2O_3 amount on thermal decomposition behaviors and burning characteristics of ammonium nitrate/ammonium perchlorate propellants. *Combust. Sci. Technol.* **2019**. [CrossRef]
3. Jos, J.; Mathew, S. Ammonium nitrate as an eco–friendly oxidizer for composite solid propellants: promises and challenges. *Crit. Rev. Solid State Mater. Sci.* **2017**, *42*, 470–498. [CrossRef]
4. Kotomin, A.A.; Dushenok, S.A.; Ilyushin, M.A. Detonation velocity of highly dispersed ammonium perchlorate and its mixtures with explosive substances. *Combust. Explos. Shock Waves* **2017**, *53*, 353–357. [CrossRef]
5. Juknelevicius, D.; Mikoliunaite, L.; Sakirzanovas, S. A Spectrophotometric Study of Red Pyrotechnic Flame Properties Using Three Classical Oxidizers: Ammonium Perchlorate, Potassium Perchlorate, Potassium Chlorate. *Z. Anorg. Allg. Chem.* **2014**, *640*, 1560–1565. [CrossRef]
6. Biessikirski, A. Analysis of morphological properties of ANFO explosives based on various types of ammonium nitrate(V) and their mixtures. *Przem. Chem.* **2018**, *97*, 1689–1692.
7. Mehilal, S.J.; Nandagopal, S.; Singh, P.P. Size and shape of ammonium perchlorate and their influence on properties of composite propellant. *Def. Sci. J.* **2009**, *59*, 294–299.
8. Htwe, Y.Z.; Denysiuk, A.P. Combustion of double-base propellants of various compositions containing ammonium nitrate. *Combust. Explos. Shock Waves* **2013**, *49*, 288–298. [CrossRef]
9. Oommen, C.; Jain, S.R. Ammonium nitrate: a promising rocket propellant oxidizer. *J. Hazard. Mater.* **1999**, *67*, 253–281. [CrossRef]
10. Kumari, A.; Mehilal; Jain, S. Nano-ammonium perchlorate: Preparation, characterization, and evaluation in composite propellant formulation. *J. Energ. Mater.* **2013**, *31*, 192–202. [CrossRef]
11. Frolov, Y.V.; Pivkina, A.N.; Zav'yalov, S.A. Physicochemical characteristics of the components of energetic condensed systems. *Combust. Explos. Shock Waves* **2010**, *4*, 916–922. [CrossRef]
12. Ma, Z.; Li, C.; Wu, R.; Chen, R.; Gu, Z. Preparation and characterization of superfine ammonium perchlorate(AP) crystals through ceramic membrane anti-solvent crystallization. *J. Cryst. Growth* **2009**, *311*, 4575–4580. [CrossRef]
13. Wu, F.; Guo, X.; Jiao, Z. Preparation and properties of ultra-fine ammonium perchlorate composite particles. *Chin. J. Energ. Mater.* **2016**, *24*, 261–268.

14. Song, J.; Guo, X.; Li, F. Preparation and property characterization of superfine spherical ammonium perchlorate. *J. Solid Rocket Technol.* **2014**, *37*, 521–524.

15. Chai, T.; Yang, X.; Wang, J. Supercritical Fluid Coating Modification of Ammonium Nitrate. China Patent ZL 200,910,074,537.1, 4 September 2009.

16. Wang, Y.; Jiang, W.; Song, D. A feature on ensuring safety of superfine explosives: the similar thermolysis characteristics between micro and nano nitroamines. *J. Therm. Anal. Calorim.* **2013**, *111*, 85–92. [CrossRef]

17. Wang, Y.; Jiang, W.; Song, X. Insensitive HMX (octahydro-1,3,5,7-tetranitro-1,3,5,7-tetrazocine) nanocrystals fabricated by high-Yield, low-cost mechanical milling. *Cent. Eur. J. Energ. Mater.* **2013**, *10*, 3–15.

18. Wang, Y.; Song, X.; Song, D. A versatile methodology using sol-gel, supercritical extraction, and etching to fabricate a nitroamine explosive: nanometer HNIW. *J. Energ. Mater.* **2013**, *31*, 49–59. [CrossRef]

19. Song, X.; Wang, Y.; An, C. Thermochemical properties of nanometer CL-20 and PETN fabricated using a mechanical milling method. *AIP Adv.* **2018**, *8*, 065009. [CrossRef]

20. Song, X.; Wang, Y.; Zhao, S. Characterization and thermal decomposition of nanometer 2,2′,4,4′,6,6′-hexanitro-stilbene and 1,3,5-triamino-2,4,6-trinitrobenzene fabricated by a mechanical milling method. *J. Energ. Mater.* **2018**, *36*, 179–190. [CrossRef]

21. Song, X.; Wang, Y.; Liu, L. Thermal decomposition performance of nano HNS fabricated by mechanical milling method. *Chin. J. Energ. Mater.* **2016**, *24*, 1188–1192.

22. Song, X.; Wang, Y.; Liu, L. Characterization of nano TATB fabricated by mechanical milling methodology. *J. Solid Rocket Technol.* **2017**, *40*, 471–475.

23. Zhao, S.; Song, X.; Wang, Y. Characterization of nanometer TAGN prepared by mechanical milling method. *Initiat. Pyrotech.* **2017**, *15*, 22–25.

24. Jacobsetc, P.W.M.; Pearson, G.S. Mechanism of the decomposition of ammonium perchlorate. *Combust. Flame* **1969**, *13*, 419–430. [CrossRef]

25. Izato, Y.-I.; Miyake, A. Thermal decomposition mechanism of ammonium nitrate and potassium chloride mixtures. *J. Therm. Anal. Calorim.* **2015**, *121*, 287–294. [CrossRef]

26. Wang, Y.; Song, X.; Li, F. Thermal behavior and decomposition mechanism of ammonium perchlorate and ammonium nitrate in presence of nanometer triaminoguanidine nitrate. *ACS Omega* **2019**, *4*, 214–225. [CrossRef] [PubMed]

27. Jain, S.R. Self-igniting fuel-oxidizer systems and hybrid rockets. *J. Sci. Ind. Res.* **2003**, *62*, 293–310.

28. Chaturvedi, S.; Dave, P.N. Review on thermal decomposition of ammonium nitrate. *J. Energ. Mater.* **2013**, *31*, 1–26. [CrossRef]

29. Sinditskii, V.P.; Egorshev, V.Y.; Levshenkov, A.I.; Serushkin, V.V. Ammonium nitrate: combustion mechanism and the role of additives. *Propellants Explos. Pyrotech.* **2005**, *30*, 269–280. [CrossRef]

30. Liu, Z.; Yin, C.; Kong, Y. The thermal decomposition of ammonium perchlorate. *Chin. J. Energ. Mater.* **2000**, *8*, 75–79.

31. Luo, T.; Wang, Y.; Huang, H. An electrospun preparation of the NC/GAP/nano-LLM-105 nanofiber and its properties. *Nanomaterials* **2019**, *9*, 854. [CrossRef] [PubMed]

32. Dobrynin, O.S.; Zharkov, M.N.; Kuchurov, I.V. Supercritical antisolvent processing of nitrocellulose: downscaling to nanosize, reducing friction sensitivity and introducing burning rate catalyst. *Nanomaterials* **2019**, *9*, 1386. [CrossRef] [PubMed]

33. Trache, D.; Tarchoun, A.F. Stabilizers for nitrate ester-based energetic materials and their mechanism of action: a state-of-the-art review. *J. Mater. Sci.* **2018**, *53*, 100–123. [CrossRef]

 *nanomaterials*

Article

Supercritical Antisolvent Processing of Nitrocellulose: Downscaling to Nanosize, Reducing Friction Sensitivity and Introducing Burning Rate Catalyst

Oleg S. Dobrynin [1], Mikhail N. Zharkov [1], Ilya V. Kuchurov [1], Igor V. Fomenkov [1], Sergey G. Zlotin [1], Konstantin A. Monogarov [2], Dmitry B. Meerov [2], Alla N. Pivkina [2] and Nikita V. Muravyev [2,*]

[1]   N.D. Zelinsky Institute of Organic Chemistry, Russian Academy of Sciences, Moscow 119991, Russia; oleg_dobrynin@list.ru (O.S.D.); m.n.zharkov@gmail.com (M.N.Z.); kuchurov@mail.ru (I.V.K.); ootx@ineos.ac.ru (I.V.F.); zlotin@ioc.ac.ru (S.G.Z.)

[2]   N.N. Semenov Federal Research Center for Chemical Physics, Russian Academy of Sciences, Moscow 119991, Russia; k.monogarov@gmail.com (K.A.M.); mmeerov@mail.ru (D.B.M.); alla_pivkina@mail.ru (A.N.P.)

*   Correspondence: n.v.muravyev@ya.ru; Tel.: +7-499-137-8203

Received: 28 August 2019; Accepted: 26 September 2019; Published: 27 September 2019

Abstract: A supercritical antisolvent process has been applied to obtain the nitrocellulose nanoparticles with an average size of 190 nm from the nitrocellulose fibers of 20 μm in diameter. Compared to the micron-sized powder, nano-nitrocellulose is characterized with a slightly lower decomposition onset, however, the friction sensitivity has been improved substantially along with the burning rate increasing from 3.8 to 4.7 mm·s^{-1} at 2 MPa. Also, the proposed approach allows the production of stable nitrocellulose composites. Thus, the addition of 1 wt.% carbon nanotubes further improves the sensitivity of the nano-nitrocellulose up to the friction-insensitive level. Moreover, the simultaneous introduction of carbon nanotubes and nanosized iron oxide catalyzes the combustion process evidenced by a high-speed filming and resulting in the 20% burning rate increasing at 12 MPa. The presented approach to the processing of energetic nanomaterials based on the supercritical fluid technology opens the way to the production of nitrocellulose-based nanopowders with improved performance.

Keywords: nitrocellulose; supercritical antisolvent process; nanoparticles; combustion

---

1. Introduction

Nitrocellulose (NC) is widely used as a base substance for conservation, adhesives, membranes and coatings [1–5] with an important application for solid rocket propellants and gunpowder production [6–8]. Long-term experience with nitrocellulose reveals some of its limitations, including certain thermal and mechanical hazards [9], inhomogeneity problems resulting from the organic nature of the cellulose precursor [10], and poor combustion performance. These challenges can be addressed by downscaling nitrocellulose to nanosized powder that should improve the burning rate, as was reported for another energetic substance, 1,3,5-trinitro-1,3,5-triazinane (RDX), whose nanoparticles demonstrate a twofold burning rate increase [11]. Also, the available data for the nanosized energetic materials show that their mechanical sensitivity is often beneficially reduced [12–14]. To date, several examples of NC with nanosized morphology have been reported. Electrostatic spinning of NC solution results in the formation of fibers with diameters from 80 to 300 nm [15,16]. The samples demonstrate the slight lowering of the decomposition onset from 198 °C to 190 °C [16], while the heat of reaction was increased [15] for nanosized NC as compared to as-received material. A complementary use of the sol-gel approach and sc-CO_2 drying allows obtaining the NC aerogel with 30–50 nm particles that also show a lower decomposition onset temperature compared to raw NC [17]. Besides, the thermolysis mechanism of this porous material is altered by its increased adsorption capacity. Zhang and Weeks explored the low temperature solvent evaporation technique to prepare nearly 500 nm-sized NC spheres (5 °C, dimethylformamide) and thin films of submicron NC (5–45 °C, MeOH or EtOH) [18]. Notably, the obtained films exhibited a more complete combustion with the rate 350% higher than that for films of ordinary nitrocellulose. From the chemical engineering prospective, there are definite complications related to the reported preparative techniques, i.e., physicochemical methods (e.g., anti-solvent [19,20], suspension dispersion [21], recrystallization [18]) are inevitably accompanied by the issues of wasted organic solvents utilization and removal of their traces trapped inside the product, whereas mechanical methods (e.g., milling [22]) are incompatible with the polymeric nature of NC. Thereby, the challenge to develop a 'greener' and more efficient approach for a NC morphology modification has not been addressed yet.

Tailoring the burning rate of the double-based propellants, i.e., nitrocellulose-based formulations, is usually performed by adding of 1–10 wt.% of catalysts or inhibitors [23,24]. The general conclusion drawn on the basis of the experience with catalysis of double-based propellants is that carbonaceous skeleton formed above the burning surface promotes the burning rate via the prolongated residence of catalyst particles and increase of heat flux to the condensed phase [24–26]. The desired carbonaceous layer can be formed either during combustion of some components, e.g., dinitrotoluene, dibutyl phthalate plasticizers, or by introduced prior to combustion additives (e.g., 0.5–3 wt.% of carbon black in [24]). Interestingly, the addition of carbon in the form of carbon nanotubes (CNTs) can increase the burning rate per se, by improving the thermal conductivity [27–29]. Thus, in the recent studies [30,31] the optimal level of multiwall carbon nanotubes additive was found to be about 1 wt.%, whereas at higher loadings the combustion is depressed. Nano-sized metal oxides render the catalytic action in diverse processes, including the decomposition and combustion of energetic materials [32–34]. Specifically, nano-Fe_2O_3 has been found to catalyze the decomposition of NC, with the governing effect of the contact surface between two components [35]. Therefore, improvement of the nitrocellulose combustion requires the composites designed to deliver the nano-sized additives to the reaction zone.

Recently, sub- and supercritical fluids media have proved valuable for modification and synthetic applications in chemical engineering [36,37]. Particularly, supercritical carbon dioxide (sc-CO_2) is most attractive due to its low critical point (P_c = 7.38 MPa and T_c = 31.0 °C [38]), non-toxicity and environmental friendliness [39,40]. Supercritical Anti-Solvent (SAS) applications of CO_2 are based on its very good miscibility with the majority of organic solvents and insolubility of the target substrate in resulting binary solvent systems (sc-CO_2/solvent) that causes the precipitation (recrystallisation) of the final material [41,42]. The SAS technology has already been applied for micronization of nitramine explosives [43,44]. Herein, we report the first application of the supercritical antisolvent process to prepare the uniform nano-sized nitrocellulose powder. Furthermore, NC-based composites with carbon nanotubes and nano-Fe_2O_3 as the burning rate modifiers have been also fabricated via the SAS processing. Thermal behavior, mechanical sensitivity, and combustion performance were evaluated showing the enhanced burning rate and reduced sensitivity for prepared nano-nitrocellulose and its composites.

2. Materials and Methods

2.1. Materials

Commercial nitrocellulose was used as a precursor for synthesis and as a reference material. The powder is constituted by fibers with a characteristic diameter of about 20 μm, as shown in Figure 1b. The elemental composition of the raw NC was found to be 26.7 ± 0.2 wt.% C, 2.9 ± 0.1% H, and 11.5 ± 0.1% N. Multi-walled carbon nanotubes ("Taunit-M" trademark by NanoTechCenter Ltd., Tambov, Russia) were used as-received. In line with the manufacturer specification [45] the inner diameter of CNTs was ca.10 nm, the outer diameter near 20 nm, and the length no less than 2 μm

(Figure S1c,d, see Supporting information for details). The laboratory sample of nanosized α-iron oxide was produced by the plasma condensation technique and appeared as spherical particles 0.05–0.3 μm in diameter (Figure S1e,f, Supporting information). High purity carbon dioxide (Russian standard TU 2114-006-05015259-2014, Linde Gas Rus, Ltd., Balashikha, Russia) and acetone (Russian standard TU 2633-012-29483781-2009, Chimmed Ltd., Moscow, Russia) were used during the study.

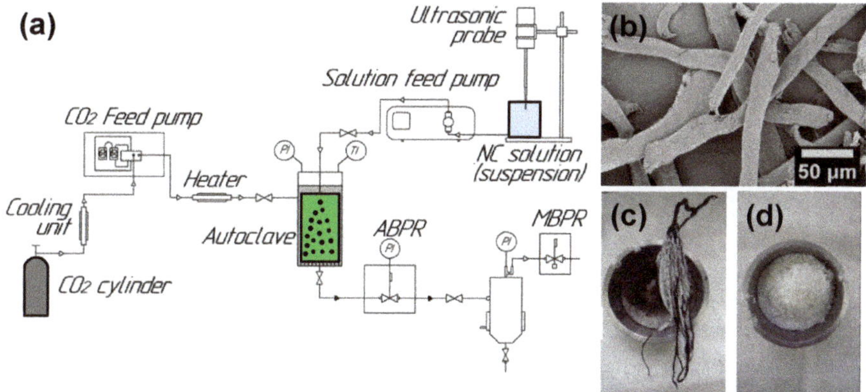

Figure 1. Supercritical anti-solvent (SAS) fabrication of nitrocellulose (NC)-based composites: (**a**) scheme of the experimental setup, (**b**) scanning electron microscopy (SEM) of raw NC fibers, (**c**) product obtained under non-optimized conditions, (**d**) pure NC after SAS treatment in optimized conditions.

2.2. Fabrication of Composites

Figure 1a illustrates the scheme of the experimental setup for the composite fabrication with the key units indicated. Cooled CO_2 was transferred from the cylinder through the high-pressure pump into the 500 mL autoclave reactor with temperature-control. The pressure inside the autoclave was maintained with the automated back pressure regulator (ABPR). The NC acetone solution (2 wt.%) or suspension of selected fillers in this solution was fed with another pump to the filled with sc-CO_2 autoclave through the capillary in the lid. The mutual diffusion of CO_2 and acetone results in the precipitation of NC or its composite particles which were collected on the metallic filter at the bottom. The binary mixture CO_2-acetone left the autoclave, went through ABPR and was divided inside the separator at the lower pressure controlled with the manual back pressure regulator (MBPR). The main units of the setup were connected with 1/8″ steel pipes.

The general procedure implies filling with CO_2 and pre-heating the system up to 40 °C. Once the desired temperature was achieved, the carbon dioxide feeding rate was set at 50 g·min^{-1}. NC solution or suspension, prepared and maintained by ultrasonic vibrations, was fed at 1–2 mL·min^{-1} rate. After the solution/suspension was fully transferred to an autoclave, and CO_2 flow pumped for an extra 30–40 min to flash off the residual acetone. Finally, the system was decompressed, and ~1 g of the target solid was taken out.

3. Results

3.1. Optimization of the Process Conditions

Preliminary runs resulted in the "herring-bone" structures instead of the desired uniform powder, as shown in Figure 1c. Several parameters of SAS process were tuned to improve the product morphology, i.e., the autoclave pressure, the ratio between the feeding rates of the solution and that of CO_2 (f/F), NC concentration in the solution, and the additive material type (Section S3, Supporting information). Pressure variation within the 9–15 MPa range specifies the optimal value to be 10–12 MPa.

At lower pressures the "herring-bone" was observed, whereas at higher pressures the uniform powder was obtained (cf. Figure 1c,d). The factor f/F controlled both the production time and CO_2 consumption; however, its increase was shown to worsen the composite morphology. Thus, the solution flow rate appeared to be no higher than 2 mL·min^{-1} at the maximum CO_2 feeding rate of 50 g·min^{-1} given by the experimental setup. The nitrocellulose concentration in the solution apparently determines the acetone and CO_2 consumption and has to be maximized. On the other hand, its increase above 2 wt.% had adverse effects on the the NC dissolution time and the pump performance (due to an increase in the mixture viscosity). When the filler, i.e., and/or nano-Fe_2O_3, was introduced into the solution, the ultrasonic processing (30% of 70 W, 20 kHz) was applied continuously to prevent settling and to maintain the suspension shelf life.

3.2. Nano-Nitrocellulose and Composites

Once the SAS process parameters were optimized, four samples were obtained: *NC(sc)* pure nitrocellulose after SAS, *NC/CNT(sc)* nitrocellulose with 1 wt.% of carbon nanotubes, *NC/Fe$_2$O$_3$(sc)* nitrocellulose with 5 wt.% of nano-iron oxide, *NC/CNT/Fe$_2$O$_3$(sc)* nitrocellulose with 1 wt.% of CNTs and 5 wt.% of nano-Fe_2O_3. The elemental composition of *NC(sc)* was unaffected by the processing: 26.7 ± 0.1 wt.% C, $2.9 \pm 0.4\%$ H, and $11.6 \pm 0.1\%$ N. In contrast to the large fibers of initial NC (Figure 1b), the product after the SAS processing with no additives appeared as nanopowder. Electron microscopy revealed that the microstructure of the powder was formed by the nano-sized spheres of 50–450 nm in diameter that were located on thin polymeric fibers (Figure 2a,b). Numerical particle size distribution gave the average diameter of 190 nm (Figure 2c, Section S4, Supporting information). Note, that transmission electron microscope (TEM) images have captured a sole fraction that is below 100 nm. Apparently, the bigger fractions were lost during the sample preparation.

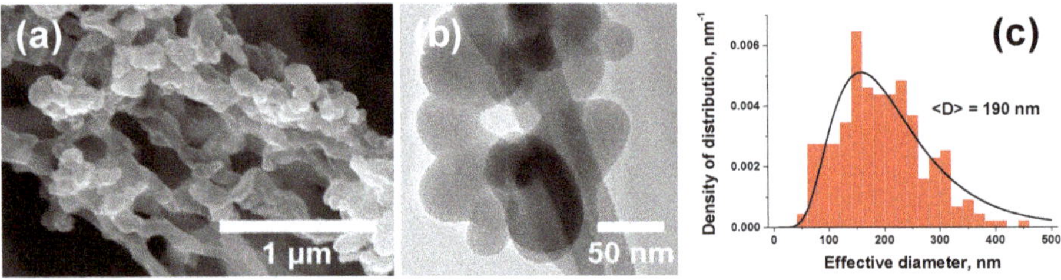

Figure 2. Morphology of *NC (sc)*: scanning (**a**), transmission (**b**) electron microscopy, and numerical particle size distribution (**c**). Note that image (**b**) shows the nanoparticles localized on transmission electron microscope (TEM) grid.

With the introduction of nanotubes the skeletal structure becomes more pronounced (Figure 3a). However, the further addition of nano-Fe_2O_3 resulted in the material with uniformly distributed nanoparticles and with no evidence of webbing (Figure 3b). These must have been the nanoparticles of iron oxide that promote the formation of NC nanospheres instead of fibers. Notably, the catalyst content within the fabricated composites was close to the target value, i.e., no material is lost during preparation (Section S5, Supporting information).

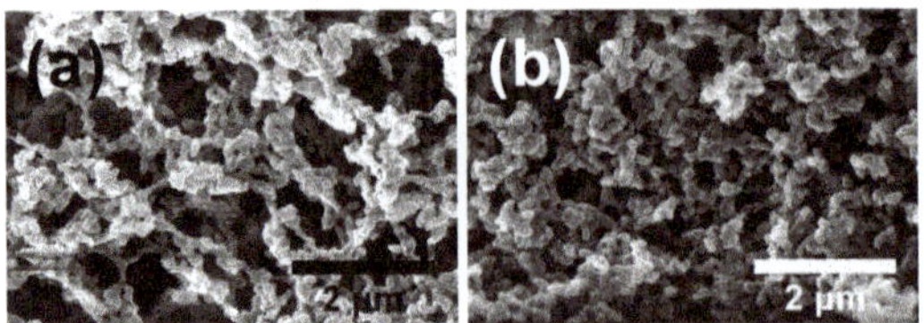

Figure 3. Electron microscopy of (**a**) NC/carbon nanotube (CNT)(sc) and (**b**) *NC/CNT/Fe$_2$O$_3$(sc)*.

3.3. Thermal Behavior

The linear heating of all samples shows the general pattern consistent with the typical nitrocellulose behavior [46,47]: a single peak of the heat release started at about 190 °C with the corresponding mass loss of nearly 80 wt.%. The heat released during thermolysis was within the range of 1800–2000 J·g^{-1} and agreed closely for all analyzed species within the experimental error. A closer examination of the four obtained nano-NC-based samples showed a lower decomposition onset temperature (see Figure 4 and Table 1), previously observed for nitrocellulose with nano-sized features [16–18]. The isoconversional Friedman analysis showed that the activation energy remained the same as for micron-NC at low- and high-conversion degrees, while at the medium range the barrier was notably lower for nano-NC (Figure 4c). A similar conclusion is drawn from the formal kinetic fit with a single reaction model in flexible reduced Sestak–Berggren form [48,49]. Kinetic parameters for nano-NC, $E_a = 186 \pm 1$ kJ·mol^{-1} and ln $A = 43.0 \pm 0.2$ s^{-1} were lower than that for raw NC, $E_a = 196 \pm 1$ kJ·mol^{-1} and ln $A = 45.7 \pm 0.3$ s^{-1} (Section S6, Supporting Information). Both pairs of the kinetic parameters were consistent with the global autocatalytic reaction proposed by Brill and Gongwer [50]. The shift of the kinetic parameters for nano-NC was apparently caused by the higher surface area available for autocatalytic agents and its increased contribution to the overall decomposition process.

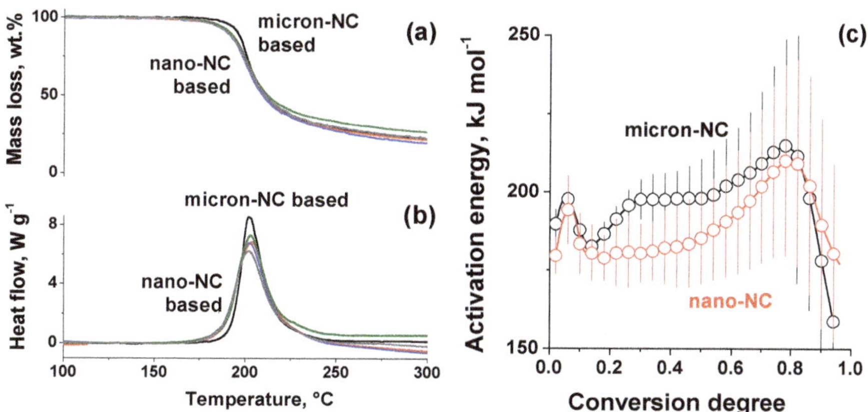

Figure 4. Thermal analysis under linear heating: mass loss (**a**), heat flow (**b**) data, and isoconversional kinetic analysis results (**c**).

Table 1. Summary of properties for raw NC and SAS-obtained composites.

Sample	Decomposition onset [1], °C	Heat Effect [1], J·g^{-1}	Friction Sensitivity, N	Burning Rate [2], mm·s^{-1}
NC	193	1880 ± 90	234 ± 21	13.8 ± 0.5
NC(sc)	189	1970 ± 30	342 ± 18	13.1 ± 0.5
NC/CNT(sc)	188	1920 ± 150	>360 (10%)	13.2 ± 0.5
NC/Fe$_2$O$_3$(sc)	186	1790 ± 150	229 ± 18	13.1 ± 0.5
NC/CNT/Fe$_2$O$_3$(sc)	189	1890 ± 150	>360 (40%)	15.8 ± 0.6
NC/CNT/Fe$_2$O$_3$	193	1590 ± 150	222 ± 31	14.6 ± 0.5

[1] Measured at 5 K·min^{-1} rate. [2] For pressed pellets burned at 12 MPa.

3.4. Friction Sensitivity

The full set of experiments according to standard STANAG 4487 [51] have been performed to derive the friction force corresponding to the 50% probability of explosion. The SAS processing decreased the pure nitrocellulose *NC(sc)* sensitivity, and the required initiation stimulus accordingly rose from 234 to 342 N. The effect must have been caused by the particle size reduction since the chemical composition of the sample was unaltered as shown above. The addition of the iron oxide sensitized the *NC/Fe$_2$O$_3$(sc)* sample down to the level of untreated raw compound. Introduction of carbon nanotubes ranked the composite *NC/CNT(sc)* as friction insensitive (10% of explosions at highest load of 360 N) and with the addition of both carbon nanotubes and iron oxide *NC/CNT/Fe$_2$O$_3$(sc)* the effect was still observed (40% of explosions at 360 N).

3.5. Combustion Tests

The energetic performance of the samples was estimated by combustion in air with the high-speed filming of the burning surface (2000 fps) and directly measured as the burning rate–pressure dependency in a Crawford bomb at elevated pressures. For the raw nitrocellulose, the observation revealed active bubble formation on the surface during combustion (Videos S1,S2, Supporting information). In this foam zone, small black particles were formed that moved inside the liquid layer to aggregate (see Figure 5a,b for sequential frames with 0.075 s time difference). Thus, the formed framework remained above the surface until burning out in a flame of the final reactions between the semiproducts of NC decomposition or until blowing away by the gas flow. Similar observations were drawn for the *NC(sc)* sample with the only exception of higher carbonaceous framework formation that remained after tests as the solid residue (Figure 5c and Figure S6, Video S2, Supporting information). The finer structure of unreacted nano-nitrocellulose seemed to either result in a more distributed localization of the seeds or in the structural modification of the formed framework to keep it intact in gas flow. Considering the burning rate values, the nano-NC exhibited a higher velocity as compared to raw NC, e.g., 3.8 against 4.7 mm·s^{-1} at 2 MPa (Figure 5g). This enhancement was apparently linked with the lowering of the activation barrier for condensed-phase decomposition as revealed by thermal analysis. The introduction of carbon nanotubes gave more seeds for carbonaceous framework (Figure 5d), and in fact a considerable porous layer grew above the surface (Figure S7, Video S3, Supporting information). However, it easily broke down into fragments, which were blown away as the combustion proceeds. The burning rate for the *NC/CNT(sc)* composite was the same as for nano-NC monopropellant showing no significant effect of the increased heat conductivity on the combustion front propagation.

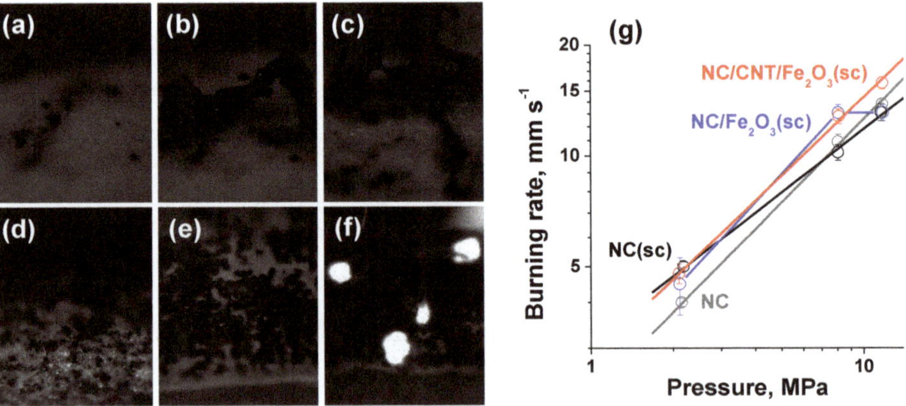

Figure 5. Combustion of the composites: still images from the high-speed filming (**a–f**) and the burning rate-pressure dependencies (**g**). Images shows the typical shots of NC (**a,b**), *NC(sc)* (**c**), *NC/CNT(sc)* (**d**), *NC/Fe$_2$O$_3$(sc)* (**e**), and *NC/CNT/Fe$_2$O$_3$(sc)* (**f**) burning. Frame width is 1.22 mm.

Figure 5e illustrates the burning process for the NC nanocomposite modified with nano-Fe$_2$O$_3$ addition. Seeds of the size bigger than that in previous samples stuck together on the surface and stayed in gas phase as a skeleton (Video S4, Supporting information). Recently, Shin et al. have proposed a facile method of nano-Fe$_2$O$_3$ transformation to Fe$_3$O$_4$ nanoparticles covered by a carbonaceous layer using a porous film made of nano-Fe_2O_3 particles filled with nitrocellulose [52]. After the burning of the NC matrix with the registered temperatures of 500–900 °C, the desired product was recovered. In line with these results, Denisyuk and Demidova [24] noticed the reduction of SnO_2 to a mixture of SnO and Sn during the combustion of an NC-based propellant. Here, we could assume the reduction of nano-Fe_2O_3 in our experiments; however the necessary high-temperature flame was not registered. As for the burning rate, we noticed the acceleration of the combustion by iron oxide addition under 8 MPa pressure (blue curve, Figure 5g).

When carbon nanotubes and nano-iron oxide are introduced together to the composite, we observe a different picture of combustion (Video S5, Supporting information). Figure 5f shows a typical shot of the burning surface covered by a crumbly carbonaceous layer with luminous spots that appeared occasionally. These luminous regions give strong evidence of the catalyzed reactions and the increased heat flux seems to be the main driver for the combustion velocity increase. Figure 5g summarizes the burning rate-pressure data showing the highest catalytic effect at high pressures, achieving a maximum increase in 20% at 12 MPa (Table 1). The simplified combustion models [53,54] assumed the pressure exponent n of the burning rate dependency $U = B \cdot P^n$ to be an indicator of the importance of gas-phase reactions, and the observed increase in it is consistent with the action of the catalyst in gas-phase zone.

To estimate the value of the SAS process in the arrangement of nanocatalyst and CNTs, we prepare the sample of the same composition $NC/CNT/Fe_2O_3$ by means of conventional dry mixing. Testing for sensitivity and combustion reveal its inferior performance as compared to the composite produced via the SAS (Table 1).

4. Conclusions

Supercritical CO_2 has been applied for the first time as anti-solvent for obtaining nano-sized nitrocellulose powder and various nitrocellulose-based composites. Thus prepared materials exhibit a significantly lower sensitivity to friction and a higher burning rate than the raw fibrous nitrocellulose, while maintaining the chemical composition and thermal stability. For the benchmark dry mixture of $NC/CNT/Fe_2O_3$ the results are notably lower than those for the SAS-produced composite. Overall, the proposed approach for introducing the burning rate modifiers into the matrices of energetic materials allows increasing the performance, satisfies the high technological and ecological requirements and may be useful for modern energetic composites production.

Supplementary Materials: The following are available online at http://www.mdpi.com/2079-4991/9/10/1386/s1: Section S1: Experimental Methods, Section S2: Starting Materials, Figure S1: Microstructure of (a, b) untreated micron-sized NC fibers, (c,d) carbon nanotubes, (e,f) nano-sized iron oxide, Section S3: Details of the SAS process, Table S1: SAS process parameters for NC modification, Section S4: Morphology of Nano-NC, Figure S2: Atomic force microscopy (AFM) image of nitrocellulose powder produced via SAS process, Table S2: Parameters of the particle-size distribution of nano-NC powder, Section S5: Determination of catalyst content in fabricated composites, Figure S3: Temperature and mass profiles during annealing of NC-based composites with Fe_2O_3 addition, Section S6: Thermokinetic analysis, Figure S4: Results of formal kinetic analysis for micron-(a) and nano-(b) sized nitrocellulose, Figure S5: Data from review and the results of the current study, Section S7: Combustion tests, Figure S6: Combustion of untreated (a–c) and processed (d–f) nitrocellulose: (a,d) view of pellet before test, (b,e) still image of the burning surface, (c,f) residue after experiment, Figure S7: Combustion nitrocellulose with additives of nano-iron oxide (a–c), carbon nanotubes (d–f) and both Fe_2O_3 and CNTs (g–i): (a,d,g) view of pellet before test, (b,e,h) still image of the burning surface, (c,f,i) residue after experiment, Table S3: Results of combustion experiments, Figure S8: Combustion velocity against pressure for all involved compositions, Videos S1–S6.

Author Contributions: Conceptualization, I.V.K. and S.G.Z.; methodology, M.N.Z., I.V.K. and K.A.M.; software, N.V.M.; validation, M.N.Z. and N.V.M.; formal analysis, I.V.K.; investigation, O.S.D., K.A.M. and D.B.M., resources, I.V.F., S.G.Z. and A.N.P.; data curation, D.B.M.; writing – original draft, N.V.M.; writing – review and editing, M.N.Z., A.N.P. and N.V.M., visualization, D.B.M. and K.A.M., project administration, I.V.K. and N.V.M.; supervision, I.V.F., S.G.Z. and A.N.P.

Funding: The authors are grateful for financial support of synthetic part and combustion behavior studies by Russian Foundation for Basic Research (Grant No. 18-29-06023) K.A.M., A.N.P., and N.V.M. acknowledge the State research project #0082-2018-0002 (CITIS #AAAA-A18-118031490034-6) to Semenov Institute of Chemical Physics for a financial support of thermal analysis.

Acknowledgments: Authors thank Ekaterina K. Kosareva for performing the atomic force microscopy and Lyubov N. Mitireva for thorough English language editing.

Conflicts of Interest: The authors declare no conflict of interest.

Nomenclature

A	Pre-exponential factor
ABPR	Automated back pressure regulator
B	Burn rate coefficient
CNTs	Carbon nanotubes
E_a	Apparent activation energy
EtOH	Ethanol
f	Feeding rate of NC-solution/suspension
F	Feeding rate of CO_2
MBPR	Manual back pressure regulator
MeOH	Methanol
micron-NC	Nitrocellulose microfibers (raw nitrocellulose)
n	Pressure exponent
nano-NC	Nitrocellulose nanoparticles
NC	Nitrocellulose
NC(sc)	Nitrocellulose particles obtained by SAS process
NC/CNT(sc)	Nitrocellulose composite with CNTs obtained by SAS process
NC/Fe$_2$O$_3$(sc)	Nitrocellulose with nano-Fe$_2$O$_3$ obtained by SAS process
NC/CNT/Fe$_2$O$_3$(sc)	Nitrocellulose with CNTs and nano-Fe$_2$O$_3$ obtained by SAS process
P	Pressure
P_c	Critical pressure of CO_2
RDX	1,3,5-trinitro-1,3,5-triazinane
SAS	Supercritical Anti-Solvent process
sc-CO_2	Supercritical carbon dioxide
SEM	Scanning electron microscopy
T_c	Critical temperature of CO_2
TEM	Transmission electron microscopy
U	Burning rate

References

1. Selwitz, C. *Cellulose Nitrate in Conservation*; Research in Conservation; Getty Conservation Institute: Marina del Rey, CA, USA, 1988; ISBN 978-0-89236-098-7.
2. Cretich, M.; Sedini, V.; Damin, F.; Pelliccia, M.; Sola, L.; Chiari, M. Coating of nitrocellulose for colorimetric DNA microarrays. *Anal. Biochem.* **2010**, *397*, 84–88. [CrossRef] [PubMed]
3. Mujawar, L.H.; Moers, A.; Norde, W.; van Amerongen, A. Rapid mastitis detection assay on porous nitrocellulose membrane slides. *Anal. Bioanal. Chem.* **2013**, *405*, 7469–7476. [CrossRef] [PubMed]
4. Holstein, C.A.; Chevalier, A.; Bennett, S.; Anderson, C.E.; Keniston, K.; Olsen, C.; Li, B.; Bales, B.; Moore, D.R.; Fu, E.; et al. Immobilizing affinity proteins to nitrocellulose: A toolbox for paper-based assay developers. *Anal. Bioanal. Chem.* **2016**, *408*, 1335–1346. [CrossRef] [PubMed]
5. Zhong, Y.; McConnell, G.C.; Ross, J.D.; DeWeerth, S.P.; Bellamkonda, R.V. A Novel Dexamethasone-releasing, Anti-inflammatory Coating for Neural Implants. In Proceedings of the 2nd International IEEE EMBS Conference on Neural Engineering, Arlington, VA, USA, 16–19 March 2005; IEEE: Piscataway, NJ, USA, 2005; pp. 522–525.
6. Kubota, N. *Propellants and Explosives*; Wiley-VCH: Weinheim, Germany, 2006; ISBN 978-3-527-61010-5.
7. Lengelle, G.; Duterque, J.; Trubert, J.-F. Physico-Chemical Mechanisms of Solid Propellant Combustion. In *Solid Propellant Chemistry, Combustion, and Motor Interior Ballistics*; Yang, V., Brill, T.B., Ren, W.Z., Eds.; Progress in Astronautics and Aeronautics; American Institute of Aeronautics and Astronautics, Inc.: Reston, VI, USA, 2000; pp. 287–334. ISBN 978-1-56347-442-2.

8.   de la Ossa, M.Á.F.; Torre, M.; García-Ruiz, C. Chapter 4. Nitrocellulose in propellants: Characteristics and thermal properties. In *Advances in Materials Science Research*; Wythers, M.C., Ed.; Nova Science Publishers, cop.: New York, NY, USA, 2012; Volume 7, pp. 201–220. ISBN 978-1-62100-045-7.

9.   Fu, G.; Wang, J.; Yan, M. Anatomy of Tianjin Port fire and explosion: Process and causes. *Process Saf. Prog.* **2016**, *35*, 216–220. [CrossRef]

10. Saunders, C.W.; Taylor, L.T. A review of the synthesis, chemistry and analysis of nitrocellulose. *J. Energy Mater.* **1990**, *8*, 149–203. [CrossRef]

11. Frolov, Y.V.; Pivkina, A.N.; Zav'yalov, S.A.; Murav'ev, N.V.; Skryleva, E.A.; Monogarov, K.A. Physicochemical characteristics of the components of energetic condensed systems. *Russ. J. Phys. Chem. B* **2010**, *4*, 916–922. [CrossRef]

12. Yang, G.; Nie, F.; Li, J.; Guo, Q.; Qiao, Z. Preparation and Characterization of Nano-NTO Explosive. *J. Energy Mater.* **2007**, *25*, 35–47. [CrossRef]

13. Bayat, Y.; Zeynali, V. Preparation and Characterization of Nano-CL-20 Explosive. *J. Energy Mater.* **2011**, *29*, 281–291. [CrossRef]

14. Lee, K.-Y.; Moore, D.S.; Asay, B.W.; Llobet, A. Submicron-Sized Gamma-HMX: 1. Preparation and Initial Characterization. *J. Energy Mater.* **2007**, *25*, 161–171. [CrossRef]

15. Xia, M.; Yunjun, L.; Hua, Y.-L. Preparation and characterization of nitrocellulose nano-fibers. *Chin. J. Energetic Mater.* **2012**, *20*, 167–171.

16. Sovizi, M.R.; Hajimirsadeghi, S.S.; Naderizadeh, B. Effect of particle size on thermal decomposition of nitrocellulose. *J. Hazard. Mater.* **2009**, *168*, 1134–1139. [CrossRef] [PubMed]

17. Jin, M.; Luo, N.; Li, G.; Luo, Y. The thermal decomposition mechanism of nitrocellulose aerogel. *J. Therm. Anal. Calorim.* **2015**, *121*, 901–908. [CrossRef]

18. Zhang, X.; Weeks, B.L. Preparation of sub-micron nitrocellulose particles for improved combustion behavior. *J. Hazard. Mater.* **2014**, *268*, 224–228. [CrossRef] [PubMed]

19. Steinitz, M.; Tamir, S. An improved method to create nitrocellulose particles suitable for the immobilization of antigen and antibody. *J. Immunol. Methods* **1995**, *187*, 171–177. [CrossRef]

20. Cox, C.D.; Liggett, T. Spherical Production of Small Particle Nitrocellulose. U.S. Patent 3671515A 1972.

21. O'Neill, J.J., Jr. Nitrocellulose Particles. U.S. Patent 2885736A 1959.

22. Plunguian, M. Process for Making Fine Particles of Nitrocellulose. U.S. Patent 3275250A 1966.

23. Androsov, A.S.; Denisyuk, A.P.; Tokarev, N.P. Laws governing the effect of lead-copper catalysts on the combustion rate of a ballistic powder. *Combust. Explos. Shock Waves* **1976**, *12*, 698–700. [CrossRef]

24. Denisyuk, A.P.; Demidova, L.A. Effect of Some Catalysts on Combustion of Double-Base Propellants. *Combust. Explos. Shock Waves* **2004**, *40*, 311–318. [CrossRef]

25. Denisyuk, A.P.; Demidova, L.A.; Shepelev, Y.G.; Baloyan, B.M.; Telepchenkov, V.E. Highly efficient low-toxicity catalysts of combustion of double-base powders. *Combust. Explos. Shock Waves* **1997**, *33*, 688–694. [CrossRef]

26. Zenin, A.A.; Finjakov, S.V.; Ibragimov, N.G.; Afiatullov, E.K. Mechanism of catalysis in combustion waves of modern ballistite propellants. *Combust. Explos. Shock Waves* **1999**, *35*, 532–542. [CrossRef]

27. Merzhanov, A.G.; Dubovitskii, F.I. On the Theory of Steady State Monopropellant. *Proc. USSR Acad. Sci.* **1959**, *129*, 153–156.

28. Kirchdoerfer, T.; Ortiz, M.; Stewart, D.S. Topology Optimization of Solid Rocket Fuel. *AIAA J.* **2019**, *57*, 1684–1690. [CrossRef]

29. He, G.; Li, X.; Dai, Y.; Yang, Z.; Zeng, C.; Lin, C.; He, S. Constructing bioinspired hierarchical structure in polymer based energetic composites with superior thermal conductivity. *Compos. Part B Eng.* **2019**, *162*, 678–684. [CrossRef]

30. Chen, S.; Tang, Y.; Yu, H.; Guan, X.; DeLuca, L.T.; Zhang, W.; Shen, R.; Ye, Y. Combustion enhancement of hydroxyl-terminated polybutadiene by doping multiwall carbon nanotubes. *Carbon* **2019**, *144*, 472–480. [CrossRef]

31. Kim, J.H.; Cho, M.H.; Kim, K.J.; Kim, S.H. Laser ignition and controlled explosion of nanoenergetic materials: The role of multi-walled carbon nanotubes. *Carbon* **2017**, *118*, 268–277. [CrossRef]

32. Yan, Q.-L.; Zhao, F.-Q.; Kuo, K.K.; Zhang, X.-H.; Zeman, S.; DeLuca, L.T. Catalytic effects of nano additives on decomposition and combustion of RDX-, HMX-, and AP-based energetic compositions. *Prog. Energy Combust. Sci.* **2016**, *57*, 75–136. [CrossRef]

33. Pivkina, A.N.; Muravyev, N.V.; Monogarov, K.A.; Fomenkov, I.V.; Schoonman, J. Catalysis of HMX Decomposition and Combustion: Defect Chemistry Approach. In *Energetic Nanomaterials*; Elsevier: Amsterdam, The Netherlands, 2016; pp. 193–230. ISBN 978-0-12-802710-3.

34. Muravyev, N.V.; Pivkina, A.N.; Schoonman, J.; Monogarov, K.A. Catalytic influence of nanosized titanium dioxide on the thermal decomposition and combustion of HMX. *Int. J. Energy Mater. Chem. Propuls.* **2014**, *13*, 211–228. [CrossRef]

35. Zhang, T.; Zhao, N.; Li, J.; Gong, H.; An, T.; Zhao, F.; Ma, H. Thermal behavior of nitrocellulose-based superthermites: Effects of nano-Fe_2O_3 with three morphologies. *RSC Adv.* **2017**, *7*, 23583–23590. [CrossRef]

36. Corr, S. 1,1,1,2-Tetrafluoroethane; from refrigerant and propellant to solvent. *J. Fluor. Chem.* **2002**, *118*, 55–67. [CrossRef]

37. *Carbon Dioxide Recovery and Utilization*; Aresta, M. (Ed.) Springer: Dordrecht, The Netherlands, 2003; ISBN 978-90-481-6335-9.

38. Lemmon, E.W.; McLinden, M.O.; Friend, D.G. Thermophysical Properties of Fluid Systems. In *NIST Chemistry WebBook*; NIST Standard Reference Database Number 69; National Institute of Standards and Technology: Gaithersburg, MD, USA, 1997.

39. Darr, J.A.; Poliakoff, M. New Directions in Inorganic and Metal-Organic Coordination Chemistry in Supercritical Fluids. *Chem. Rev.* **1999**, *99*, 495–542. [CrossRef] [PubMed]

40. Sihvonen, M. Advances in supercritical carbon dioxide technologies. *Trends Food Sci. Technol.* **1999**, *10*, 217–222. [CrossRef]

41. Jung, J.; Perrut, M. Particle design using supercritical fluids: Literature and patent survey. *J. Supercrit. Fluids* **2001**, *20*, 179–219. [CrossRef]

42. Pourmortazavi, S.M.; Hajimirsadeghi, S.S. Application of Supercritical Carbon Dioxide in Energetic Materials Processes: A Review. *Ind. Eng. Chem. Res.* **2005**, *44*, 6523–6533. [CrossRef]

43. Seo, B.; Kim, T.; Park, H.J.; Kim, J.-Y.; Lee, K.D.; Lee, J.M.; Lee, Y.-W. Extension of the Hansen solubility parameter concept to the micronization of cyclotrimethylenetrinitramine crystals by supercritical anti-solvent process. *J. Supercrit. Fluids* **2016**, *111*, 112–120. [CrossRef]

44. Lee, B.-M.; Jeong, J.-S.; Lee, Y.-H.; Lee, B.-C.; Kim, H.-S.; Kim, H.; Lee, Y.-W. Supercritical Antisolvent Micronization of Cyclotrimethylenetrinitramin: Influence of the Organic Solvent. *Ind. Eng. Chem. Res.* **2009**, *48*, 11162–11167. [CrossRef]

45. *CNTs "Taunit" and Synthesis Thereof*; NanoTechCenter Ltd.: Tambov, Russia, 2019.

46. Eisenreich, N.; Pfeil, A. Non-linear least-squares fit of non-isothermal thermoanalytical curves. reinvestigation of the kinetics of the autocatalytic decomposition of nitrated cellulose. *Thermochim. Acta* **1983**, *61*, 13–21. [CrossRef]

47. Binke, N.; Rong, L.; Xianqi, C.; Yuan, W.; Rongzu, H.; Qingsen, Y. Study on the Melting Process of Nitrocellulose by Thermal Analysis Method. *J. Therm. Anal. Calorim.* **1999**, *58*, 249–256. [CrossRef]

48. Šesták, J.; Berggren, G. Study of the kinetics of the mechanism of solid-state reactions at increasing temperatures. *Thermochim. Acta* **1971**, *3*, 1–12. [CrossRef]

49. Burnham, A.K. Application of the Šesták-Berggren Equation to Organic and Inorganic Materials of Practical Interest. *J. Therm. Anal. Calorim.* **2000**, *60*, 895–908. [CrossRef]

50. Brill, T.B.; Gongwer, P.E. Thermal Decomposition of Energetic Materials 69. Analysis of the kinetics of nitrocellulose at 50–500 °C. *Propellants Explos. Pyrotech.* **1997**, *22*, 38–44. [CrossRef]

51. STANAG 4487. *Explosives, Friction Sensitivity Tests*; NATO: Brussels, Belgium, 2002; p. 1.

52. Shin, J.; Lee, K.Y.; Yeo, T.; Choi, W. Facile One-pot Transformation of Iron Oxides from Fe_2O_3 Nanoparticles to Nanostructured Fe_3O_4@C Core-Shell Composites via Combustion Waves. *Sci. Rep.* **2016**, *6*. [CrossRef] [PubMed]

53. Zeldovich, Y.B. Theory of combustion of propellants and explosives Zh. *Eksperimentalnoy i Teor. Fiz.* **1942**, *12*, 498–524.

54. Sinditskii, V.P.; Egorshev, V.Y.; Berezin, M.V.; Serushkin, V.V. Mechanism of HMX combustion in a wide range of pressures. *Combust. Explos. Shock Waves* **2009**, *45*, 461–477. [CrossRef]

nanomaterials

MDPI

Article

Preparation and Characterization of Silicon-Metal Fluoride Reactive Composites

Siva Kumar Valluri [1], Mirko Schoenitz [1] and Edward Dreizin [1,2,*

[1] O.H. York Department of Chemical and Materials Engineering, New Jersey Institute of Technology, Newark, NJ 07102, USA; sv476@njit.edu (S.K.V.); schoenit@njit.edu (M.S.)
[2] High-Energy and Special Materials Research Laboratory, Tomsk State University, 634050 Tomsk, Russia
* Correspondence: dreizin@njit.edu

Received: 16 October 2020; Accepted: 25 November 2020; Published: 28 November 2020

Abstract: Fuel-rich composite powders combining elemental Si with the metal fluoride oxidizers BiF_3 and CoF_2 were prepared by arrested reactive milling. Reactivity of the composite powders was assessed using thermoanalytical measurements in both inert (Ar) and oxidizing (Ar/O_2) environments. Powders were ignited using an electrically heated filament; particle combustion experiments were performed in room air using a CO_2 laser as an ignition source. Both composites showed accelerated oxidation of Si when heated in oxidizing environments and ignited readily using the heated filament. Elemental Si, used as a reference, did not exhibit appreciable oxidation when heated under the same conditions and could not be ignited using either a heated filament or laser. Lower-temperature Si fluoride formation and oxidation were observed for the composites with BiF_3; respectively, the ignition temperature for these composite powders was also lower. Particle combustion experiments were successful with the Si/BiF_3 composite. The statistical distribution of the measured particle burn times was correlated with the measured particle size distribution to establish the effect of particle sizes on their burn times. The measured burn times were close to those measured for similar composites with Al and B serving as fuels.

Keywords: reactive materials; nanocomposite; metal combustion; ignition; thermal analysis

1. Introduction

Silicon serves as a fuel in several pyrotechnic compositions [1]. Its high energy density, abundance in the lithosphere [2] and cheap manufacturing make it an attractive fuel for a broad range of applications, including propellants and explosives. However, it is known to be relatively difficult to burn [3,4]. Silicon has a high boiling point (3538 K) compared to its oxide (2503 K) and hence burns heterogeneously. Further, the nascent refractory oxide layer on silicon particles limits the inward diffusion of oxygen at temperatures below 1273 K [5]. The direct oxidation of silicon from gaseous oxygen [6] and nitrogen [7] is only accelerated at higher temperatures, when diffusion rates increase.

To make silicon more reactive, previous work focused on employing aggressive condensed phase oxidizers. Transition metal oxides [4,8–11], alkali metal nitrates [4,9–11], perchlorates [9] and manganates [9] were explored as oxidizers for micron-sized and nanometric silicon powders. These compositions, though favorably reactive, do not fully exploit the potential of silicon as they yield largely condensed phase combustion products. Respective silicon-condensed oxidizer composites have reduced energy content compared to elemental silicon reacting with oxygen gas; thus, the reactions have reduced adiabatic flame temperatures. Use of high surface area, porous nanosilicon has attracted attention [9,12–14]; however, such materials may be difficult to handle and their applications are mostly limited to pyrotechnics.

In comparison to oxidation, fluorination of silicon is preferred in fuel-limited systems because the negative heat of formation of silicon tetrafluoride, SiF_4, (−1615 kJ/mol) is substantially lower than that of SiO_2 (-911 kJ/mol), making the reaction more exothermic. Silicon fluorides (and most oxyfluorides [15]) are gases under standard conditions.

Combustion-related work on fluorination of silicon was mostly limited to the use of fluoropolymers, e.g., polytetrafluoroethylene (PTFE) as oxidizers. Silicon-PTFE compositions are predicted to have a higher adiabatic combustion temperature and produce more gaseous products compared to combinations of silicon with oxide-based oxidizers [9]. Silicon-PTFE milled composites with a range of compositions (10–40 wt.% silicon) have been explored [16–18].

In recent studies it was shown that metal fluorides, such as CoF_2, NiF_2 and BiF_3, could be effective oxidizers with metal fuels, e.g., aluminum. Reduced ignition temperatures and higher burn rates than for pure aluminum (in air) were reported for respective composites [19,20]. Further, metal fluorides were combined with boron, known to be more difficult to ignite than aluminum. A marked improvement in the ignition and combustion characteristics for boron/metal fluoride composites was observed with as little as 10 wt.% of BiF_3 [21,22]. The addition of fluorides altered combustion of boron particles: it occurred in a single stage with shortened burn times [21–23]. In constant volume explosion experiments, the 10 wt.% BiF_3 coated boron powders exhibited a higher rate of pressurization and higher maximum pressure compared to fine aluminum [22]. Experiments with aluminum and boron-based reactive composites containing metal fluorides as oxidizers show consistently significant improvements in kinetics of reactions leading to ignition [24,25].

Despite the increased reactivity, the fluoride composites made with aluminum and boron were found to be safer to handle, as they were insensitive to electrostatic stimulus unlike oxide based compositions [24,25]. It should also be noted that the gas-phase fluorinated reaction products could be aggressive; thus care must be taken when such products are released.

The use of metal fluorides as oxidizers for silicon is explored in the current work. The fluorides used in this study, BiF_3 and CoF_2 were selected because they are less hygroscopic, easy to handle [25], and contain metals that interact differently with metal fuels: Bi does not alloy with most metals or metalloids while Co forms alloys readily. The phase of the reduced metal (pure element or alloy) could affect further oxidation and combustion reactions. Based only on gravimetric and volumetric heats of reaction shown in Table 1, replacing oxides of Co and Bi with respective fluorides does not lead to tangible advantages. The heats of reactions are nearly matching for the metal-rich composites (with 50 wt.% of Si) reacting in presence of the external oxidizer, so that excess of Si and the metals obtained from the reduced fluorides can further be oxidized. Thus, the anticipated advantages of using metal fluorides are not based on the expected theoretical energy release but are based on the expected accelerated reaction rates and formation of gas phase products.

Table 1. Estimated heats of reactions involving Si with oxides and fluorides of Co and Bi for stoichiometric and fuel-rich (50/50 wt.%) composites reacting in inert and oxidizing environments.

Composite Type			Heat of Reaction	
			kJ/g	**kJ/cm^3**
Si with Bi_2O_3 and BiF_3	Stoichiometric	$3Si + 2Bi_2O_3 \rightarrow 3SiO_2 + 4Bi$	-1.56	-11.25
		$3Si + 4BiF_3 \rightarrow 3SiF_4 + 4Bi$	-1.05	-5.12
	50/50 wt.% (fuel rich) *	$Si + Bi_2O_3 \rightarrow Si + SiO_2 + Bi$	-0.85	-3.14
		$Si + BiF_3 \rightarrow Si + SiF_4 + Bi$	-0.57	-1.84
		Added ambient oxidizer $Si + Bi_2O_3 + O_2 \rightarrow SiO_2 + Bi_2O_3$	-15.6	-57.6
		Added ambient oxidizer $Si + BiF_3 + O_2 \rightarrow SiF_4 + SiO_2 + Bi_2O_3$	-15.5	-50.22

Table 1. *Cont.*

Composite Type			Heat of Reaction	
			kJ/g	kJ/cm^3
Si with CoO and CoF$_2$	Stoichiometric	Si + 2CoO → SiO$_2$ + 2Co	−2.45	−12.32
		Si + 2CoF$_2$ → SiF$_4$ + 2Co	−1.28	−5.11
	50/50 wt.% (fuel-rich) *	Si + CoO → Si + SiO$_2$ + Co	−1.45	−4.97
		Si + CoF$_2$ → Si + SiF$_4$ + Co	−0.60	−1.82
		Added ambient oxidizer Si + CoO + O$_2$ → SiO$_2$ + CoO	−14.63	−50.05
		Added ambient oxidizer Si + CoF$_2$ + O$_2$ → SiF$_4$ + SiO$_2$ + CoO	−14.46	−44.26

* For clarity of the concept, reactions are written without balancing reactants and products.

Both BiF$_3$ and CoF$_2$ were incorporated into silicon-metal fluoride composites prepared by arrested reactive milling [26].

2. Experimental

2.1. Materials Preparation

The silicon-metal fluoride composite powders were prepared in a SPEX Certiprep 8000 series shaker mill (by SPEX CertiPrep, Metuchen, NJ, USA). The constituent silicon (−325 mesh crystalline Si, 99%) and anhydrous fluoride powders, bismuth (III) fluoride (BiF$_3$, 98%) and cobalt (II) fluoride (CoF$_2$, 99%) were sourced from Alfa Aesar, Ward Hill, MA, USA. Preliminary inspection using electron microscopy showed that BiF$_3$ contained crystalline particle agglomerates, mostly in the 10–15 μm range. For CoF$_2$, the powder appeared as finer powder, with most particles in the single micron range. The compositions targeted 30 and 50 wt.% of silicon with the remaining mass of bismuth or cobalt fluoride. The materials were prepared in 5 g batches in steel milling vials. The powders were milled using steel milling media with hexane as the process control agent. The ball to powder mass ratio was fixed at 10.

In an initial, exploratory experiment, 50 wt.% silicon and 50 wt.% bismuth fluoride were milled for 60 min using 9.25-mm (3/8 in) diameter steel balls with 7 mL of hexane. This material is referred to as 1-stage 50Si·50BiF$_3$. Examination of 1-stage 50Si·50BiF$_3$ samples showed the presence of large Si particles that were not mixed with BiF$_3$. To improve homogeneity, milling was therefore carried out it two steps for all other prepared composites. Silicon was milled separately in the first milling step to refine it. This step lasted 60 min and used 14 mL of hexane and 5-mm steel balls. This size-reduced silicon was also used as reference material for thermal analysis and other experiments.

In the second step, the size-reduced (premilled) silicon was milled with bismuth fluoride or cobalt fluoride for 60 min using 4.76-mm steel balls with 7 mL of hexane. These composites are referred to as 50Si·50BiF$_3$, 50Si·50CoF$_2$, 30Si·70BiF$_3$ and 30Si·70CoF$_2$, with respective mass percentages of components used as identifiers. Table 2 presents characteristics of the prepared materials. The equivalence ratios describe reactions.

$$Si + \frac{4}{x}MeF_x \rightarrow SiF_4 \uparrow + \frac{4}{x}Me \tag{1}$$

where Me stands for Bi or Co. Reactions (1) reduce the respective metal fluorides and yield SiF$_4$. All prepared composites are fuel-rich, making them potential fuel candidates for applications where oxygenated gas environments are available.

Table 2. Compositions of prepared silicon-metal fluoride reactive composites.

Milling Protocol	Material	Fluoride Volume (%)	Equivalence Ratio ϕ	Mass of Fluorine (%)	
				Prepared	Stoichiometric
Single stage	1-stage 50Si·50BiF$_3$	30.4	12.6	10.7	19.9
Two stages	50Si·50BiF$_3$				
	30Si·70 BiF$_3$	50.5	5.4	15	
	50Si·50CoF$_2$	34.3	6.9	19.6	34.2
	30Si·70CoF$_2$	54.9	3.0	27.4	

The prepared materials were passivated in an argon-filled glovebox, where oxygen was present at a low partial pressure. The passivation lasted 48 h. Passivated materials were covered by hexane and stored in the laboratory for the duration of this project.

Reactivity was initially assessed by taking 100–200 mg of powder on a ventilated filter paper and igniting the paper by a butane lighter inside a fume hood. The 30 wt.% silicon powders burned quickly with a bright noiseless flash; the 50 wt.% silicon powders were comparatively slower and generated particle streaks jetting from the burning powder bulk. For the 50 wt.% silicon compositions, despite a coarser scale of mixing and poor homogeneity, 1-stage 50Si·50BiF$_3$ was visibly more reactive than two-stage milled 50Si·50BiF$_3$. There was no apparent difference between composites with BiF$_3$ and CoF$_2$ serving as oxidizers.

2.2. Material Characterization

The surface morphology and homogeneity of mixing between components in the prepared powders was examined using a JSM-7900 field emission scanning electron microscope (SEM) by JEOL, Tokyo, Japan. The imaging was performed by back-scattered electrons to achieve compositional contrast. To identify material composition and impurities incorporated by milling, energy dispersive X-ray spectroscopy (EDX) was used. The prepared powders were also analyzed using X-ray diffraction (XRD) on a PANalytical Empyrean multipurpose research diffractometer by Malvern Panalytical, Malvern, UK, operated at 45 kV and 40 mA using filtered Cu K$_\alpha$ radiation.

Reactivity of powders containing 50 wt.% silicon was tested by thermogravimetric analysis (TG) coupled with differential thermal analysis (DTA)/differential scanning calorimetry (DSC) on a STA409PG by Netzsch, Selb, Germany. The powders with a composition of 30 wt.% silicon were not tested to avoid releasing large amounts of aggressive fluorinated gases in the instrument.

Samples were heated in open alumina DTA crucibles only up to 973 K (700 °C) to limit release of volatile fluorinated species. Samples were tested in argon (99.998%) flowing at 50 mL/min as an inert environment and in an oxidizing environment, with an additional 50 mL/min flow of oxygen (99.994%). Both gases were sourced from Airgas. For each case, a baseline was obtained by heating an empty crucible under identical conditions. The sample mass was fixed at 5 mg, except for a few runs in argon performed with larger sample masses of 10 and 24 mg. For the oxidizing environment, heating rates 1, 2 and 5 K/min were used. Experiments in argon were performed with heating rates of 2, 5 and 10 K/min. To probe the progress of reactions, samples were recovered from specific intermediate temperatures and analyzed using XRD.

2.3. Heated Filament Ignition

The ignition temperatures for the prepared powders were obtained by coating a sample onto an electrically heated 24 gauge (0.5105 mm) nickel-chromium wire. The time of ignition was determined from a high-speed video and the wire temperature was measured optically, using a calibrated photo diode. The basic experimental setup and procedure have been presented earlier [27].

The photodiode-based pyrometer has been described in Ref. [24]. Briefly, the wire was coated with a suspension of the powder in hexane. Once the hexane dried, the wire was electrically heated using a variable set of rechargeable large cell batteries (7238K57 McMaster Carr. Elmhurst, IL, USA) and a 1 Ω rheostat connected in series to achieve targeted heating rates in the range of 10^3–10^4 K/s. A fiber optics cable was focused at a section of the wire that was uncoated and fed into germanium photodiode (PDA30B2 by Thorlabs, Newton, NJ, USA). The photodiode was calibrated against a black body emission source (BB4A by Omega Engineering, Bridgeport, NJ, USA) to function as a high-speed pyrometer. To correlate the time of ignition to the temperature of the wire, the ignition was recorded using a MotionPro 500 camera by Redlake, San Diego, CA, USA, synchronized with the ignition circuit. At least 5 runs were performed for each targeted heating rate.

2.4. Particle Combustion

The particle combustion was studied by aerosolizing the materials in a room-temperature air stream fed into the focal spot of a sealed CO_2 laser (Evolution 125 by Synrad, Mukilteo, WA, USA). The optical emission of the ignited particles was captured by filtered photomultiplier tubes (PMT) to obtain burn times. Each burning particle produced an emission pulse; the pulse width taken at 10% of its peak value was interpreted as the burn time.

The powder feeder described elsewhere [28] consisted of a screw driven by a DC motor, with 30 threads coated by 0.12 g of powder. The powder lodged between the threads was blown off the screw and aerosolized by a carrier gas, air flown at 0.68 L/min. The aerosolized particles were fed via a 2.39 mm internal diameter brass tube. The upper end of the tube was placed 2 mm below the focal spot of the laser. The laser beam was focused by a ZnS lens to a spot of about 250 μm diameter. The laser was operated at 37.5 W, or ca. 30% of its maximum power, which is much higher than the ignition threshold for the materials tested. Thus, particles passing through the laser beam ignited consistently. Further experimental details are available elsewhere [29].

A fiber optics bundle trained to capture the particle emissions was placed 2 cm above and 16 cm away from the brass feeder tube. The fiber was split to feed to two R3896-03 PMTs by Hamamatsu, Hamamatsu City, Japan, filtered at 700 and 800 nm. The signal from the PMTs was acquired through a 16 bit PCI-6123 board by National Instruments, Austin, TX, USA, at 100,000 samples per second and processed using LabView software (Labview 16 by National Instruments, August 2016). Signal was collected for periods of 8 s at a time. Several such 8 s runs were required to acquire over 800 pulses produced by individual burning particles for each material. Overlapping, or closely spaced pulses were dismissed. The widths of the collected individual particle pulses, interpreted as burn times, were developed into respective statistical distributions employing a smoothing kernel density function [21].

2.5. Aerosolized Particle Collection and Sizing

The aerosolization of the composite powders may cause size classification. Thus, the size distributions for the particles observed to burn in the present experiments could differ from those of the prepared powders. To account for this, samples of powders passed through the feeder were collected and analyzed. A flat aluminum substrate covered with a double-sided adhesive carbon tape was placed 2.5 cm above the brass feeder. Particles were trapped onto the tape for 10–15 s using the feeder operated at the same gas flowrate as in the combustion tests; however, with the laser beam turned off. The collected particles were imaged by the SEM and their sizes were obtained using ImageJ software [30]. From a table of all particles of one material, size distributions were generated using a kernel density function. The particle size distributions were correlated with the burn time distributions to identify the effect of particle size on the burn time. Such correlations broadly assume that larger particles burn longer.

3. Results

3.1. Particle Morphology and Composition

SEM images of the prepared 50 wt.% silicon- 50 wt.% metal fluoride composites are presented in Figure 1. The grey crystalline phase was silicon, and the brightest phase was the fluoride. The carbon tape in the background appeared darkest. The single stage milled 1-stage $50Si·50BiF_3$ powder is shown in Figure 1A,B. It is apparent from Figure 1A that the fluoride was not uniformly distributed in the material. Many relatively large, ca. 20–40 μm, Si particles and some unattached BiF_3 clusters were observed. Part of the image in Figure 1A with a relatively well-homogenized composite particle was magnified and shown as Figure 1B. Individual Si and BiF_3 phases were interspersed on a submicron scale. Such particles were few with a majority of silicon particles having little to no fluoride attached to them.

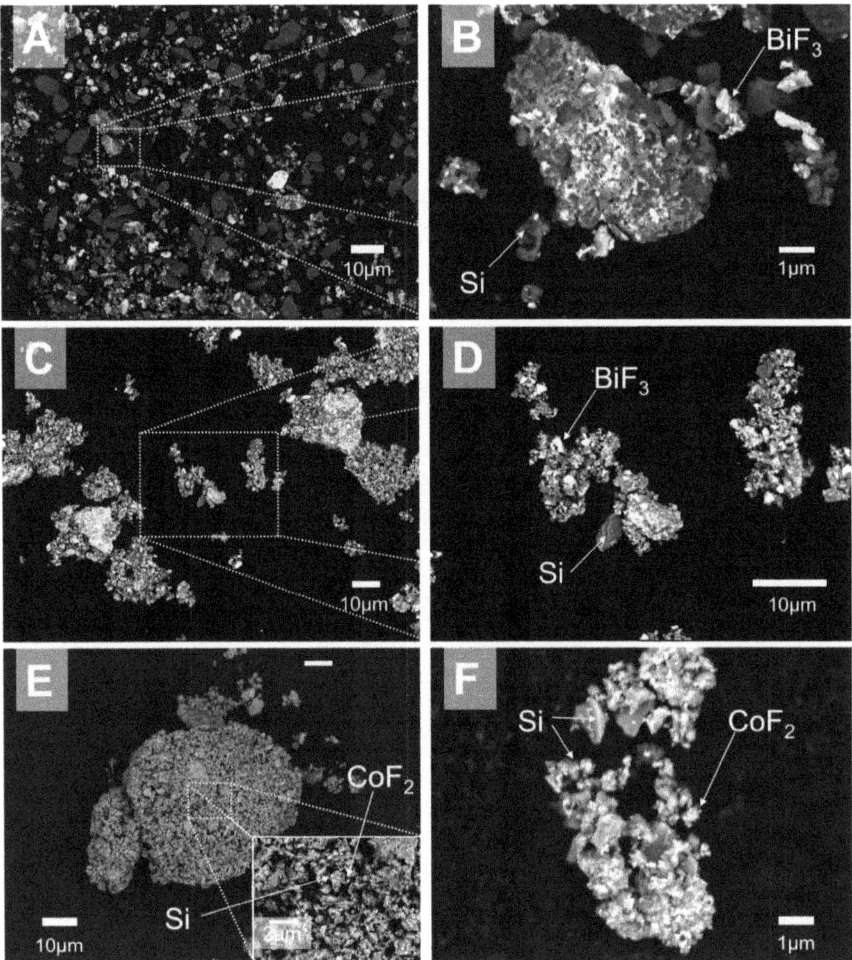

Figure 1. Backscattered electron images of as-milled 50 wt.% silicon-50 wt.% fluoride composites: (**A,B**) 1-stage $50Si·50BiF_3$, (**C,D**) $50Si·50BiF_3$ and (**E,F**) $50Si·50CoF_2$.

A composite prepared using the two-stage protocol, 50Si·50BiF₃ is shown in Figure 1C,D. All particles show good mixing of the fluoride and silicon and reduced particle sizes in Figure 1C. A magnified image of a representative particle in Figure 1D shows angular silicon crystals decorated by bright particles of BiF_3.

The images of the two-stage milled 50Si·50CoF₂ powder are presented in Figure 1E,F. In Figure 1E, a large, agglomerated particle of 60 µm, with no discernable difference in phase contrast is seen. Such agglomerates were common. The inset in Figure 1E shows that the particle surface had distinct angular platelets and smaller particles. Through EDX, the larger platelets were confirmed to be Si and the smaller particles were found to be CoF_2. A smaller particle of 50Si·50CoF₂ is shown in Figure 1F. Distinct morphology enables one to identify both silicon and CoF_2 phases mixed rather uniformly on the nanoscale.

SEM images of powders with 30 wt.% silicon and 70 wt.% metal fluoride are presented in Figure 2. These images were obtained using secondary electrons to highlight the surface morphology, while losing on the phase contrast. General particle morphology for 30Si·70BiF₃ is illustrated in Figure 2A,B. Similarly, the morphology for 30Si·70CoF₂ is illustrated by Figure 2C,D.

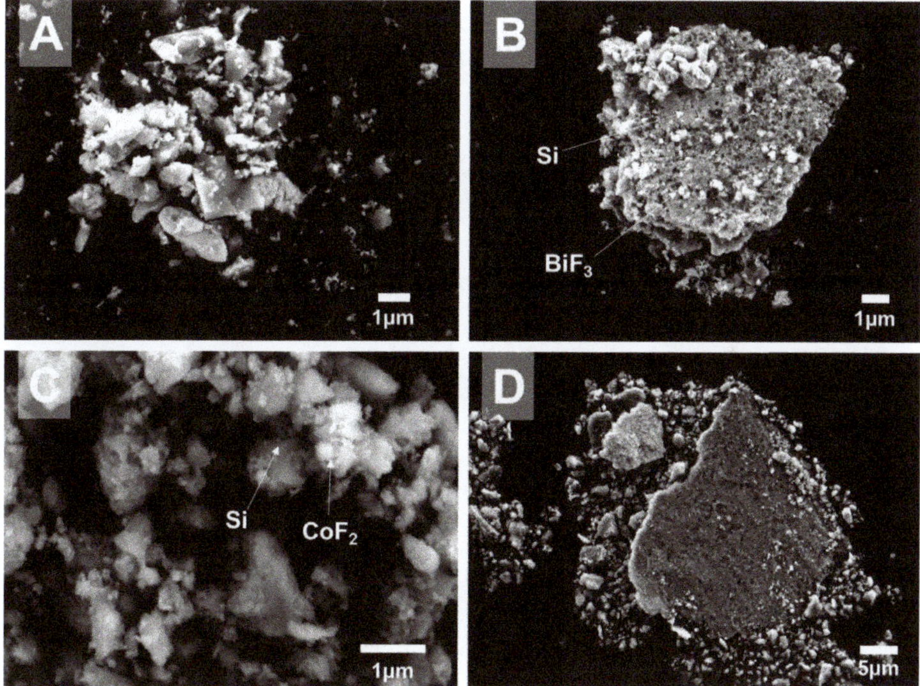

Figure 2. Secondary electron images of as milled 30 wt.% silicon-70 wt.% fluoride composites: (**A,B**) 30Si·70BiF₃ and (**C,D**) 30Si·70CoF₂.

A large amount of unattached BiF_3 particles and small crystalline silicon particles were observed in Figure 2A. Some relatively large BiF_3 particles decorated agglomerates of silicon crystals (Figure 2B); although the coating did not appear to be uniform.

In Figure 2C, cobalt fluoride was observed as small fuzzy particles covering jagged crystals of silicon. No unattached CoF_2 was observed. Generally, 30Si·70CoF₂ composite contained silicon particles with a broad size distribution and all Si surfaces were rather uniformly coated by cobalt fluoride, unlike 30Si·70BiF₃. Occasional large particles (>10 µm) were observed in both materials, as shown in Figure 2B,D. A flat surface, typically formed by milling balls pressing into a softer material, was commonly observed for such particles. SEM images show that such large particles consisted of fine silicon particles embedded in a fluoride matrix.

The XRD patterns of milled composites are presented in Figure 3. Peaks corresponding to BiF_3 and $Bi_7F_{11}O_5$ [31] were observed in all Bi-containing composites. Detectable amounts of reduced Bi were also observed. In the two-stage milled composites, the fluoride and reduced Bi peak intensities scale with the loaded BiF_3 concentration; stronger fluoride peaks were observed for $30Si\cdot70BiF_3$ as compared to $50Si\cdot50BiF_3$. Results of quantitative composition analysis for BiF_3-containing composites based on the whole pattern refinement performed using the X-pert Highscore software package [32] are shown in Table 3. The data suggest that $50Si\cdot50BiF_3$ retained only ca. 64% of the initial fluorine. On the other hand, the poorly refined 1-stage $50Si\cdot50BiF_3$ exhibited the weakest bismuth peaks and strongest $BiF_3/Bi_7F_{11}O_5$ peaks among the three $Si\cdot BiF_3$ composites; it retained approximately 71% if the initial fluorine.

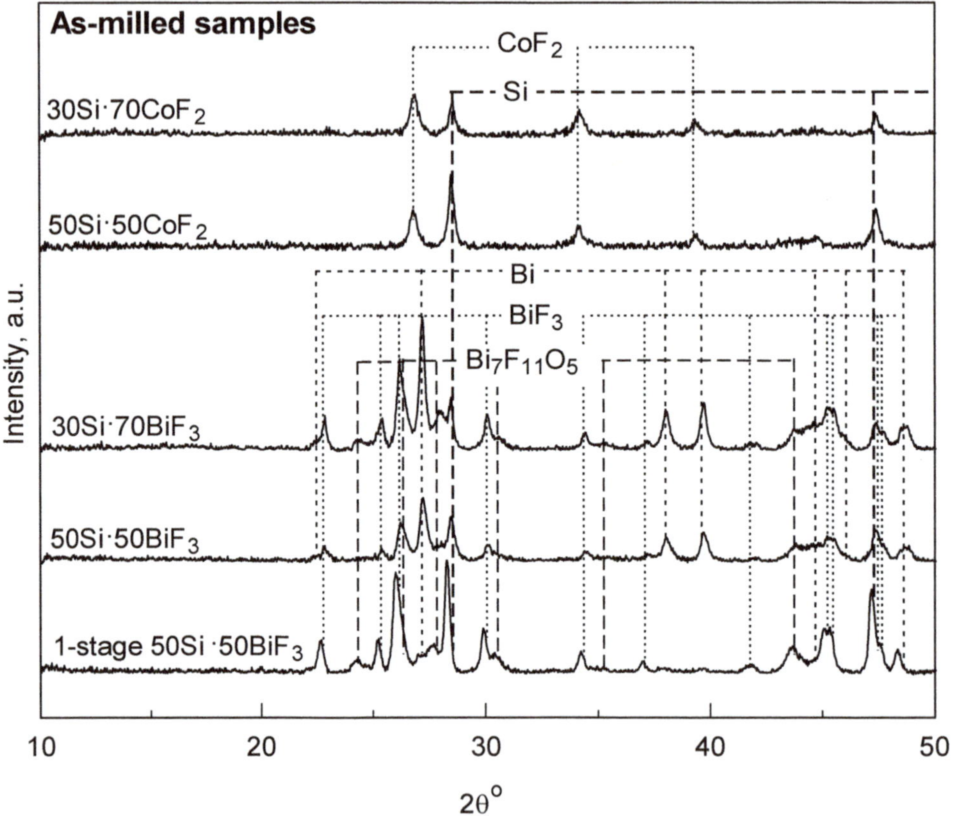

Figure 3. XRD patterns of the prepared composite powders.

The XRD pattern of the composites with CoF_2 shows peaks corresponding to silicon and cobalt fluoride. No reduced cobalt is observed. The scaling of the fluoride peak intensity as a function of the loaded fluoride concentration was less apparent.

Table 3. Compositions of the prepared Si·BiF$_3$ composites obtained from the whole pattern refinement of the XRD traces.

Sample	Mass (%)				
	Si	Bi	BiF$_3$	BiF$_{1.57}$O$_{0.71}$	BiF$_{1.98}$O$_{0.51}$
1-stage-50Si·50BiF$_3$	55.2	1.7	22.4	9.7	11.0
50Si·50BiF$_3$	45.6	12.0	16.6	17.3	8.5
30Si·70BiF$_3$	26.6	16.4	24.9	20.8	11.3

3.2. Thermal Analysis

The TG and their corresponding DTA traces for 50Si·50CoF$_2$ heated in both oxidizing and inert environments are shown in Figure 4. The DTA traces for the oxidizing and inert runs used separate vertical scales because of difference in the heat effects observed. The TG trace measured in Ar (Figure 4A) shows a single mass loss event with the onset below 673 K (400 °C); the total observed mass loss was close to 23% of the initial mass. The respective DTA trace (Figure 4B) shows weak features, which can be interpreted at two broad exothermic humps (peaks at 498 and 893 K (or 225 and 620 °C)) or a very broad exothermic feature overlaid with an endothermic event. To support the latter interpretation, much of the endothermic feature correlated with the observed mass loss.

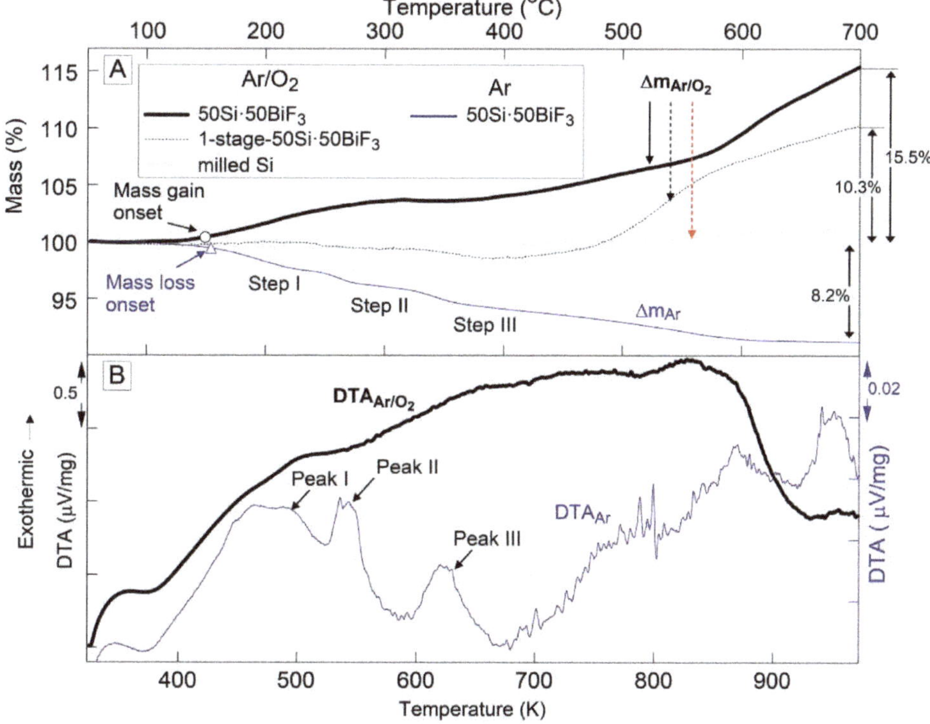

Figure 4. (**A**) Thermogravimetric analysis (TG) and corresponding (**B**) differential thermal analysis (DTA) traces for 50Si·50CoF$_2$ heated at 5 K/min in different environments.

For TG in Ar/O$_2$, the onset for a gradual mass gain was at 548 K (275 °C). The mass gain was reversed at 723 K (450 °C), and the maximum mass loss of ca. 8.3% was observed by 798 K (525 °C). After the minimum, the mass began increasing again. The corresponding DTA trace exhibited a rising baseline (despite the correction applied using an empty crucible). A small exothermic peak was noted at 280 °C, corresponding to the onset of the mass gain. A stronger exotherm was observed at 773 K (500 °C) correlating with the mass loss.

The TG and DTA plots for 50Si·50BiF$_3$ heated in different environments are presented in Figure 5. In Figure 5A, the oxidative TG runs for the milled/size-reduced Si and 1-stage 50Si·50BiF$_3$ were presented for comparison. No mass change is observed for the Si reference.

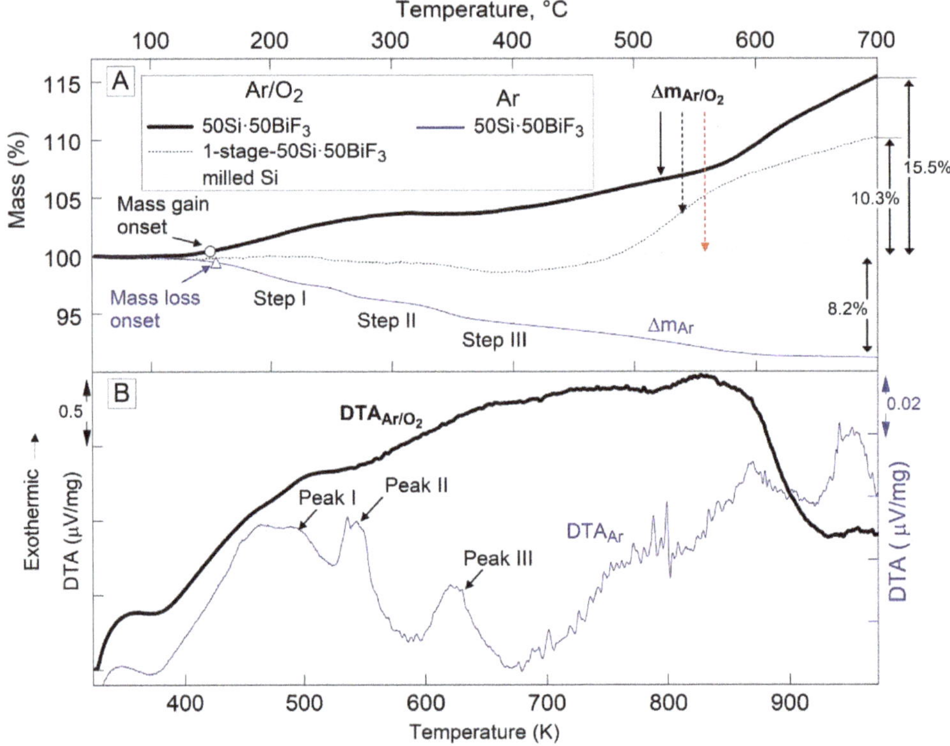

Figure 5. (**A**) TG and corresponding (**B**) DTA traces for 50Si·50BiF$_3$ in heated at 5 K/min in oxidizing and inert environments. The oxidative TG runs for 1-stage 50Si·50BiF$_3$ and milled Si are also shown.

The TG trace for 50Si·50BiF$_3$ in Ar shows a gradual mass loss of 8.2%. It begins around 423 K (150 °C) and ends by 873 K (600 °C). Weak but reproducible step-like features are noted in the TG trace; the onset for the mass loss are marked by a filled circle as labeled in Figure 5A. The corresponding DTA plot in Figure 5B, shows multiple weak peaks. It is possible to interpret the trace as showing a broad endothermic peak, similar to that in Figure 4. Additionally, several exothermic features are noted. The first three of these features, labeled in Figure 5B, correlated with the mass loss steps. The second exotherm, at ca. 543 K (270 °C; marked Peak II), had a small but reproducible superimposed endothermic feature.

For 50Si·50BiF$_3$ experiments in both Ar and Ar/O$_2$ were performed at different heating rates. The shift of the respective mass loss and mass gain points to higher temperatures at greater heating rates was used to consider relevance of the respective reactions to ignition, as discussed below.

In Ar/O$_2$, 1-stage milled 50Si·50BiF$_3$, exhibited a small mass loss (ca. 1.5%, onset at 583 K (310 °C)) and then mass gain, which was not complete by the end of the run. For the two-stage milled composite, 50Si·50BiF$_3$, heated in Ar/O$_2$, the mass began to increase at ca. 423 K (150 °C) as marked in Figure 5A. The mass was stabilized by about 593 K (320 °C). It began increasing again at 658 K (385 °C); the rate of mass gain increased around 873 K (600 °C). The final measured mass gain was 15.5%, it was not complete by the end of the run. The corresponding DTA curve for 50Si·50BiF$_3$ seen in Figure 5B exhibited a rising baseline indicating a broad exothermic process initiated at least from 393 K (120 °C). The exothermic feature ended around 923 K (650 °C).

For both oxidizing and inert environments runs for 50Si·50BiF$_3$, partially reacted samples were recovered from 473, 673 and 973 K (200, 400 and 700 °C).

To better resolve weak processes occurring in Ar, runs with the powder mass increasing from 5 to 10 and to 23 mg were performed for 50Si·50BiF$_3$. The respective TG traces are presented in Figure 6. The step-like features observed during mass loss were prominent for the lowest sample loading of 5 mg, for which a mass gain was also noted around 543 K (270 °C), which was close to the melting temperature of bismuth (marked in the Figure 6). The mass gain could have been caused by oxidation of the sample reacting with trace oxygen remaining in the furnace despite it being flushed with Ar. Melting of Bi could have accelerated such oxidation. When the sample was heated to a higher temperature of 1473 K (1200 °C), cf. 23 mg trace, a strong mass loss of 33% was observed with an onset at 1173 K (900 °C), which could only be associated with the loss of Bi.

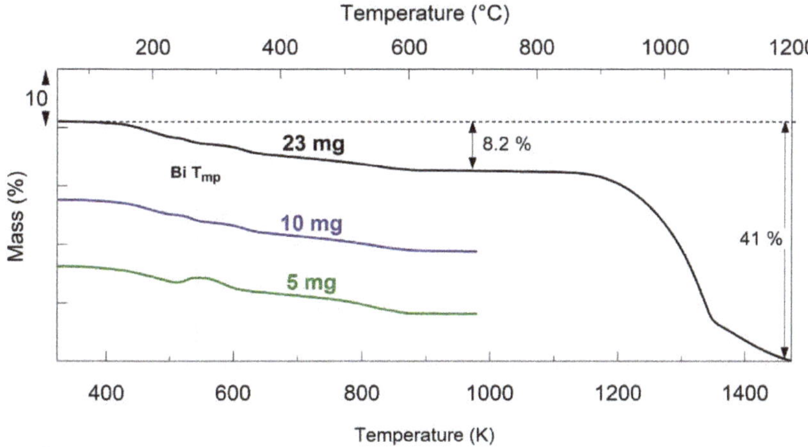

Figure 6. TG traces for different mass samples of 50Si·50BiF$_3$ heated in Ar at 5 K/min.

SEM images of partially reacted 50Si·50BiF$_3$ recovered from different temperatures in inert and oxidizing environments are presented in Figure 7. The powder heated to 700 °C in Ar (Figure 7A) shows bright spherical bismuth particles studded on darker silicon particles. The dispersion of bismuth across silicon particles was largely preserved. For the sample heated to 700 °C in Ar/O$_2$ (Figure 7B), elongated crystals, appearing to be Bi$_2$SiO$_5$ from EDX analysis, are observed; see the inset for details. Minor bright spheroidal particles rich in Bi were found decorating the surface of the darker Si-rich matrix. EDX shows oxygen present in all phases.

Figure 7C shows a characteristic particle heated to 1473 K (1200 °C) in Ar. It is porous and consists of several fused parts. Figure 7D provides a higher magnification image, presenting the surface morphology with multiple small spherical submicron bismuth-rich particles and lamellae of silicon crystal. In materials heated in both Ar and Ar/O$_2$, EDX detects iron and chromium; both contaminants came from the steel milling balls.

Figure 7. Backscattered electron images of the two-stage milled composite 50Si·50BiF$_3$ heated at 5 K/min and recovered from 973 K (700 °C): (**A**) in Ar, (**B**) in Ar/O$_2$ and recovered from 1473 K (1200 °C) in Ar (**C,D**).

XRD patterns for 50Si·50BiF$_3$ and 50Si·50CoF$_2$ heated to different temperatures in Ar and recovered for analysis are presented in Figure 8, along with the reference pattern for the as-milled material. The specific peaks of oxyfluoride Bi$_7$F$_{11}$O$_5$ in the as milled sample have not been presented in the Figure 8 due to space constraints.

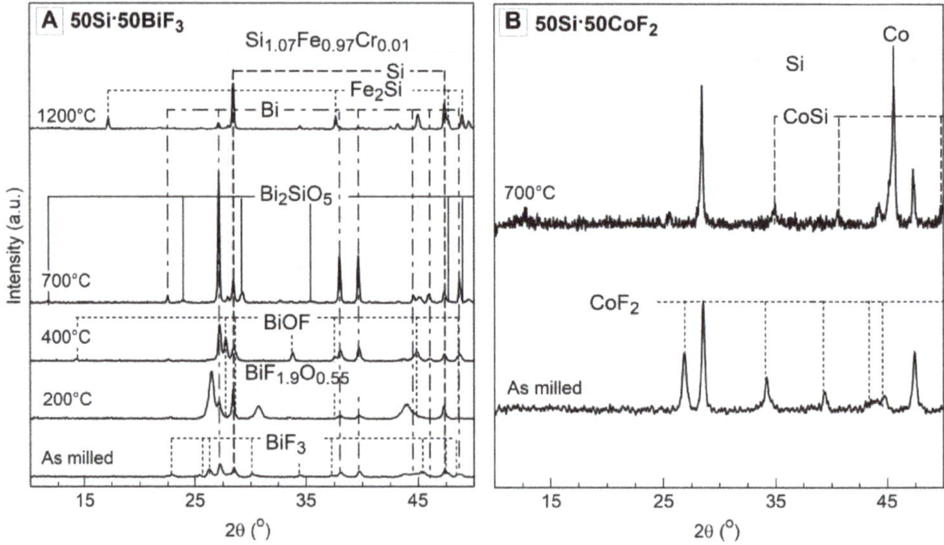

Figure 8. The XRD traces of the two-stage milled composites: (**A**) 50Si·50BiF$_3$ and (**B**) 50Si·50CoF$_2$ heated at 5 K/min in Ar and quenched at different temperatures.

Despite Ar purge, traces of oxygen were expected in these runs. For 50Si·50BiF$_3$ (Figure 8A) peaks associated to BiF$_3$ are depressed by 473 K (200 °C). Simultaneously, BiF$_{1.9}$O$_{0.55}$ peaks [33] became dominant and $Bi_7F_{11}O_5$ peaks from starting powder disappeared. At 673 K (400 °C), strong peaks of Bi, Si and the fluorine containing BiOF became predominant. By 973 K (700 °C), peaks corresponding to fluorinated species disappeared, remaining were peaks of Bi and Si and minor peaks attributed to Bi_2SiO_5 and Fe_2Si. At 1473 K (1200 °C), Bi peaks diminished, and peaks of Si became stronger. Patterns attributed to iron and chromium impurities, such as Fe_2Si and $Si_{1.07}Fe_{0.97}Cr_{0.01}$ were also noted.

For 50Si·50CoF$_2$, the XRD patterns of the starting powder and of the material heated to 700 °C in Ar are shown in Figure 8B. CoF_2 peaks disappeared for the heated material; aside from elemental Co and Si, a pattern for CoSi was observed.

XRD patterns for the powders heated to different temperatures in Ar/O$_2$ for 50Si·50BiF$_3$ and 50Si·50CoF$_2$ are presented in Figure 9. As seen in Figure 9A, the as milled 50Si·50BiF$_3$ had prominent peaks of BiF_3, Si, Bi and weaker peaks of $Bi_7F_{11}O_5$. The material recovered from 473 K (200 °C) exhibited weaker BiF_3 peaks while Bi and the oxyfluoride $BiF_{1.9}O_{0.55}$ peaks became stronger. By 673 K (400 °C), the BiF_3 peaks were absent, Si peaks diminished appreciably, while a new oxyfluoride species, BiOF was observed. By 973 K (700 °C), oxidized products, Bi_2SiO_5 and $Bi_6Cr_2O_{15}$ were the primary species. Minor peaks corresponding to Si were observed as well.

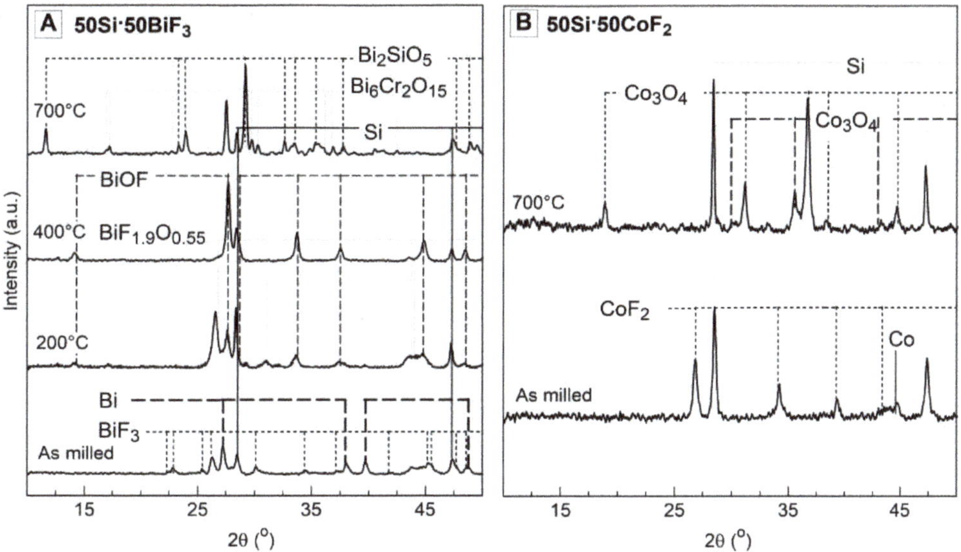

Figure 9. XRD patterns of the two-stage milled composites (**A**) 50Si·50BiF$_3$ and (**B**) 50Si·50CoF$_2$ heated at 5 K/min in Ar/O$_2$ and recovered from different temperatures.

The XRD patterns for as-milled 50Si·50CoF$_2$ and that quenched at 973 K (700 °C) in Ar/O$_2$ are presented in Figure 9B. Strong peaks of the oxide of cobalt, Co_3O_4 and unreacted Si were observed.

3.3. Heated Filament Ignition

High-speed video frames for all materials igniting on a heated filament are presented in Figure 10. The frames are labeled with the temperature of the filament.

All prepared materials exhibited ignition at low temperatures well before the filament turns incandescent. The incandescence of particles marks the ignition in fuel-rich 50 wt.% silicon compositions as indicated by arrows in the respective first frames. The 1-stage 50Si·50BiF$_3$ composite exhibited less bright, short-lived particle streaks within an incandescent plume. At higher temperatures (see 857 K) a weak secondary ignition event was observed as marked in the frame.

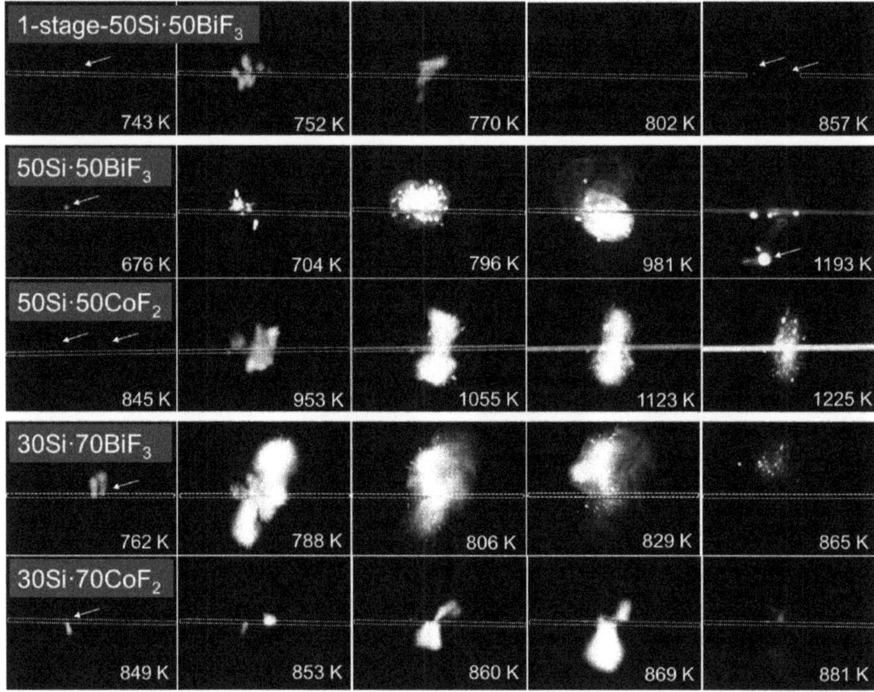

Figure 10. High-speed video frames of the ignition of 1-stage 50Si·50BiF₃ (heated at 2100 K/s) and two-stage milled materials; 50Si·50BiF₃ (heated at 3800 K/s), 50Si·50CoF₂ (heated at 2100 K/s), 30Si·70BiF₃ (heated at 2200 K/s) and 30Si·70CoF₂ (heated at 2200 K/s), on an electrically heated nichrome wire. The wire temperatures are indicated.

For 50Si·50BiF₃, bright smoke enveloped the burning particles flying off the filament, as seen in frames captured above 796 K. The 50Si·50CoF₂ material ignited producing a bright plume with few particle streaks, which became prominent at higher wire temperatures (>1123 K).

For materials with higher fluoride content, 30Si·70BiF₃ and 30Si·70CoF₂, ignition caused luminescent plume flares. For 30Si·70BiF₃, particle streaks and smoke were observable above 806 K. The 30Si·70CoF₂ composite generated a bright plume and no detectable burning particles.

The measured ignition temperatures of all prepared materials are shown as a function of heating rates in Figure 11. The error bars represent the scatter in 5 runs performed for each targeted heating rate. All ignition temperatures increase with the heating rate.

Lower ignition temperatures were measured for the materials with BiF₃ compared to those with CoF₂. The composites with 50 wt.% Si ignited consistently at lower temperatures compared to the composites with 30 wt.% Si. The 1-stage 50Si·50BiF₃ had a higher ignition temperature comparable to its two-stage milled analog, 50Si·50BiF₃. The secondary ignition for 1-stage 50Si·50BiF₃ could only be clearly detected at lower heating rates. At higher heating rates, this secondary ignition could have occurred at much higher temperatures and was not discernable due to the bright filament emission.

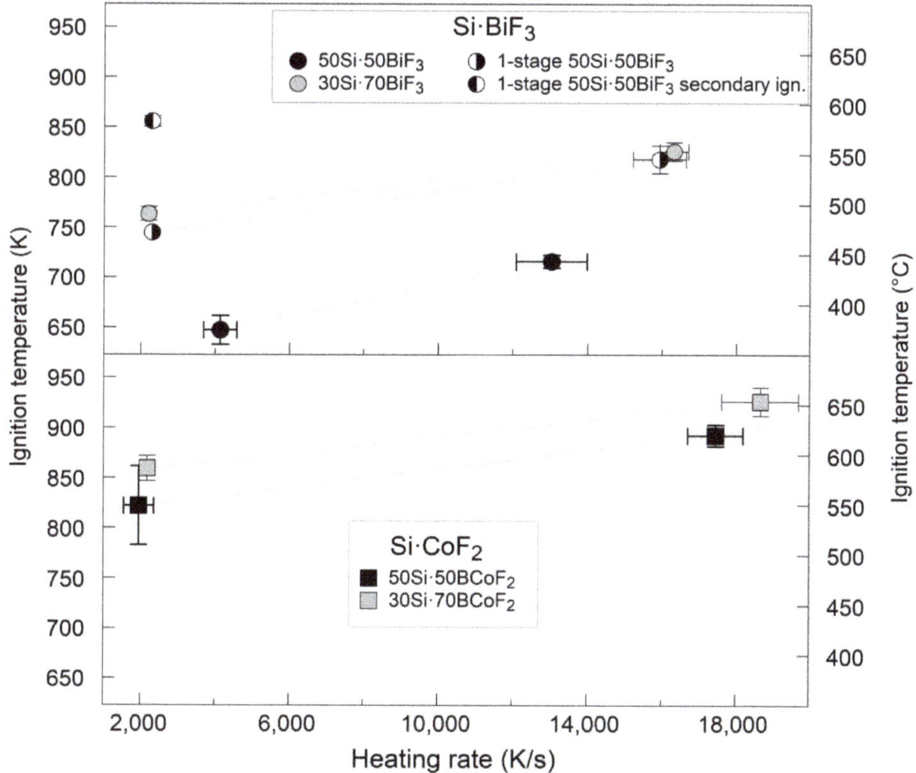

Figure 11. Ignition temperature in air of all prepared materials as a function of the heating rate.

3.4. Particle Combustion

In experiments employing the laser ignition of composite particles, 50Si·50CoF$_2$ powder could not be fed consistently. Conversely, 50Si·50BiF$_3$ powder was easy to feed; thus, only the experimental results with this powder are presented. Attempts to ignite milled Si powder, which could be fed into the laser beam, were not successful. Additionally, condensed combustion products could not be collected for analysis, likely because of the substantial amounts of gaseous combustion products.

A representative emission pulse of a laser-ignited 50Si·50BiF$_3$ particle burning in air is shown in Figure 12. All pulses observed had single peaks with minor oscillatory features. A distribution of the particle burn times is shown in Figure 13. It peaked at 0.43 ms with a tail trailing towards longer times.

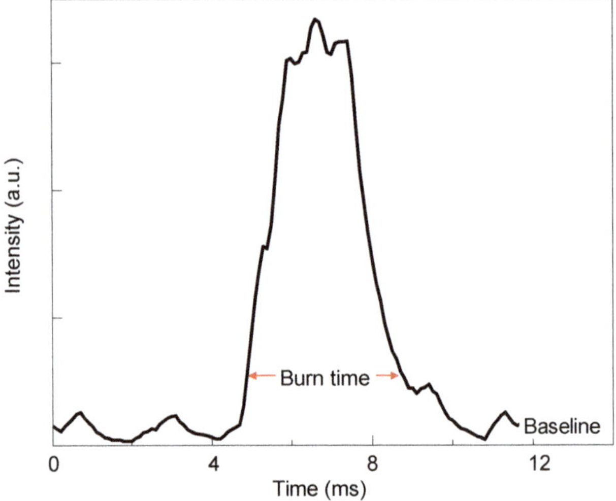

Figure 12. A representative emission pulse produced by a 50Si·50BiF$_3$ particle burning in air.

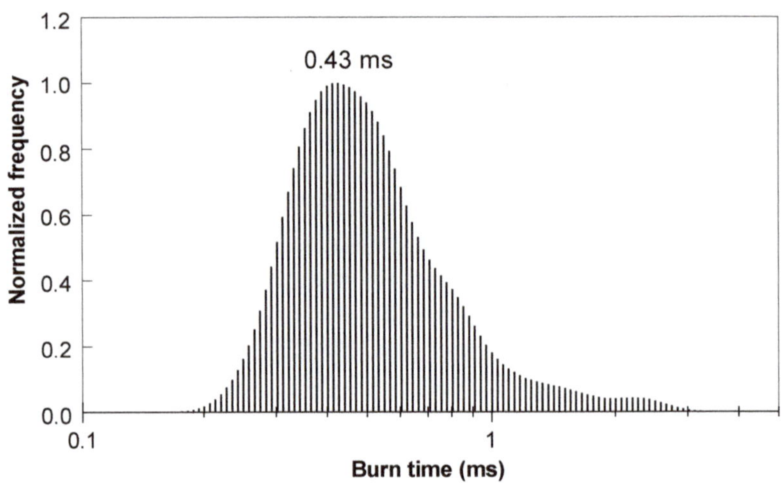

Figure 13. The burn time distribution for particles of 50Si·50BiF$_3$ burning in air.

The size distribution of the aerosolized 50Si·50BiF$_3$ particles passed through the feeder is shown in Figure 14. The distribution was nearly symmetrical, with a mode of 3.09 μm.

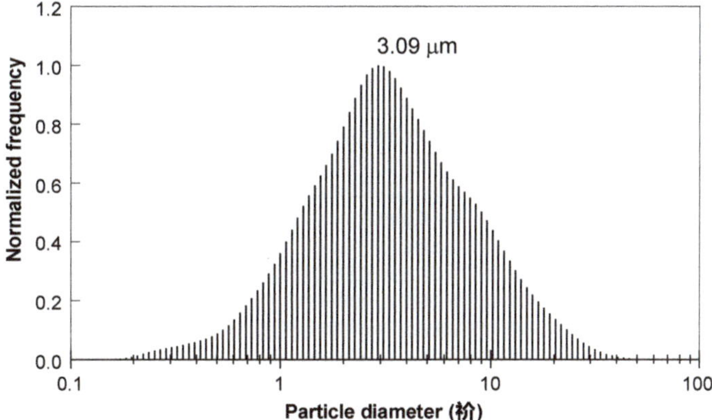

Figure 14. The particle size distribution for aerosolized 50Si·50BiF$_3$ powder passed through the feeder.

The burn time distribution was correlated with the particle size distribution for 50Si·50BiF$_3$; the correlation plot is presented in Figure 15 along with earlier results for other metal-metal fluoride composites obtained using the same data processing method [20,23]. Like Al and B-based composites using BiF$_3$ as an oxidizer, 50Si·50BiF$_3$ composite powder particles burn rapidly, with most burn times under 1 ms. These times were shorter than reported burn times for Al particles of the same sizes.

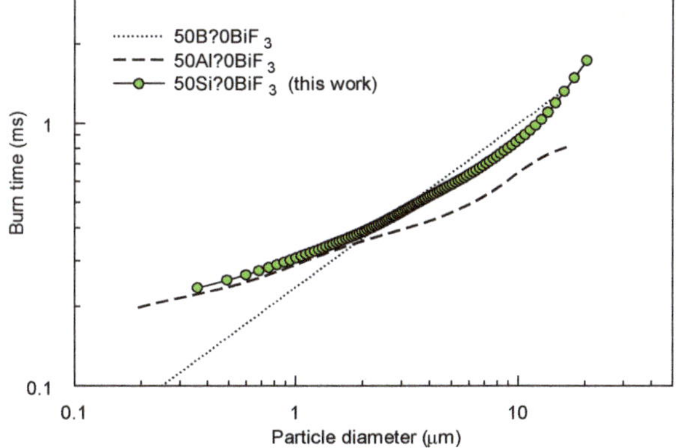

Figure 15. Burn time–particle size correlation for 50Si·50BiF$_3$ powder ignited in air by a laser. For comparison, similar correlations are shown for powders of 50Al·50BiF$_3$ [20] and 50B·50BiF$_3$ [23] obtained by arrested reactive milling.

The adiabatic flame temperature and predicted mole fractions of the combustion products of 50Si·50CoF$_2$ burning in air calculated by NASA CEA code [34] are presented in the Supplementary Section Figure S1. The particle surface temperature during combustion is expected to be close to 2750 K (predicted to be the adiabatic flame temperature), sufficiently high for gasifying a significant fraction of the main combustion product SiO$_2$.

4. Discussion

4.1. Material Preparation

A single-stage milling protocol does not effectively mix the physically different fluoride and silicon powders. Conversely, using premilled silicon for the second step milling with metal fluorides yields well-homogenized nanocomposite powders.

Materials with CoF_2 have distinct morphological features, where the fluoride formed a coating on the larger silicon particles. Even with increased fluoride content, no loose fluoride or uncoated silicon particles were observed in $30Si \cdot 70CoF_2$. It is surprising that such a fine scale of mixing was achieved without detectable mechanochemical reaction, e.g., lack of reduced Co detected by XRD of as-milled powders (see Figure 3). The XRD results were consistent with the thermal analysis, Figure 4A. The complete reduction of CoF_2 by silicon to form gaseous SiF_4 resulted in a mass loss of 26%. This was close to the mass loss of 23 wt.% measured by TG.

Based on XRD of the as-milled $50Si \cdot 50BiF_3$ showing the presence of reduced bismuth and bismuth oxyfluorides, it was estimated that BiF_3 made up only 28.5–30 wt.%, with the remaining 20–21.5 wt.% of BiF_3 lost before or during milling. Accounting for this initial amount of BiF_3 available in the as-milled powder, it is expected that the complete reaction of Si with the remaining F in the material would result in an observable mass loss of close to 8.9%. This agreed reasonably with the mass loss of 8.3% observed in TG (Figure 7A). While a smaller fraction of the fluoride reacted and was lost for the 1-stage $50Si \cdot 50BiF_3$, the mixing scale and homogeneity for that composite material were inadequate.

4.2. Low-Temperature Reactions

The TG for $50Si \cdot 50BiF_3$ and $50Si \cdot 50CoF_2$ in Ar shows a mass loss due to the evolution of SiF_4, described by reaction (1). As seen in Figure 4, SiF_4 formation in $50Si \cdot 50CoF_2$ occurred in a single step between 661 and 873 K (388 and 600 °C). Conversely, for $50Si \cdot 50BiF_3$, the mass decreased gradually, with discernable stages starting from ca. 423 K (150 °C) and extending to 873 K (600 °C).

In Ar/O_2 oxidation of the reduced metal and oxidation of silicon occur parallel to fluorination of silicon. Both oxide formation reactions form condensed products and thus increase the mass. Assuming for simplicity that the fluorination occurs similarly in both Ar and Ar/O_2 environments, the mass gain due to oxidation, Δm_{ox}, can be estimated from the mass difference between measurements in Ar/O_2 and in Ar:

$$\Delta m_{ox} = \Delta m_{Ar/O_2} - \Delta m_{Ar} \tag{2}$$

This change of mass is shown in Figure 16. The mass gain due to oxidation consists of two components, caused by oxidation of silicon, Δm_{ox}^{Si} and that of the reduced metal, Δm_{ox}^{Me}. The value of Δm_{ox}^{Me} can be estimated from the mass measured in Ar, assuming Δm_{Ar} is entirely due to evolution of SiF_4 and assuming that all the reduced metal oxidizes immediately when the reaction occurs in Ar/O_2:

$$\Delta m_{ox}^{Me} = \Delta m_{Ar}^{Me} = \frac{M_{Me}}{M_{SiF_4}} \cdot \frac{x}{4} \Delta m_{Ar} \tag{3}$$

Subtracting this mass from Δm_{ox} gives:

$$\Delta m_{ox}^{Si} = \Delta m_{ox} - \Delta m_{ox}^{Me} \tag{4}$$

Finally, the rate of Si oxidation can be obtained dividing the derivative $d\left(\Delta m_{ox}^{Si}\right)/dt$ by the mass of Si remaining in the material accounting for SiF_4 leaving as gas (Equation (1)), also based on the TG measured in Ar.

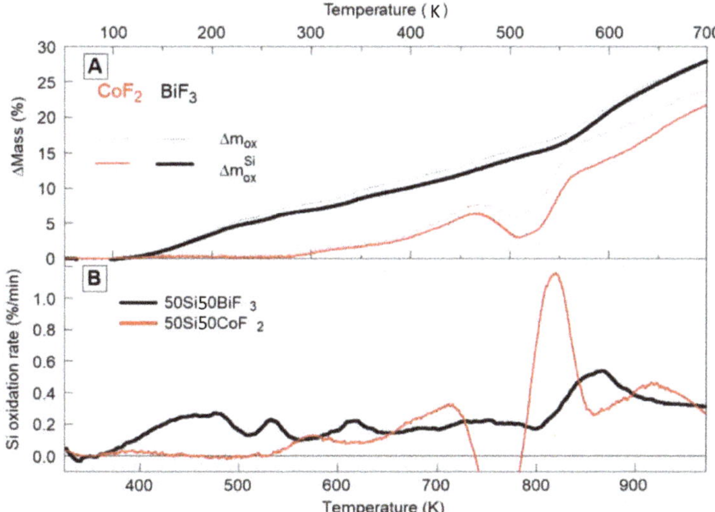

Figure 16. The (**A**) oxidative mass gain and corresponding (**B**) silicon oxidation rate of two-stage milled materials 50Si·50BiF$_3$ and 50Si·50CoF$_2$.

Processing experimental traces using Equations (2) and (4) yields traces for Δm_{ox}, Δm_{ox}^{Si}, and the rate, $\frac{1}{m_{Si}} \frac{d\left(\Delta m_{ox}^{Si}\right)}{dt}$ shown in Figure 16. Figure 16B shows that the rate of silicon oxidation parallels the observed rate of oxidation in the Ar/O$_2$ environment. Since only as much Me oxide can form as MeF$_x$ has been reduced, the bulk of the observed mass gain represents the oxidation of Si. The apparent negative Si oxidation rate seen for 50Si·50CoF$_2$ at between 450 and 500 °C suggests that the assumption that SiF$_4$ formation is identical in Ar and Ar/O$_2$ environments is not strictly true.

For 50Si·50CoF$_2$, Figure 16 suggests that oxidation of Si began at a higher temperature than the corresponding SiF$_4$ formation (cf. Figure 4); it also began at a higher temperature than for 50Si·50BiF$_3$, for which oxidation began simultaneously with fluorination (cf. Figure 5). For both composites, the oxidation rate of Si increased initially and then became relatively stable over a range of temperatures. For both composites the low-temperature rates of heat release associated with the observed initial rate of oxidation of Si could be added to the rates of heat release caused by the Si fluorination occurring simultaneously (e.g., recovered from data in Figures 4 and 5 for 50Si·50CoF$_2$ and 50Si·50BiF$_3$, respectively) to describe the exothermic reactions leading to ignition of such composite materials.

Aside of relatively small changes in the reaction rate for 50Si·50BiF$_3$, a stepwise increase in the oxidation rate is observed between 803 and 973 K (530 and 600 °C). As discussed below, this accelerated oxidation at relatively high temperatures was unlikely to affect ignition of the prepared composite. For 50Si·50CoF$_2$ the apparent oxidation rate of Si dropped around 723 K (450 °C). However, based on Figure 5, it can be concluded that this effect was superficial and simply means that the fluorination of Si accelerated rapidly at such temperatures, with the associated mass loss becoming stronger than mass gain due to oxidation of both Si and reduced Co. Indeed, after a while, the oxidation rate of Si returned to about the same level it was at just below 723 K.

4.3. Reactions Leading to Ignition

The onsets of TG features observed in both inert and oxidizing environment runs for both 50Si·50BiF$_3$ and 50Si·50CoF$_2$ were plotted along with their respective heated filament ignition temperatures in the Kissinger plot shown in Figure 17. The onsets of the mass loss observed in Ar and mass gain in Ar/O$_2$ for 50Si·50BiF$_3$ nearly coincided with each other for different heating rates. Extrapolating the kinetic trend from the ignition data points down to the range of low heating rates used in thermal analysis pointed clearly to the onsets of the observed mass gain (in Ar/O_2) and loss (in Ar) as reactions associated with ignition. As discussed above, for 50Si·50BiF$_3$, both fluorination and oxidation of Si occurred nearly simultaneously with both exothermic reactions contributing to igniting the material.

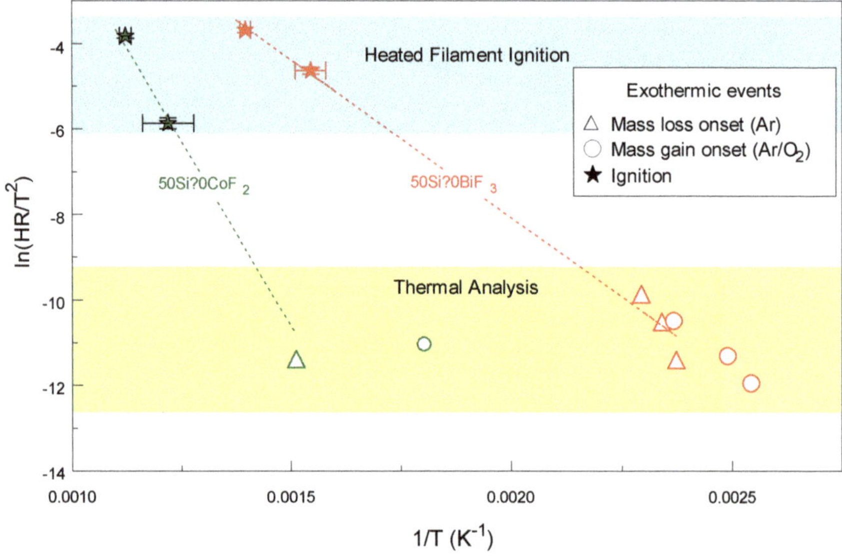

Figure 17. Kissinger plot combining TG and DTA features and heated filament ignition data for two-stage milled 50Si·50BiF$_3$ and 50Si·50CoF$_2$.

The onset of mass loss observed in Ar for 50Si·50CoF$_2$ (for which the measurement was only made at one heating rate) was shifted from that of the mass gain observed for this material heated in Ar/O_2 at a lower temperature. The kinetic trend implied by the respective ignition temperatures lined up better with the mass loss observed in Ar, and thus it was likely that the ignition was mostly associated with SiF$_4$ formation for this composite. It is nonetheless expected that the low-temperature oxidation of Si, once started, assisted ignition. The apparent activation energy that can describe ignition was roughly evaluated from the slopes of line fits shown in Figure 17. These activation energies were around 62 and 140 kJ/mol for 50Si·50BiF$_3$ and 50Si·50CoF$_2$, respectively. For 50Si·50BiF$_3$, the inferred activation energy was comparable to that reported for fluorination of silicon wafers by elemental fluorine at reduced fluorine partial pressures [35].

The higher temperature reactions are not shown but would be shifted far left in Figure 17 and thus would be clearly irrelevant for ignition.

4.4. Particle Combustion

The micron-sized silicon particles could not be initiated in the laser-assisted combustion experiment. Conversely, the prepared 50Si·50BiF$_3$ composite could be readily initiated and burned in the time frame comparable to other reactive metal-fluoride composite powders. Most likely, initial fluorination of Si made it easier to initiate the reaction; however, it is expected that the laser energy was sufficient to rapidly heat Si particles, which did not sustain combustion once exposed to the room temperature air. It is thus suggested that the presence of reduced or partially reduced Bi accelerates both low- and high-temperature heterogeneous oxidation of Si, assisting in establishing and sustaining the particle flame.

As estimated by the CEA code (Figure S1), the high adiabatic flame temperatures volatilize a significant mass fraction of refractory silicon oxides formed during combustion of $50Si\cdot50CoF_2$. In the case of $50Si\cdot50BiF_3$, the mass fraction of volatile species might be higher because both bismuth and bismuth oxide vaporize easily. However, once bismuth is oxidized in the presence of a more reactive metal fuel, it is likely to be immediately reduced, serving as an oxygen shuttle. Thus, presence of Bi can effectively accelerate oxidation of Si, similarly to how it was suggested to accelerate oxidation of Al [20] and B [23]. When BiF_3 serves as an oxidizer with different fuels, including Al, B and Si, gas phase products (including fluorides and oxyfluorides of the respective fuels) are favored a the high combustion temperatures. At the same time, the surface reaction may be governed by oxidation of Bi. Respectively, similar rates of combustion, with heterogeneous surface reactions yielding volatile products, are expected for all such composite materials. The rate of such reactions is expected to be controlled by diffusion of the oxidizer (oxygen) to the particle surface in all cases. This reasoning can explain comparable burn times observed for different composites, $50Al\cdot50BiF_3$, $50B\cdot50BiF_3$ and $50Si\cdot50BiF_3$ for similarly sized particles (Figure 15)

5. Conclusions

Preparing Si-based composites with BiF_3 and CoF_2 as oxidizers required a premilling step to refine the starting commercial Si powder in order to achieve nanoscale mixing between the material components. Preparation of well homogenized composites of Si with BiF_3 could not be achieved without partial reaction of BiF_3; however, no such reaction of CoF_2 was observed. Upon heating, oxidation of Si occurs at lower temperatures than its fluorination when the added oxidizer is CoF_2. Conversely, both oxidation and fluorination occurred nearly simultaneously and at a lower temperature with BiF_3 as an oxidizer. The presence of reduced Co and Bi accelerated oxidation of Si upon its heating in an oxygen-containing environment.

Unlike elemental Si, prepared composite powders ignited readily when placed as a coating on an electrically heated filament. Ignition of the composite with BiF_3 as an oxidizer occurred at consistently lower temperatures than that of the composite with CoF_2. Results suggest that low-temperature fluorination and oxidation caused ignition for the composites with BiF_3 as an oxidizer; apparent activation energy for reactions leading to ignition was close to 62 kJ/mol. For the composites with CoF_2 as an oxidizer, it is likely that ignition is associated with Si fluorination.

Prepared composite powder with the BiF_3 oxidizer ignited readily when passed through a CO_2 laser beam in room temperature air. The measured particle burn times were nearly the same as those of similar composites combining BiF_3 with such fuels as Al and B. Combustion products were mostly gaseous making it impossible to collect and analyze produced condensed species.

Supplementary Materials: The following are available online at http://www.mdpi.com/2079-4991/10/12/2367/s1, Figure S1: The calculated adiabatic flame temperature and the combustion products presented as mole fractions at a range of equivalence ratios for $50Si\cdot50CoF2$.

Author Contributions: S.K.V. performed interpreted experiments, prepared first draft of the manuscript; M.S. discussed experimental results and correlated material properties with the reactions observed; E.D. set objectives of the effort, analyzed results, and edited the manuscript. All authors have read and agreed to the published version of the manuscript.

Funding: This research was funded in parts by the Office of Naval Research, Grant N00014-19-1-2048 and by the Defense Threat Reduction Agency, Grant HDTRA1-17-1-0044.

Conflicts of Interest: The authors declare no conflict of interest.

References

1. Koch, E.; Clément, D. Special Materials in Pyrotechnics: VI. Silicon—An Old Fuel with New Perspectives. *Propellants Explos. Pyrotech.* **2007**, *32*, 205–212. [CrossRef]
2. Yaroshevsky, A.A. Abundances of chemical elements in the Earth's crust. *Geochem. Int.* **2006**, *44*, 48–55. [CrossRef]

3.   Gelain, C.; Cassuto, A.; Le Goff, P. Kinetics and mechanism of low-pressure, high-temperature oxidation of silicon-II. *Oxid. Met.* **1971**, *3*, 139–151. [CrossRef]

4.   Rugunanan, R.A.; Brown, M.E. Reactions of powdered silicon with some pyrotechnic oxidants. *J. Therm. Anal. Calorim.* **1991**, *37*, 1193–1211. [CrossRef]

5.   Deal, B.E.; Grove, A.S. General Relationship for the Thermal Oxidation of Silicon. *J. Appl. Phys.* **1965**, *36*, 3770–3778. [CrossRef]

6.   Cabrera, N.; Mott, N.F. Theory of the oxidation of metals. *Rep. Prog. Phys.* **1949**, *12*, 163–184. [CrossRef]

7.   Braun, G.; Boden, G.; Henkel, K.; Rossbach, H. Thermal analysis of the direct nitridation of silicon to Si3N4. *J. Therm. Anal. Calorim.* **1988**, *33*, 479–485. [CrossRef]

8.   Rugunanan, R.A.; Brown, M.E. Combustion of Binary and Ternary Silicon/Oxidant Pyrotechnic Systems, Part I: Binary Systems with Fe_2O_3 and SnO_2 as Oxidants. *Combust. Sci. Technol.* **1993**, *95*, 61–83. [CrossRef]

9.   Mason, B.A.; Groven, L.; Son, S.; Yetter, R.A. Combustion Performance of Several Nanosilicon-Based Nanoenergetics. *J. Propuls. Power* **2013**, *29*, 1435–1444. [CrossRef]

10.   Anil Rugunanan, R.; Brown, M.E. Combustion of Binary and Ternary Silicon/Oxidant Pyrotechnic Systems, Part III: Ternary Systems. *Combust. Sci. Technol.* **1993**, *95*, 101–115. [CrossRef]

11.   Anil Rugunanan, R.; Brown, M.E. Combustion of Binary and Ternary Silicon/Oxidant Pyrotechnic Systems, Part IV: Kinetic Aspects. *Combust. Sci. Technol.* **1993**, *95*, 117–138. [CrossRef]

12.   Piekiel, N.W.; Churaman, W.A.; Morris, C.J.; Currano, L.J. Combustion and material characterization of porous silicon nanoenergetics. In Proceedings of the 2013 IEEE 26th International Conference on Micro Electro Mechanical Systems (MEMS), Taipei, Taiwan, 20–24 January 2013; pp. 449–452.

13.   Piekiel, N.W.; Morris, C.J.; Currano, L.J.; Lunking, D.M.; Isaacson, B.; Churaman, W.A. Enhancement of on-chip combustion via nanoporous silicon microchannels. *Combust. Flame* **2014**, *161*, 1417–1424. [CrossRef]

14.   Plummer, A.; Kuznetsov, V.; Gascooke, J.R.; Shapter, J.G.; Voelcker, N.H. Laser shock ignition of porous silicon based nano-energetic films. *J. Appl. Phys.* **2014**, *116*, 54912. [CrossRef]

15.   Kasi, S.R.; Liehr, M.; Cohen, S. Chemistry of fluorine in the oxidation of silicon. *Appl. Phys. Lett.* **1991**, *58*, 2975–2977. [CrossRef]

16.   Yarrington, C.D.; Son, S.; Foley, T.J. Combustion of Silicon/Teflon/Viton and Aluminum/Teflon/Viton Energetic Composites. *J. Propuls. Power* **2010**, *26*, 734–743. [CrossRef]

17.   Terry, B.C.; Son, S.; Groven, L.J. Altering combustion of silicon/polytetrafluoroethylene with two-step mechanical activation. *Combust. Flame* **2015**, *162*, 1350–1357. [CrossRef]

18.   Terry, B.C.; Lin, Y.-C.; Manukyan, K.; Mukasyan, A.; Son, S.; Groven, L.J. The Effect of Silicon Powder Characteristics on the Combustion of Silicon/Teflon/Viton Nanoenergetics. *Propellants Explos. Pyrotech.* **2014**, *39*, 337–347. [CrossRef]

19.   Valluri, S.K.; Bushiri, D.; Schoenitz, M.; Dreizin, E.L. Fuel-rich aluminum–nickel fluoride reactive composites. *Combust. Flame* **2019**, *210*, 439–453. [CrossRef]

20.   Valluri, S.K.; Schoenitz, M.; Dreizin, E. Combustion of Aluminum-Metal Fluoride Reactive Composites in Different Environments. *Propellants Explos. Pyrotech.* **2019**, *44*, 1327–1336. [CrossRef]

21.   Valluri, S.K.; Ravi, K.K.; Schoenitz, M.; Dreizin, E.L. Effect of boron content in B·BiF_3 and B·Bi composites on their ignition and combustion. *Combust. Flame* **2020**, *215*, 78–85. [CrossRef]

22.   Valluri, S.K.; Schoenitz, M.; Dreizin, E. Bismuth fluoride-coated boron powders as enhanced fuels. *Combust. Flame* **2020**, *221*, 1–10. [CrossRef]

23.   Valluri, S.K.; Schoenitz, M.; Dreizin, E. Combustion of Composites of Boron with Bismuth and Cobalt Fluorides in Different Environments. *Combust. Sci. Technol.* **2019**, 1–16. [CrossRef]

24.   Valluri, S.K.; Schoenitz, M.; Dreizin, E. Boron-Metal Fluoride Reactive Composites: Preparation and Reactions Leading to Their Ignition. *J. Propuls. Power* **2019**, *35*, 802–810. [CrossRef]

25.   Valluri, S.K.; Monk, I.; Schoenitz, M.; Dreizin, E.L. Fuel-Rich Aluminum-Metal Fluoride Thermites. *Int. J. Energetic Mater. Chem. Propuls.* **2017**, *16*, 81–101. [CrossRef]

26.   Badiola, C.; Zhu, X.; Schoenitz, M.; Dreizin, E. Aluminum Rich Al-CuO Nanocomposite Materials Prepared by Arrested Reactive Milling at Cryogenic and Room Temperature. In Proceedings of the 47th AIAA Aerospace Sciences Meeting including the New Horizons Forum and Aerospace Exposition, Orlando, FL, USA, 5–8 January 2009. [CrossRef]

27.   Ward, T.S.; Trunov, M.A.; Schoenitz, M.; Dreizin, E.L. Experimental methodology and heat transfer model for identification of ignition kinetics of powdered fuels. *Int. J. Heat Mass Transf.* **2006**, *49*, 4943–4954. [CrossRef]

28. Corcoran, A.; Mercati, S.; Nie, H.; Milani, M.; Montorsi, L.; Dreizin, E. Combustion of fine aluminum and magnesium powders in water. *Combust. Flame* **2013**, *160*, 2242–2250. [CrossRef]

29. Mohan, S.; Trunov, M.A.; Dreizin, E. Heating and Ignition of Metallic Particles by a CO_2 Laser. *J. Propuls. Power* **2008**, *24*, 199–205. [CrossRef]

30. Chintersingh, K.-L.; Schoenitz, M.; Dreizin, E.L. Combustion of boron and boron–iron composite particles in different oxidizers. *Combust. Flame* **2018**, *192*, 44–58. [CrossRef]

31. Laval, J.P.; Champarnaudmesjard, J.C.; Frit, B.; Britel, A.; Mikou, A. BI7F11O5—A new ordered anion-excess fluorite-related structure withcolumnar clusters. *Eur. J. Solid State Inorg. Chem.* **1994**, *31*, 943–956.

32. Degen, T.; Sadki, M.; Bron, E.; König, U.; Nénert, G. The HighScore suite. *Powder Diffr.* **2014**, *29*, S13–S18. [CrossRef]

33. Laval, J.; Champarnaud-Mesjard, J.; Britel, A.; Mikou, A. $Bi(F,O)_{2.45}$: An Anion-Excess Fluorite Defect Structure Deriving from Rhombohedral LnFO Type. *J. Solid State Chem.* **1999**, *146*, 51–59. [CrossRef]

34. McBride, B.J.; Gordon, S. *Computer Program for Calculation of Complex Chemical Equilibrium Compositions and Applications II. Users Manual and Program Description*; NASA RP 1311; NASA Glenn Research Center: Cleveland, OH, USA, 1996.

35. Kuriakose, A.K.; Margrave, J.L. Kinetics of Reaction of Elemental Fluorine. III. Fluorination of Silicon and Boron1. *J. Phys. Chem.* **1964**, *68*, 2671–2675. [CrossRef]

Publisher's Note: MDPI stays neutral with regard to jurisdictional claims in published maps and institutional affiliations.

 *nanomaterials*

Communication

Layered Al/CuO Thin Films for Tunable Ignition and Actuations

Ludovic Salvagnac, Sandrine Assie-Souleille and Carole Rossi *

CNRS, LAAS, University of Toulouse, 7 Avenue du Colonel Roche, F-31400 Toulouse, France;
salvagnac@laas.fr (L.S.); souleille@laas.fr (S.A.-S.)
* Correspondence: rossi@laas.fr

Received: 20 August 2020; Accepted: 8 October 2020; Published: 12 October 2020

Abstract: Sputter-deposited Al/CuO multilayers are used to manufacture tunable igniters and actuators, with applications in various fields such as defense, space and infrastructure safety. This paper describes the technology of deposition and the characteristics of Al/CuO multilayers, followed by some examples of the applications of these energetic layers.

Keywords: nanothermite; pyroMEMS; nanoenergetics; reactive thin film

1. Introduction

Energetic materials are widely used to suddenly generate high amounts of thermal or mechanical energy under an electrical, mechanical or thermal stimulus. As typical energetic materials, nanothermites, which contain Al and oxide, have attracted much attention as they exhibit not only better combustion efficiencies and better ignitability compared to traditional explosives, but also their reaction outputs can be tuned, thanks to the selections of fuels, oxidizers, architectures and reactant size, allowing multiple actions. Among nanothermites, sputter-deposited Al/CuO multilayers represent interesting energetic thin film nanomaterials for tunable ignition, to replace old hot-wire ignitors.

In that context, our research group demonstrated several miniature pyrotechnical ignition devices integrating Al/CuO multilayers within microelectromechanical systems (MEMS) based microheaters, called pyroMEMS, for civilian and military applications such as triggering the inflation of airbags, micro-propulsion systems, and arm and fire devices.

This paper briefly reviews the technological process, focusing on the sputter-deposition technique, presents the properties of Al/CuO multilayered films, details the fabrication process flow of a micro-igniter, and present examples of applications of pyroMEMS.

2. Al/CuO Multilayered Nanothermites

Two decades ago, the introduction of nanotechnologies enabled the emergence of a new class of energetic materials called nanothermites that could play a great role in future society. Nanothermites use raw materials that can be found in abundance, are low-cost and non-polluting (green materials). They also exhibit better combustion efficiencies and better ignitability compared to typical CHNO energetic mixtures, while being safer. Importantly, the reactions and ability to trigger specific actions can be tailored by varying the size and composition of the oxide, allowing multiple applications. Focusing on thermite multilayers, the overwhelming majority of works concern Al/CuO systems, as they feature an exceptionally high-energy release with gaseous production.

2.1. Multilayers Deposition

Al/CuO thermite multilayers are mainly produced by the sputter-deposition technique [1–5], as this provides excellent control over the layering and stoichiometry. Cupric oxide thin film is synthesized using the direct current (DC) reactive magnetron sputtering technique in an oxygen-enriched environment. To produce Al/CuO multilayered films, layers of Al and CuO are deposited on top of each other in an alternating fashion (see Figure 1) without venting the chamber. However, after each layer of deposition, the sample stage is cooled at ambient temperature for 600 s. At each interface between the cupric oxide and Al layers, an intermixing layer of 4–8 nm is present that forms during the deposition itself.

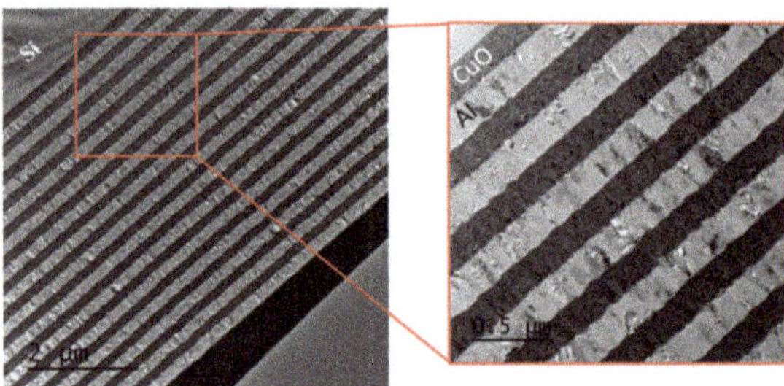

Figure 1. Cross-sectional transmission electron micrographs of the Al/CuO multilayer obtained by magnetron sputtering.

Depositions parameters have been published several times and can be found in [6,7].

2.2. Ignition and Reaction Properties

Applying an external source of energy locally on the multilayer results in an increase in the temperature of the multilayer section directly in contact with the heated surface. This can be done by electrostatic discharge [8], mechanical impact [9], laser irradiation [10], electrical heating (spark) [11] or thermal hot points [12], the latter of which is the widest used. After being ignited, the multilayers react by a self-sustained propagating reaction following the chemistry of Equation (1).

$$2Al + 3CuO \rightarrow Al_2O_3 + 3Cu\ (+ \Delta H) \tag{1}$$

In other words, after the temperature is raised locally and rapidly to a characteristic onset reaction temperature, the reaction enthalpy (ΔH) spreads into neighboring unreacted portions of the film, leading to a self-sustained reaction wave, which is highly luminous.

As the exothermic chemical reactions are controlled by the outward migration of oxygen atoms from the CuO matrix towards the aluminum layers, the reaction rate strongly depends on the reactant spacing (bilayer thickness). The sustained combustion rate is commonly characterized with a standard high-speed camera under ambient pressure [5,7].

The combustion velocity increases for thinner bilayers, with a maximum at ~90 m/s for free-standing foils (see Figure 2a). However, the reaction stops suddenly for extremely small bilayer thicknesses, which is due to the presence of interfacial layers, as described earlier, which reduces the amount of stored energy, i.e., reaction enthalpy, ΔH.

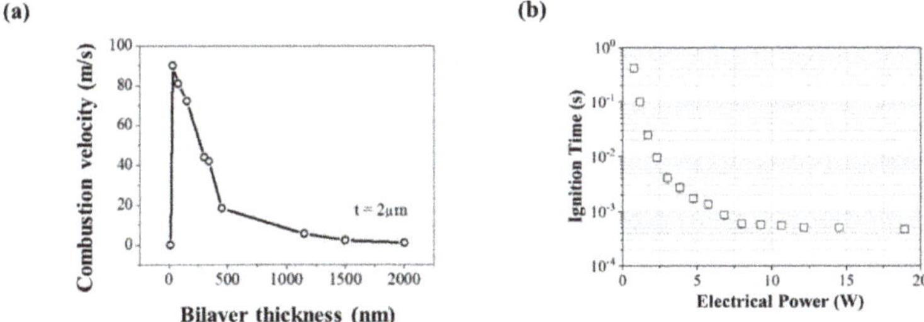

Figure 2. (**a**) Sustained combustion velocity in air of free-standing multilayers as a function of bilayer thickness (total thickness t = 2 μm); (**b**) ignition time as a function of the electrical power sent through the micro-heater for 5 bilayers of Al/CuO.

3. PyroMEMS Integration and Characterization

When integrated onto a device/substrate, the widest and simplest ignition method for igniting an Al/CuO multilayered thin film is using hot-wire, which functions via local metallic thin film resistance. A pyroMEMS is mainly composed of a substrate, thin metallic resistance and the Al/CuO nanothermites (see Figure 3a).

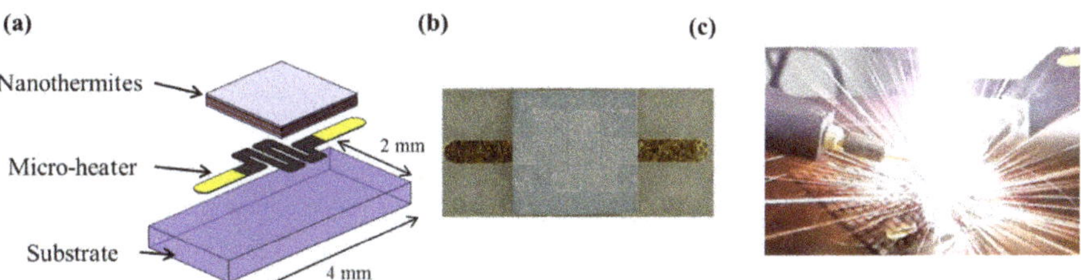

Figure 3. (**a**) 3D schematic representation of a pyroMEMS; (**b**) photo of the pyroMEMS; (**c**) photo during the nanothermite reaction.

To minimize the ignition energy, the substrate must be a good thermal insulator. We chose 500 μm thick 4-inches glass substrates AF32 (Schott AG, Mainz, Germany). After being cleaned in oxygen plasma to remove surface contaminants, 300 nm thick titanium followed by 300 nm thick gold layers are evaporated onto the glass substrate, and the filament is patterned using the lift-off technique. Then the gold is chemically etched and patterned to define the Ti resistance and Au electrical pads. Finally, Al/CuO multilayers are sputter-deposited through a silicon shadow mask. A photo of a pyroMEMS thus manufactured is presented in Figure 3b.

We characterized that, for any heating surface areas (micro-heater sizes), the ignition time (delay between the time of application of the electrical power and the emission of the light) rapidly decreases when the ignition power density increases, until an asymptotic value, defining the minimum response ignition time, which is a characteristic of the multilayered film itself.

As an illustration, Figure 2b provides the ignition time versus the ignition power for a thermite made of 15 300 nm thick 1:1 Al/CuO bilayers. Below a certain electrical power, no ignition occurs, regardless of the duration of application of the power. This is simply explained by the fact that, below this ignition power threshold, the electrical energy supplied to the micro-heater is not sufficient to compensate for the energy lost by conduction through the substrate, by radiation from the exposed surface, or by convection in the air. The multilayer cannot reach its ignition temperature, also known as the onset reaction temperature.

This threshold highly depends on the heating surface area, and detailed results can be found in [12].

4. Examples of Applications

To date, igniters are the widest investigated applications and are widely used to trigger the inflation of airbags, micro-propulsion systems, and the arm and fire devices used in missiles, rockets and any other ordnance systems. Traditionally, igniter technology consists of a metallic hot-wire (Ni/Cr, et al.) or bridge wire in contact with a secondary or primary explosive. The fabrication and integration of explosives in igniters requires extreme precaution due to their high sensitivity to electrostatic discharge, friction or shock.

Al/CuO multilayers can efficiently replace the hot-wire and primary explosive. Other advantages of replacing traditional hot-wire with pyroMEMS are as follows: (1) The overall integration is easier, requiring only the electrical connections of the pyroMEMS, whereas traditional pyrotechnical igniters require at least two steps (hot-wire and primary pyrotechnic composition deposition). (2) The Al/CuO multilayers are safe and less sensitive to the environment. (3) Ignition threshold and reaction output (flame temperature and gaseous products) can be easily tuned by changing the Al/CuO bilayer thickness and stoichiometry (bilayer thickness ratio).

Taton et al. [13] first reported the design, realization and characterization of hot-wire ignition integrating Al/CuO multilayers (see Figure 4) for use in pyrotechnical systems for space applications. The reactive Al/CuO multilayered thin film resides on a 100 μm thick epoxy/polyethyleneterephtalate (PET) membrane to insulate the reactive layer from the bulk substrate. When current is supplied, Al/CuO reacts and the products of the reaction produce sparks that can ignite any secondary energetic composition, such as RDX (hexogen). The authors demonstrated a 100% success of ignition over a 0.25–4 A firing current range, corresponding to 80–244 μJ, and with response times ranging from 2 to 260 μs. Then, Glavier et al. [14] used this igniter to cut and propel a thin metallic foil and ignite RDX in a detonation. Authors showed that a stainless-steel flyer of 40 mg can be properly cut and propelled at velocities calculated from 665 to 1083 m/s, as a function of the RDX's extent of compaction and ignition charge. The impact of the flyer can directly initiate the detonation of an RDX explosive, which is very promising in terms of removing primary explosives from detonators. A schematic view and photo of one miniature detonator is given in Figure 4.

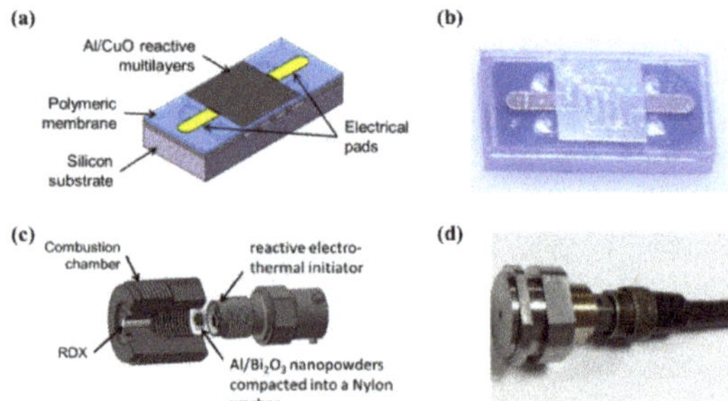

Figure 4. (**a**) Schematic and (**b**) photo of the reactive electro-thermal initiator on Epoxy/PET membrane. Chip dimension is 4 mm × 2 mm. (**c**) Schematic and (**d**) photo of one miniature detonator integrating Al/CuO multilayers as the initiator.

More recently, Nicollet et al. [12] developed the Al/CuO multilayer igniter concept for several substrates, and simplified the fabrication process to adapt it to airbag initiation. Additionally, our team has demonstrated miniature one-shot circuit breakers [15], based on the combustion of a nanothermite. Each device is simply made from two assembled printed board circuits (PCBs) to define a hermetic cavity in which an Al/CuO multilayer initiator chip ignites—in less than 100 μs—a few milligrams of nanothermite, to cut a thick copper connection (see Figure 5). The authors demonstrated the good operation (100% success rate) with a response time of 0.57 ms, which is much lower than the response time of classical mechanical circuit breakers (>ms).

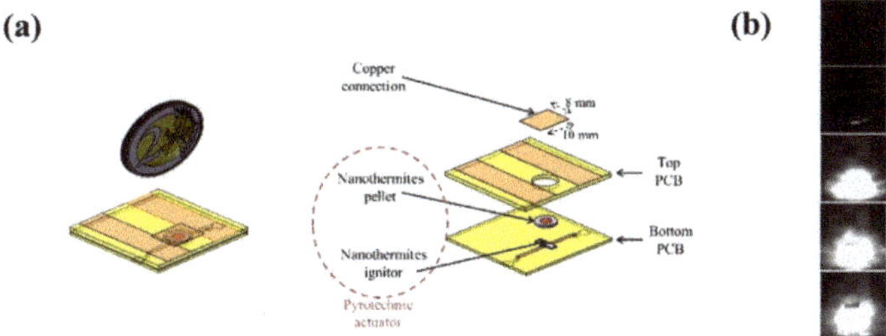

Figure 5. (**a**) 3D schematic of the nanothermite-based circuit breaker with exploded views and (**b**) snapshots of high speed images taken for the tested circuit breaker. The time between each picture is 100 µs. Taken from [15].

5. Conclusions

Sputter-deposited Al/CuO multilayers, composed of alternating Al and CuO nanolayers with total thicknesses ranging from 1 to tens of µm, present an opportunity for tunable ignition and actuation. The interest in and integration of igniter applications have been demonstrated for different applications, such as in inflators, aerospace or actuators.

Author Contributions: L.S. developed the sputter-deposition process and assisted in pyroMEMS development. S.A.-S. developed all the experimental benches and assisted in the characterization part. C.R. supervised the research and wrote the paper. All authors have read and agreed to the published version of the manuscript.

Funding: C.R. received funding from the European Research Council (ERC) under the European Union's Horizon 2020 research and innovation program (grant agreement No 832889—PyroSafe). L.S. received funding for sputter-deposition equipment from FEDER/Region Occitanie program Grant THERMIE.

Acknowledgments: The authors acknowledge support from the European Research Council (H2020 Excellent Science) Researcher Award (grant 832889—PyroSafe) and the Occitanie Region/European Union for their FEDER support (THERMIE grant). The authors thank Séverine Vivies from LAAS-CNRS and Teresa Hungria-Hernandez from "Centre de Micro Caractérisation Raymond Castaing (UMS 3623)" for their great support with the sample preparation and TEM experiments.

Conflicts of Interest: The authors declare no conflict of interests.

References

1. Petrantoni, M.; Rossi, C.; Salvagnac, L.; Conédéra, V.; Esteve, A.; Tenailleau, C.; Alphonse, P.; Chabal, Y.J. Multilayered Al/CuO thermite formation by reactive magnetron sputtering: Nano versus micro. *J. Appl. Phys.* **2010**, *108*, 084323. [CrossRef]
2. Adams, D. Reactive multilayers fabricated by vapor deposition: A critical review. *Thin Solid Films* **2015**, *576*, 98–128. [CrossRef]
3. Fu, S.; Shen, R.; Zhu, P.; Ye, Y. Metal–interlayer–metal structured initiator containing Al/CuO reactive multilayer films that exhibits improved ignition properties. *Sens. Actuators A Phys.* **2019**, *292*, 198–204. [CrossRef]
4. Manesh, N.A.; Basu, S.; Kumar, R. Experimental flame speed in multi-layered nano-energetic materials. *Combust. Flame* **2010**, *157*, 476–480. [CrossRef]
5. Rossi, C. Engineering of Al/CuO Reactive Multilayer Thin Films for Tunable Initiation and Actuation. *Propellants Explos. Pyrotech.* **2018**, *44*, 94–108. [CrossRef]
6. Lahiner, G.; Zapata, J.; Cure, J.; Richard, N.; Djafari-Rouhani, M.; Estève, A.; Rossi, C. A redox reaction model for self-heating and aging prediction of Al/CuO multilayers. *Combust. Theory Model.* **2019**, *23*, 700–715. [CrossRef]

7.   Zapata, J.; Nicollet, A.; Julien, B.; Lahiner, G.; Esteve, A.; Rossi, C. Self-propagating combustion of sputter-deposited Al/CuO nanolaminates. *Combust. Flame* **2019**, *205*, 389–396. [CrossRef]

8.   Weir, C.; Pantoya, M.L.; Ramachandran, G.; Dallas, T.; Prentice, D.; Daniels, M. Electrostatic discharge sensitivity and electrical conductivity of composite energetic materials. *J. Electrost.* **2013**, *71*, 77–83. [CrossRef]

9.   Cheng, J.; Hng, H.H.; Lee, Y.; Du, S.; Thadhani, N. Kinetic study of thermal- and impact-initiated reactions in Al–Fe2O3 nanothermite. *Combust. Flame* **2010**, *157*, 2241–2249. [CrossRef]

10.   Damm, D.; Maiorov, M. Thermal and radiative transport analysis of laser ignition of energetic materials. *Opt. Eng. Appl.* **2010**, *7795*, 779502. [CrossRef]

11.   Hadjiafxenti, A.; Gunduz, I.; Doumanidis, C.; Rebholz, C. Spark ignitable ball milled powders of Al and Ni at NiAl composition. *Vacuum* **2014**, *101*, 275–278. [CrossRef]

12.   Nicollet, A.; Lahiner, G.; Belisario, A.; Assié-Souleille, S.; Rouhani, M.D.; Estève, A.; Rossi, C. Investigation of Al/CuO multilayered thermite ignition. *J. Appl. Phys.* **2017**, *121*, 034503. [CrossRef]

13.   Taton, G.; Lagrange, D.; Conedera, V.; Renaud, L.; Rossi, C. Micro-chip initiator realized by integrating Al/CuO multilayer nanothermite on polymeric membrane. *J. Micromech. Microeng.* **2013**, *23*, 105009. [CrossRef]

14.   Glavier, L.; Nicollet, A.; Jouot, F.; Martin, B.; Barberon, J.; Renaud, L.; Rossi, C. Nanothermite/RDX-Based Miniature Device for Impact Ignition of High Explosives. *Propellants Explos. Pyrotech.* **2017**, *42*, 308–317. [CrossRef]

15.   Nicollet, A.; Salvagnac, L.; Baijot, V.; Estève, A.; Rossi, C. Fast circuit breaker based on integration of Al/CuO nanothermites. *Sens. Actuators A Phys.* **2018**, *273*, 249–255. [CrossRef]

 *nanomaterials*

Article

How Thermal Aging Affects Ignition and Combustion Properties of Reactive Al/CuO Nanolaminates: A Joint Theoretical/Experimental Study

A. Estève [1], G. Lahiner [1], B. Julien [1], S. Vivies [1], N. Richard [2] and C. Rossi [1,*]

[1] LAAS-CNRS, University of Toulouse, 7 Avenue du colonel Roche, 31077 Toulouse, France;
    aesteve@laas.fr (A.E.); glahiner@laas.fr (G.L.); bjulien@laas.fr (B.J.); svivies@laas.fr (S.V.)
[2] CEA-DAM, DIF, 91297 Arpajon, France; nicolas.richard@cea.fr
* Correspondence: rossi@laas.fr

Received: 9 September 2020; Accepted: 19 October 2020; Published: 21 October 2020

Abstract: The paper reports a joint experimental/theoretical study on the aging of reactive Al/CuO nanolaminates, investigating both structural modifications and combustion properties of aged systems. We first show theoretically that the long-term storage (over several decades) in ambient temperature marginally affects nanolaminates structural properties with an increase in an interfacial layer of only 0.3 nm after 30 years. Then, we observe that the first thermal aging step occurs after 14 days at 200 °C, which corresponds to the replacement of the natural Al/CuO interfaces by a proper ~11 nm thick amorphous alumina. We show that this aging step does impact the nanolaminates structure, leading, for thin bilayer thicknesses, to a substantial loss of the energetic reservoir: considering a stoichiometric Al/CuO stack, the heat of reaction can be reduced by 6–40% depending on the bilayer thickness ranging from 150 nm (40%) to 1 µm (6%). The impact of such thermal aging (14 days at 200 °C) and interfacial modification on the initiation and combustion properties have been evaluated experimentally and theoretically. Varying Al to CuO ratio of nanolaminates from 1 to 3, we show that ignition time of aged systems does not increase over 10% at initiation power densities superior to 15 W·mm^{-2}. In contrast, burn rate can be greatly impacted depending on the bilayer thickness: annealing a stoichiometric nanolaminates with a bilayer thickness of 300 nm at 200 °C for 14 days lowers its burn rate by ~25%, whereas annealing a fuel rich nanolaminates with the same bilayer thickness under the same thermal conditions leads to a burn rate decrease of 20%. When bilayer thickness is greater than 500 nm, the burn rate is not really affected by the thermal aging. Finally, this paper also proposes a time–temperature diagram to perform accelerated thermal aging.

Keywords: nanothermite; Al; CuO; aging; combustion; initiation

1. Introduction

Al-based thermite materials have attracted great attention for decades because of their high energy content and overall stability compared to CHNO energetic mixtures. In addition, ignition and combustion properties, which are easily tuned by modifying the mixture structure and fuel to oxidizer ratio, make these metal-based energetic materials very interesting for many different types of applications: additives in propellants, explosives and pyrotechnics [1–3], customized heat sources [4,5], microenergetics [6,7], rapid fuses and microinitiators [8–13], brazing of materials, as well as use as a pressure generator as for molecular delivery (such as biological neutralization) [14–16]. The potential spreading of nanothermite materials, including reactive thermite nanolaminates [9,17–22], in a variety of applications in both defense and civilian domains, absolutely requires a clear statement of how these nanomaterials do behave with time and storage conditions. Despite the active experimental research dedicated to nanothermites over the two last decades, the literature specifically dedicated to aging issues is rare and exclusively concerns nanothermites in the form of mixed powders [23,24]. Wang et al. [24] investigated the short-term storage stability of Al/CuO mixed powder prepared by electrospray method, and observed modifications of the mixture performance after 13 months aging at the ambient temperature. They mentioned a slight lowering of the pressure generation from 686 down to 626 kPa, and an increase in the initiation temperature (from 754 to 775 °C) after observing a slight decrease at mid-term aging. Counter-intuitively, thermal analyses of aged systems indicate a continuous lowering of the onset temperature along storage time.

It is reasonable to expect that the aging kinetics of sputter-deposited Al/CuO having specific chemical compositions of interfacial layer will qualitatively and quantitatively differ from those of Al/CuO mixed powders with identical stoichiometry, dimensional features, and composition. Along this line, a very recent and thorough experimental study showed that under slow heating rates, the release of gaseous oxygen from the CuO occurring well below reaction onset (i.e., at ~250 °C) initiates the Al oxidation process at the vicinity of native interfaces until a proper Al_2O_3 is formed at ~400 °C [25]. Based on these findings, Lahiner et al. proposed a redox reaction model suitable for investigating slow heating rates and low temperature experiments [26]. The model permits describing the kinetics of the reactant and product layer thicknesses (including Al, CuO, Cu_2O, Al_2O_3) of reacting sputter-deposited Al/CuO nanolaminates upon annealing at low temperature (ambient to 500 °C) [26]. This preceding work offers a unique opportunity to investigate Al/CuO nanolaminates aging in an original manner, allowing for efficient in silico screening of structures modification and ensuing performances alteration when the Al/CuO thermite nanolaminates are subject to long-term storage at the ambient temperature and upon annealing as well.

In this respect, representative types of Al/CuO nanolaminates structures are considered, and their aging analyzed, notably in terms of interfacial layer thickness evolution and energetic reservoir consumption. This aging effect (temperature out of any other chemistry associated with the storage environment) on both initiation time and combustion performances (burn rate) of various structures is then detailed. In complement to simulations, a set of Al/CuO nanolaminates structures was fabricated, and some of them were annealed as prescribed by the proposed aging model to support the theoretical predictions.

This work represents important progress for the community of thermites, not only thanks to the data collected on the aging of Al/CuO nanolaminates, but also as it presents an original methodology that can be applied to other thermite materials. Indeed, whereas the traditional methods in aging investigations are only based on empirical model-free approaches, not supported by physical considerations, our methodology is totally guided by the physico-chemistry of all elementary mechanisms occurring in sputter-deposited Al/CuO multilayers upon annealings. Hence, the elementary mechanism(s) that is (are) activated during an aging process inducing a modification in the thermite performances is known which opens the possibility to the material designers to find solutions in order to remediate.

2. Materials and Methods

2.1. Computational Details

A set of elementary diffusion/reaction mechanisms activated at low temperature (oxygen diffusion, structural transformations, and polymorphic phase changes) has been determined by coupling calorimetric characterization with electronic microscopy imaging on Al/CuO nanolaminates [25]. The main finding is that, below Al melting, the rate of reaction is limited by the transport of oxygen species across the growing interfacial layer separating Al and CuO layers. Three main reaction steps were observed to describe the nanolaminate self-heating as depicted in Figure 1: (i) the release of gaseous oxygen from the CuO identified as the first reaction step starting at ~250 °C, this gas accumulates at interfaces, (ii) between 300 and 350 °C, the migration of oxygen species is thermally activated, giving birth to a sudden enhancement of the Al oxidation rate and the subsequent formation of a well-defined amorphous alumina oxide in replacement of the native ill-defined interface. The formation of this amorphous layer hinders further diffusion of oxidizing species, (iii) between 500 and 600 °C, the amorphous alumina turns into a γ-alumina structure that stops further migration of oxygen species.

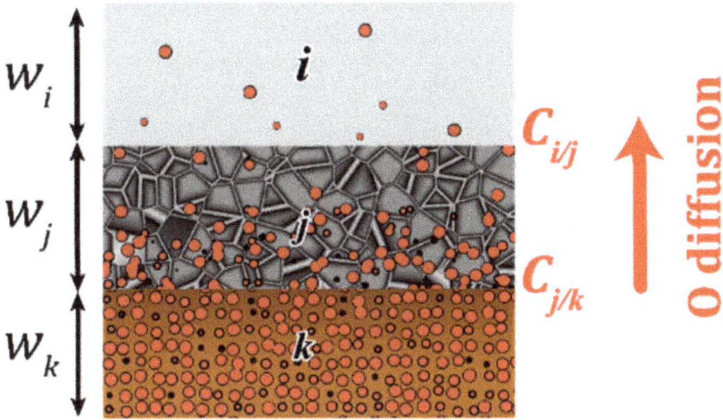

Figure 1. Schematic of the model-system: copper oxide layer W_k (in orange), alumina barrier layer W_j (in dark grey) and aluminum W_i (in light grey). Oxygen diffusion across the interface is schematized with red balls.

Based on these experimental findings, a heterogeneous reaction model was developed, coupling the basic thermal equation with a diffusion/reaction scheme as proposed by Deal and Grove [27]. In brief, for each layer, Al, CuO, Cu_2O, or Al_2O_3, the time evolution of their thicknesses w_i is expressed by:

$$\frac{dw_i}{dt} = \varnothing(t)\frac{M_i}{3\rho_i} \tag{1}$$

M_i and ρ_i is the molar mass (in $kg \cdot mol^{-1}$) and volumetric density ($kg \cdot m^{-3}$) of the material i, i representing Al, CuO, Cu_2O, or Al_2O_3. The oxygen flux $\varnothing(t)$ is defined by the general formulation:

$$\varnothing(t) = \frac{D_j\left(C(t)_{\frac{i}{j}} - C(t)_{\frac{j}{k}}\right)}{w(t)_j} \tag{2}$$

where D_j is the oxygen diffusivity in the material j that obeys a Fickian process; $C_{i/j}$, $C_{j/k}$ are the oxygen concentrations at interfaces between materials i and j, j and k. w_i, w_i, w_k, … correspond to the thicknesses of materials i, j, k, all representing Al, CuO, Cu_2O or Al_2O_3. The oxygen diffusivity D_j follows a single Arrhenius dependence on temperature calibrated from experiments.

Then, Equation (1) is coupled via a source term related to the energy release during the exothermic reaction.

$$P_{DSC} = \frac{dH(t)}{dt} - nS\phi(t) \times Q \tag{3}$$

P_{DSC} is the differential external power (in W) as measured in Differential Scanning Calorymetry (DSC) experiments. The total enthalpy H (in $J \cdot g^{-1}$) is expressed as: $H(t) = H_a + \sum\limits_{t,T_t > T_a} M_t h_t \theta(T(t) - T_t)$. With H_a, the total enthalpy of the bilayer at ambient temperature (T_a) and h_t the molar enthalpy corresponding to the various phase transitions t that may occur (such as Al melting) at temperature T_t. M_t is the number of moles involved in the phase transition. θ is the Heaviside step function: $\theta = 0$ for $T < T_t$, and $\theta = 1$ for $T > T_t$. $nS\phi(t) \times Q$ is the source term representing the exothermal Al oxidation, with Q corresponding to the value of the redox reaction $2Al + 3CuO \rightarrow Al_2O_3 + 3Cu$ given in [28]. For more details about the model implementation, see [26]. Table 1 summarizes the main model parameters used in the following aging study. A more detailed table that includes all main thermodynamic parameters is provided in the Supporting Information file, Table S1.

Table 1. Model parameters used for aging simulations of Al/CuO nanolaminates. Thermo-physical parameters are given in Supporting Information file, **Table S1**.

Mechanisms	Arrhenius Parameters	Source
Oxygen diffusion through natural $Al_xCu_yO_z$	$D_0 = 1 \times 10^{-7}$ m$^2\cdot$s^{-1} $E_a = 100$ kJ$\cdot$mol^{-1}	Calibrated from DSC experiments, see Ref. [26]
Oxygen diffusion through amorphous Al_2O_3	$D_0 = 1.67 \times 10^{-11}$ m$^2\cdot$s^{-1} $E_a = 120$ kJ$\cdot$mol^{-1}	Isotopic labelling, see Ref. [29]

2.2. Nanolaminate Fabrication

A few configurations of Al/CuO nanolaminates were selected to perform experimental aging. Each sample was characterized by the Al to CuO equivalence ratio ξ, defined as the ratio of the actual Al/CuO ratio to the stoichiometric Al/CuO ratio and bilayer thickness, w. Note that in a ξ 1 sample, the aluminum thickness is half the CuO thickness (stoichiometric stack) whereas $\xi > 1$ corresponds to a fuel rich situation with thicker aluminum layers. The details of the magnetron sputtering of Al/CuO nanolaminates can be found in previous studies (see [30,31]). Throughout the paper, including figures and captions, the following notation was adopted to characterize the Al/CuO nanolaminates: ξ Al/CuO w_{Al}/w_{CuO}, w_{Al} and w_{CuO} being the Al and CuO layer thicknesses, in nm, respectively.

2.3. Experiments

For each produced nanolaminate, thermal analysis, initiation time and burn rate measurements were performed as deposited and after selected aging conditions.

2.3.1. Initiation Time Experiments

The sustained reactions in CuO/Al nanolaminates were initiated via a localized titanium resistive filament, resulting in self-propagating combustion fronts. For the ignition experiments, 700 nm thick Ti film was patterned on glass wafers. Around 12 mm^2 Al/CuO nanolaminates were sputter-deposited in contact with the Ti filament. The initiation time of the nanolaminates was characterized using a photodiode (Vishay Semiconductors, Heilbronn, Germany, reference BPV10) placed at a few inches distance from the thermite to record the optical emission.

2.3.2. Burn-Rate Experiments

Nanolaminates were sputter-deposited along 25 × 2 mm lines, onto a glass substrate. Resistive titanium filaments were patterned at both ends underneath the thermite line to ignite it. The self-propagating reaction velocity, called burn rate, was characterized with a SA3 Photron high-speed camera (West Wycombe, UK) at a framerate of 20,000 fps placed at a few inches distance from the thermite line to record the flame propagation.

2.3.3. Thermal Analysis

Thermal analyses using a NETZSCH DSC 404 F3 Pegasus (Selb, Germany) were performed under constant heating rates (20 °C$\cdot$min^{-1}) in Ar atmosphere (99.998% pure) at a flow rate of 20 mL$\cdot$min^{-1}. The device, equipped with a DSC-Cp sensor type S and a Platinum furnace, recorded DSC signals in a temperature range from room temperature to 1000 °C. The DSC traces are always normalized by the mass of nanothermite material (around 5 mg).

3. Results and Discussion

3.1. Modelling the Long-Term Aging at the Ambient Temperature

The redox model is first used to predict how long-term storage (up to 30 years) at the ambient temperature (300 K) affects the nanolaminate structures and enthalpy of reaction. We perform the simulations on three sets of Al/CuO nanolaminate systems having three different aluminum thicknesses: 75, 100, and 150 nm. CuO thickness is adjusted to obtain a stoichiometric ratio ξ ranging from 1 to 3. Table 2 provides results in term of the energy loss in percentage (EL%) after 30 years of storage at the ambient temperature. EL% is expressed as $EL\% = \frac{H_f - H_a}{H_a}$, with H_a and H_f being the total enthalpy of the bilayer at the ambient temperature after deposition and after the 30 years storage, respectively. Note that the number of bilayers does not affect the results.

Table 2. Interfacial alumina thickness and Energy Loss (EL%) after 30 years of storage at the ambient temperature for 9 sets of Al/CuO nanolaminates (3 stoichiometries and, for each, 3 bilayer thicknesses).

Nanolaminate Configuration	Grown Interfacial Alumina Thickness	Energy Loss in %
ξ 1 Al/CuO 75/150		0.023%
ξ 1 Al/CuO 100/200		0.017%
ξ 1 Al/CuO 150/300		0.008%
ξ 2 Al/CuO 75/75		0.048%
ξ 2 Al/CuO 100/100	0.3 nm	0.036%
ξ 2 Al/CuO 150/150		0.018%
ξ 3 Al/CuO 75/50		0.072%
ξ 3 Al/CuO 100/67		0.054%
ξ 3 Al/CuO 150/100		0.027%

We observe that, after a 30 years storage, the EL% remains negligible (<0.1%) clearly demonstrating that Al/CuO nanolaminates are stable for decades under the ambient temperature. Note that the unique physico-chemical process activated at the ambient temperature is the growth of an amorphous alumina layer in the immediate contact of the as deposited natural interface, a process that only consumes a negligible part of the Al reservoir. What is herein called "natural interfaces" corresponds to the intermixed layers formed between sputtered Al and CuO during the deposition. The interface formed upon the deposition of CuO onto Al is a flat $Cu_xAl_yO_z$ layer (~4 nm in thickness), whereas the interface formed upon the deposition of CuO onto Al is ill-defined over a thickness of ~5 ± 5 nm, as detailed in [25]. For simplification in the model, we took equivalent CuO/Al and Al/CuO interfacial layers: 4 nm of $Al_xCu_yO_z$, see Table 1.

Typically, after 30 years at the ambient temperature, the interface grows marginally (sub nanometer range, not observable experimentally) for all considered systems, even though the results qualitatively show a tendency for the nanolaminates to be more sensitive to aging when considering thin bilayer thicknesses and fuel rich stoichiometry.

Storing a nanolaminate for long periods of time, such as 10, 20, and 30 years, to investigate the performance evolution is not reasonable. Therefore, thermal accelerated aging experiments are the most usual way to investigate long periods of natural aging. Next, we calculate the temperature–time couple that is equivalent to aging at the ambient temperature for 10, 20, and 30 years. For that purpose, we have adopted an equivalence in terms of the energy loss EL%, i.e., we systematically calculated and identified time–temperature couples leading to the energy loss associated with 10, 20, and 30 years of aging at the ambient temperature, respectively. Figure 2 plots the obtained curves for a fuel rich material corresponding to ξ 2 Al/CuO 150/150 sample. Results show that, for the shortest duration

(10 years at the ambient temperature), even though the EL% is nearly zero, a constant temperature of roughly 100 °C applied for one day is required to achieve the EL% equivalence. For 30 years aging at the ambient, applying a 100 °C temperature requires about a week duration to obtain the EL% equivalence. This illustrates the model capability to anticipate any accelerated aging protocol. Note that, as the interfacial layer growth is almost zero, the aging diagram such as the one presented in Figure 2 will not vary when modifying stoichiometry or bilayer thickness.

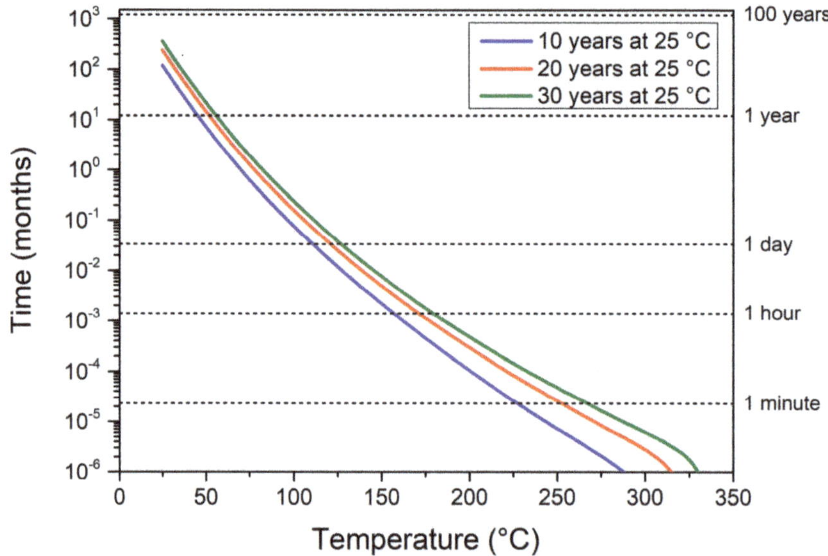

Figure 2. Time–temperature couples giving equivalence to 10, 20, and 30 years natural aging at the ambient temperature for a fuel rich ξ 2 Al/CuO 150/150.

3.2. Accelerated Thermal Aging

Here, as a first aging step, we consider the activation of the mechanism, exhibiting the lower energy barrier, which was introduced in the Table 1 and labelled as Oxygen diffusion through natural $Al_xCu_yO_z$. The complete activation of this mechanism corresponds to the replacement of the native ill-defined interface by an amorphous alumina barrier layer that further stops the oxygen transport until ~500 °C [25]. In Figure 3, we plot the couple annealing duration/temperature needed to fully activate the first aging process i.e., the loss of the first DSC peak after aging. We observe two regimes apart from a transition couple: 200 °C/14 days. Below 200 °C, the curve is almost a vertical line with respect to the time plotted in a logarithmic scale. This indicates that temperature is no longer affecting aging time that quickly expands from a few tens of years to centuries. In contrast, above 200 °C, the slope reduces drastically, indicating that the annealing time quickly drops with increasing temperature. Above 350 °C, the time to replace the natural interface by an amorphous alumina falls below the hour duration.

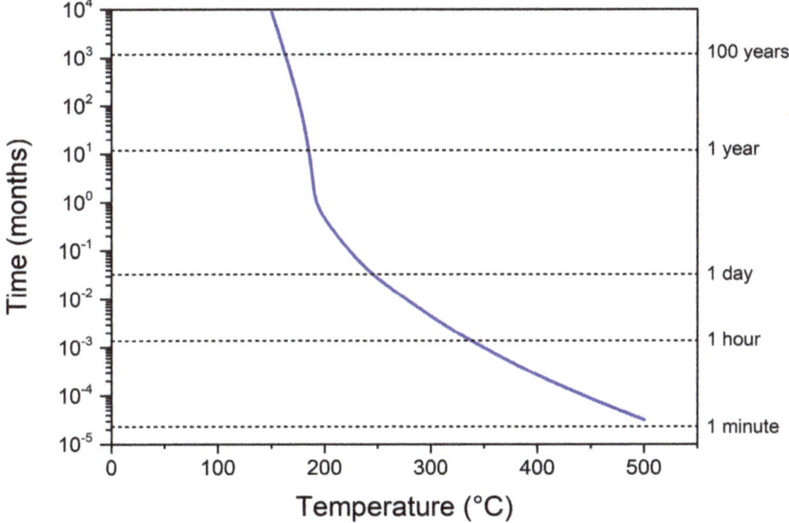

Figure 3. Simulated aging predictions of Al/CuO nanolaminates, depending on time and temperature, to reach an aging leading to the complete conversion of the natural $Al_xCu_yO_z$ interface into a well-defined amorphous alumina layer.

Interestingly, 200 °C corresponds to the activation of the release of gaseous oxygen from the columnar CuO layer, starting its decomposition. Below 200 °C, the effect of temperature is therefore negligible on nanolaminate properties, as no gaseous oxygen is available in the structure. Note that, for all systems considered in this study, in terms of stoichiometric ratio and bilayer thicknesses, the aging time–temperature diagram will be equivalent to the one shown in Figure 3. This is due to the kinetics of thermal growth: the modification of the interfacial region is equivalent whatever the stoichiometry and bilayer thickness, as soon as the nanolaminate is deposited following the same sputtering conditions.

Seeking for an experimental validation of the aging process, we selected the time–temperature aging couple corresponding to the transition point, i.e., sample annealed at 200 °C for 14 days. Figure 4 compares experimental and simulated DSC curves obtained for stoichiometric and fuel rich Al/CuO nanolaminates collected at 20 °C·min^{-1} after deposition (as-deposited) and after being annealed at 200 °C for 14 days (aged). Overall, the simulated DSC results show good agreement with the experimental curves with the three main reaction steps at 440, 578, and close to 800 °C. The first experimental exothermal event between 350 and 540 °C, corresponding to the oxygen diffusion through the native interfaces, is broad showing a two peaks contribution. This is probably due to the difference in nature and thickness of the two natural interfacial layers associated respectively with the deposition of Al onto CuO and CuO onto Al. Instead, the virtual DSC traces feature a single exothermic peak contribution, as we assumed the two Al/CuO and CuO/Al interfaces identical. Most importantly, the first DSC peak totally disappears for both experimental and virtual DSC traces after the annealing at 200 °C for 14 days: this represents a loss of ~20% of the initial energetic reservoir. Note that the model also predicts the theoretical melting of Al is set to 660 °C (endotherm), which is the bulk melting point, whereas it is not always seen experimentally, depending on layer thicknesses, stoichiometry, and heating ramp.

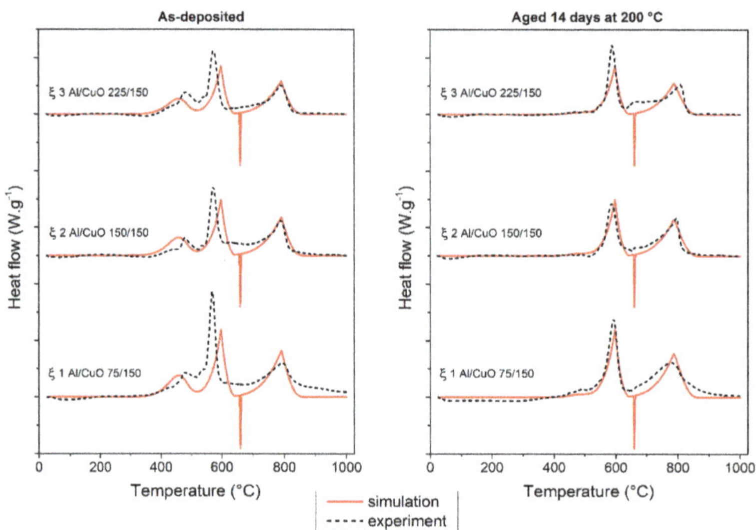

Figure 4. Experimental and theoretical DSC curves of Al/CuO nanolaminates: (left column) as-deposited and (right column) being aged 14 days at 200 °C.

3.3. Modelling the Structural Modifications and Associated Loss of Energy Reservoir Upon Thermal Aging

The structural changes due to thermal aging, when modifying stoichiometric ratio and bilayer thicknesses, may impact the Al to oxygen ratio of aged materials. The kinetics of these changes are theoretically examined for different bilayer thicknesses (w = 150–375 nm) and stoichiometric ratio (ξ 1–3). Considering a ξ 2 Al/CuO 150/150 nanolaminates at 200 °C, Figure 5 shows the prediction of the interfacial alumina growth and subsequent energy loss percentage (EL%) evolution with annealing time.

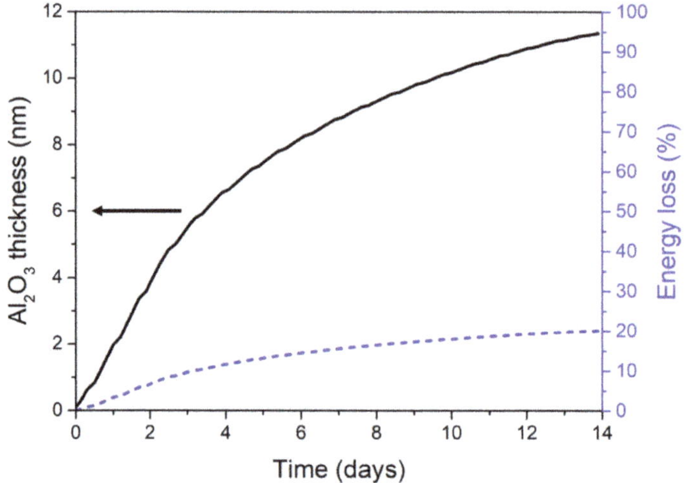

Figure 5. Time evolution of both the amorphous alumina thickness (black line) and Energy loss EL% (dashed blue line), for a ξ 2 Al/CuO 150/150 nanolaminate, aged 14 days at 200 °C.

Modelling results show that at 200 °C, a ~11 nm thick amorphous alumina is grown after 14 days leading to an energy loss of 20%. Figure 6a,b shows how the stoichiometry and the bilayer thicknesses affect the energy loss keeping the same aging conditions (200 °C/14 days), respectively. In Figure 6a, the bilayer thickness is kept constant and stoichiometric ratio varies. We observe that EL% decreases when the stoichiometry increases, which is explained by the fact that the reaction enthalpy (H_a) is maximum at ξ 1 and decreases for fuel rich conditions.

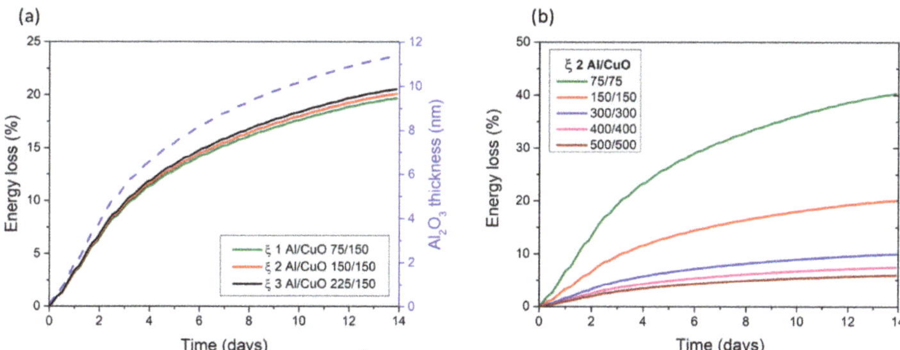

Figure 6. Time evolution of the energy loss EL% upon 200 °C aging conditions for, (**a**) three different stoichiometries (ξ 1–3), (**b**) ξ 2 stack with five bilayer thicknesses w = 150, 300, 600, 800, and 1000 nm.

Still considering a fuel rich nanolaminate (ξ 2), we observe a strong variation of the energy loss, from 5 to 40% with the decrease in the bilayer thicknesses (Figure 6b). EL% is of 10 and 8% for bilayer thicknesses of 300 and 400 nm, respectively. This can be explained by the fact that the quantity of Al consumed to grow the amorphous alumina interface remains constant whatever the bilayer thickness, whereas the total quantity of Al increases with the bilayer thickness increase. Hence, increasing the Al quantity in the nanolaminate diminishes the impact of interface evolution and aging. We see that increasing bilayer thickness to the submicron range clearly allows lowering the aging effect, with an energy loss of nearly 6% when reaching w = 1 μm. In contrast, when downscaling the bilayers below 300 nm, the effect of aging is impactful, with energy losses largely exceeding 20% of the initial energetic reservoir; a nanolaminate with a bilayer thickness of 150 nm loses 20% of its energetic reservoir after 14 days at 200 °C.

As summary of this first part, we theoretically demonstrated the stability of Al/CuO sputter-deposited nanolaminates at ambient temperature over a few tens of years. We also showed that the full activation of the first aging mechanism, which is the replacement of the native interface by an amorphous alumina, reduces the energy reservoir up to 40% when bilayer thickness is below 300 nm. In order to reduce the impact of interfacial modification upon heating, nanolaminates must be designed with bilayer thicker than 500 nm. In any case, we anticipate that the ignition and reaction performances will be affected as initiation and further reaction are function of the nature and thickness of the layers. This is assessed and discussed in the next section.

3.4. Impact of Aging on Nanolaminate Performance

This section discusses the effect of nanolaminate storage on initiation and burn rate. In this section we consider the aged material when its native interface is replaced by the amorphous alumina, which we have shown earlier to require 14 days at 200 °C. Note that all simulations of initiation are performed using the redox model presented in Section 2.1 coupled with heat transport equation to model the propagation. Only the thermal losses and the initiation procedure is different as described in ref. [32]. The multilayer ignition is triggered by imposing a constant heat power density at the left end of the film over a width of 750 μm (details in ref. [32]).

3.4.1. Initiation

Figure 7 plots the initiation time calculated for a ξ 2 Al/CuO 150/150 nanolaminate initiated at different power densities. ξ 2 Al/CuO 150/150 was chosen as this configuration is the widest use in applications related to MEMS based initiators [15]. Overall, the simulation points show good agreement with experimental points. Two regimes can be observed: at low power density, i.e., below 20 W· mm^{-2}, the initiation time corresponds to the time needed to heat up the material to its initiation temperature (heating time). Initiation times are in the millisecond to few second range. Above 20 W·mm^{-2}, the heating time requested to heat the material to its initiation temperature is very low (sub-millisecond regime). Then the nanolaminate starts reacting until the self-sustained reaction is reached: this reaction time takes from a few µs to hundreds of µs depending on the bilayer thickness. Therefore, above 20 W·mm^2, the ignition delay is only controlled by the reaction time, explaining the curve flattening at high initiation power densities.

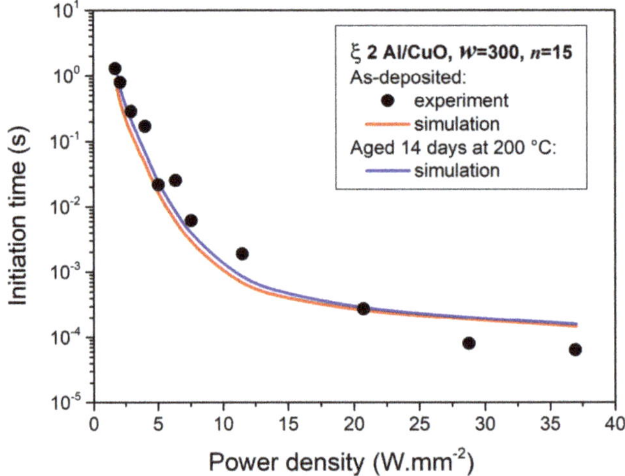

Figure 7. Initiation time as a function of applied power density on a ξ 2 Al/CuO 150/150 nanolaminates composed of black solid circles are experimentally measured for as-deposited samples. The red and blue simulated curves correspond to an as-deposited nanolaminate (red curve) and after being aged 14 days at 200 °C (blue curve), respectively.

Simulations of initiation time using power densities inferior to 20 W·mm^{-2} better fit the experimental points, while systematic overestimation is obtained beyond this value. This might be explained by the fact that "thermal chocks" may be imposed to the nanolaminate which are not considered in the model, while being observed experimentally [11]. Comparing now aged and non-aged simulation points, initiation is more affected by nanolaminate aging at low power densities (<5 W·mm^{-2}): initiation time is 34% longer for aged nanolaminates compared to fresh ones (ξ 2 and $w = 300$ nm). When increasing the power density, this delay drops quickly. Beyond 10–15 W·mm^{-2}, the initiation delay is slightly increased by ~7% which can be considered negligible.

Unfortunately it is impossible to characterize the initiation of aged samples because of the concomitant aging of the Ti thin film resistor in contact with nanolaminate.

3.4.2. Burn Rate

To predict the nanolaminates burn rates, we use a home-built propagation model specifically dedicated to the simulation of the self-propagating combustion. This model detailed in [32] has been calibrated empirically to account for high temperature processes driving the flame front propagation.

Before exploring how aging affects the burn rate on a vast number of nanolaminate configurations (Figure 8), we start first with a validation step on a reduced set of experimentations (Table 3).

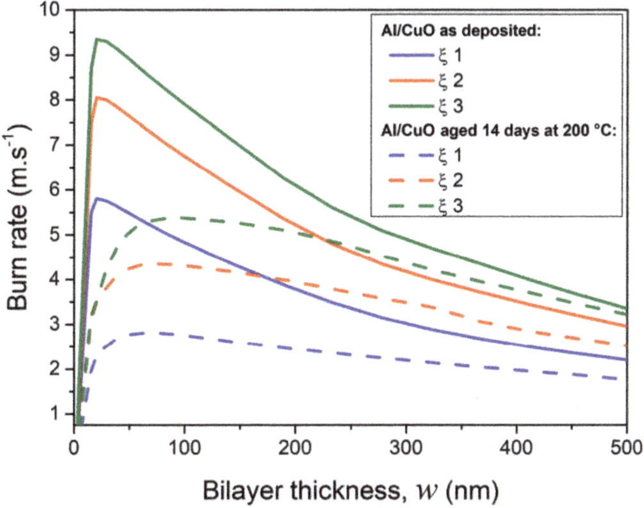

Figure 8. Simulated burn rates of Al/CuO nanolaminates as a function of bilayer thickness for three stoichiometries: ξ 1–3. The total thickness of the nanolaminate as well as heat losses are kept constant by varying the number of bilayers. Solid and dashed curves correspond to as deposited and aged (14 days at 200 °C) nanolaminates, respectively.

Table 3. Experimental burn rates obtained on various sputter-deposited Al/CuO nanothermites produced by our team and comparison with theoretical results obtained with a home-built propagation model as detailed in [32].

	Experiments Macroscopic Burn Rate m/s	From Model [32] m/s
ξ 1 Al/CuO 75/150		
15 bilayers		
As-deposited	4.6 ± 0.5	4.2
Aged 14 days at 200 °C	2.7 ± 0.5	3.0
ξ 2 Al/CuO 150/150		
11 bilayers		
As-deposited	11.7 ± 0.5	4.6
Aged 14 days at 200 °C	5.5 ± 0.5	3.8
ξ 3 Al/CuO 225/150		
9 bilayers		
As-deposited	10.1 ± 0.5	4.7
Aged 14 days at 200 °C	3.3 ± 0.5	3.7

Table 3 assembles a list of unpublished and recently published experiments [31,33] performed on sputter-deposited Al/CuO nanolaminates produced by our team and characterized in combustion.

For both experimental and theoretical data, we have considered fresh nanolaminates, i.e., characterized just after sputter deposition, and aged nanolaminates, i.e., being annealed at 200 °C for 14 days.

The first observation is that the model qualitatively reproduces the dependency of the burn rate on dimensions and stoichiometries in studied nanolaminates. Although the model is able to predict the experimental macroscopic burn rate in the ξ 1 Al/CuO 75/150 system (discrepancy between modelled and experimental data ~9%), it fails in predicting the experimental macroscopic burn rate for fuel rich stacks. To explain this, one has to note that the macroscopic burn rate corresponds to the global measurement of the reaction front velocity in the Al/CuO system. As discussed in [33], it corresponds to measuring the microscopic (local) burn rates at the reaction front multiplied by the corrugation of the flame front, which is the ratio of the total geometrical length of the flame to its width. While the exact origin of corrugation is still elusive, it is clearly the result of mechanisms different from the pure condensed phase combustion that fundamentally operates in our propagation model. The corrugation factor is ~1.3 for stoichiometric stacks leading to a macroscopic burn rate close to the microscopic one. However, the flame corrugation factor is of ~3 for fuel rich nanolaminates. The significant corrugation for fuel rich cases leads to a larger burning surface area ($\times$ ~3) and, consequently, faster observed burn rate (macroscopic) in fuel rich samples (details in [33]). Hence, considering microscopic burn rate, measured by microscale high speed imaging (not shown in this article but detailed in [33]) and with respect to fresh samples, fair agreement is found with the experimental values, being of 4 and 3.5 m·s^{-1} for ξ 2 and ξ 3 stacks. For aged samples, comparison of fuel rich samples is not straightforward, as the corrugation and microscopic burn rate were not characterized.

As summary, for fuel rich systems, the model is able to predict microscopic burn rate, which is explained by the fact that the model is exclusively based on thermal conduction and does not take into consideration advection and convection mechanisms which both greatly contribute to the flame propagation of fuel rich samples.

Figure 8 plots the burn rates of as-deposited and aged nanolaminates having bilayer thicknesses ranging from 25 to 500 nm and for three different stoichiometric ratios (ξ 1, 2, and 3). Whatever ξ value, aging effect on the burn rate increases with the bilayer thickness decrease. For a bilayer thicknesses inferior to 200 nm, the burn rate is reduced by 30% after an aging at 200 °C/14 days. For bilayer thickness equal to 300 nm, the burn rate drops from 3.4 m·s^{-1} to 2.6 m·s^{-1} and from 4.6 m·s^{-1} to 3.8 m·s^{-1} for ξ = 1 and 2, respectively. For thick bilayer thicknesses (w > 500 nm), the burn rate remains unchanged (1 to 2 m·s^{-1} depending on stoichiometry) after aging. This result was expected since the grown oxide at the nanolaminate interface becomes marginal for thick bilayers.

4. Conclusions

This paper investigated in detail the long-term storage and thermal aging of Al/CuO sputter deposited nanolaminates. The main findings are that: (1) Al/CuO nanolaminates are stable over decades at the ambient temperature. The energy reservoir is not affected after 30 years of storage under ambient conditions. (2) An aging transition is reached after annealing at 200 °C for 14 days: this corresponds to the replacement of the "ill-defined" $Al_xCu_yO_z$ native interfacial layer by an amorphous alumina. This mechanism affects the energetic reservoir and can therefore modify the initiation time and burn rate. (3) The importance of this interfacial modification on the energetic performance (initiation and burn rate) greatly depends on the stack dimensional characteristics, i.e., bilayer thickness and stoichiometric ratio: a Al/CuO stack having a bilayer thickness of 150 nm sees its energetic reservoir and burn rate dropping of 40% and 20%, respectively. The same Al/CuO stack with a bilayer thickness of 1 μm has an energetic reservoir and burn rate dropping of 6% and 16%, respectively. Finally, based on a redox reaction model specifically calibrated on our Al/CuO sputter deposited nanolaminates, we provided temperature/time diagram allowing to perform accelerated aging, which can be of great interest for future applications.

Nanomaterials **2020**, *10*, 2087

Supplementary Materials: The following are available online at http://www.mdpi.com/2079-4991/10/10/2087/s1, Table S1: Physical and thermodynamic parameters used in the model.

Author Contributions: G.L. built the redox model and assisted A.E. for aging simulations and paper preparation. S.V. fabricated and prepared the samples and B.J. performed all the aging and performances characterizations. A.E. developed the redox model with G.L. and prepared the manuscript with C.R. C.R. provided support for the manuscript preparation and supervised the research. N.R. provided support for the aging study conceptualization, notably giving some aging specifications and provided support to edit the manuscript. All authors have read and agreed to the published version of the manuscript.

Funding: C. Rossi received funding from the European Research Council (ERC) under the European Union's Horizon 2020 research and innovation program (grant agreement No. 832889–PyroSafe). This study was also funded by the CEA-DAM, DIF and CNRS.

Acknowledgments: The authors thank Sylvain Pelloquin for his assistance in samples preparation.

Conflicts of Interest: The authors declare no conflict of interest.

References

1. Bezmelnitsyn, A.; Thiruvengadathan, R.; Barizuddin, S.; Tappmeyer, D.; Apperson, S.; Gangopadhyay, K.; Gangopadhyay, K.; Redner, P.; Donadio, M.; Kapoor, D.; et al. Modified Nanoenergetic Composites with Tunable Combustion Characteristics for Propellant Applications. *Propellants Explos. Pyrotech.* **2010**, *35*, 384–394. [CrossRef]

2. Staley, C.S.; Raymond, K.E.; Thiruvengadathan, R.; Apperson, S.J.; Gangopadhyay, K.; Swaszek, S.M.; Taylor, R.J.; Gangopadhyay, S. Fast-Impulse Nanothermite Solid-Propellant Miniaturized Thrusters. *J. Propuls. Power* **2013**, *29*, 1400–1409. [CrossRef]

3. Zhou, X.; Torabi, M.; Lu, J.; Shen, R.; Zhang, K. Nanostructured Energetic Composites: Synthesis, Ignition/Combustion Modeling, and Applications. *ACS Appl. Mater. Interfaces* **2014**, *6*, 3058–3074. [CrossRef] [PubMed]

4. Stamatis, D.; Dreizin, E.L.; Higa, K. Thermal Initiation of Al-MoO$_3$ Nanocomposite Materials Prepared by Different Methods. *J. Propuls. Power* **2011**, *27*, 1079–1087. [CrossRef]

5. Duckham, A.; Spey, S.J.; Wang, J.; Reiss, M.E.; Weihs, T.; Besnoin, E.; Knio, O.M. Reactive nanostructured foil used as a heat source for joining titanium. *J. Appl. Phys.* **2004**, *96*, 2336–2342. [CrossRef]

6. Rossi, C.; Zhang, K.; Esteve, D.; Alphonse, P.; Tailhades, P.; Vahlas, C. Nanoenergetic Materials for MEMS: A Review. *J. Microelectromech. Syst.* **2007**, *16*, 919–931. [CrossRef]

7. Xu, J.; Shen, Y.; Wang, C.; Dai, J.; Tai, Y.; Ye, Y.; Shen, R.; Wang, H.; Zachariah, M.R. Controlling the energetic characteristics of micro energy storage device by in situ deposition Al/MoO$_3$ nanolaminates with varying internal structure. *Chem. Eng. J.* **2019**, *373*, 345–354. [CrossRef]

8. Zhou, X.; Shen, R.; Ye, Y.; Zhu, P.; Hu, Y.; Wu, L. Influence of Al/CuO reactive multilayer films additives on exploding foil initiator. *J. Appl. Phys.* **2011**, *110*, 094505. [CrossRef]

9. Zhu, P.; Shen, R.; Ye, Y.; Fu, S.; Li, D. Characterization of Al/CuO nanoenergetic multilayer films integrated with semiconductor bridge for initiator applications. *J. Appl. Phys.* **2013**, *113*, 184505. [CrossRef]

10. Zhu, P.; Shen, R.; Fiadosenka, N.N.; Ye, Y.; Hu, Y. Dielectric structure pyrotechnic initiator realized by integrating Ti/CuO-based reactive multilayer films. *J. Appl. Phys.* **2011**, *109*, 084523. [CrossRef]

11. Nicollet, A.; Lahiner, G.; Belisario, A.; Assié-Souleille, S.; Rouhani, M.D.; Estève, A.; Rossi, C. Investigation of Al/CuO multilayered thermite ignition. *J. Appl. Phys.* **2017**, *121*, 034503. [CrossRef]

12. Taton, G.; Lagrange, D.; Conedera, V.; Renaud, L.; Rossi, C. Micro-chip initiator realized by integrating Al/CuO multilayer nanothermite on polymeric membrane. *J. Micromech. Microeng.* **2013**, *23*, 105009. [CrossRef]

13. Fu, S.; Shen, R.; Zhu, P.; Ye, Y. Metal–interlayer–metal structured initiator containing Al/CuO reactive multilayer films that exhibits improved ignition properties. *Sens. Actuators A Phys.* **2019**, *292*, 198–204. [CrossRef]

14. Korampally, M.; Apperson, S.J.; Staley, C.S.; Castorena, J.A.; Thiruvengadathan, R.; Gangopadhyay, K.; Mohan, R.R.; Ghosh, A.; Polo-Parada, L.; Gangopadhyay, S. Transient pressure mediated intranuclear delivery of FITC-Dextran into chicken cardiomyocytes by MEMS-based nanothermite reaction actuator. *Sens. Actuators B Chem.* **2012**, *171*, 1292–1296. [CrossRef]

15. Sullivan, K.T.; Piekiel, N.W.; Chowdhury, S.; Wu, C.; Zachariah, M.R.; Johnson, C.E. Ignition and Combustion Characteristics of Nanoscale Al/AgIO$_3$: A Potential Energetic Biocidal System. *Combust. Sci. Technol.* **2010**, *183*, 285–302. [CrossRef]

16. Sullivan, K.T.; Wu, C.; Piekiel, N.W.; Gaskell, K.; Zachariah, M.R. Synthesis and reactivity of nano-Ag$_2$O as an oxidizer for energetic systems yielding antimicrobial products. *Combust. Flame* **2013**, *160*, 438–446. [CrossRef]

17. Rossi, C. Engineering of Al/CuO Reactive Multilayer Thin Films for Tunable Initiation and Actuation. *Propellants Explos. Pyrotech.* **2018**, *44*, 94–108. [CrossRef]

18. Dreizin, E.L. Metal-based reactive nanomaterials. *Prog. Energy Combust. Sci.* **2009**, *35*, 141–167. [CrossRef]

19. Egan, G.C.; Mily, E.J.; Maria, J.-P.; Zachariah, M.R. Probing the Reaction Dynamics of Thermite Nanolaminates. *J. Phys. Chem. C* **2015**, *119*, 20401–20408. [CrossRef]

20. Manesh, N.A.; Basu, S.; Kumar, R. Experimental flame speed in multi-layered nano-energetic materials. *Combust. Flame* **2010**, *157*, 476–480. [CrossRef]

21. Mily, E.; Oni, A.; Lebeau, J.; Liu, Y.; Brown-Shaklee, H.; Ihlefeld, J.F.; Maria, J.-P. The role of terminal oxide structure and properties in nanothermite reactions. *Thin Solid Films* **2014**, *562*, 405–410. [CrossRef]

22. Zhu, P.; Shen, R.Q.; Ye, Y.H.; Zhou, X.; Hu, Y.; Wu, L.Z. Energetic Igniters Based on Al/CuO/B/Ti Reactive Multilayer Films. Theory and Practice of Energetic Materials (Vol Ix). In Proceedings of the 2011 International Autumn Seminar on Propellants, Explosives and Pyrotechnics, Nanjing, China, 20–23 September 2011; pp. 756–760.

23. Nie, H.; Chan, H.Y.; Pisharath, S.; Hng, H.H. Combustion characteristic and aging behavior of bimetal thermite powders. *Def. Technol.* **2020**. [CrossRef]

24. Wang, C.-A.; Xu, J.-B.; Shen, Y.; Wang, Y.-T.; Yang, T.-L.; Zhang, Z.-H.; Li, F.-W.; Shen, R.-Q.; Ye, Y. Thermodynamics and performance of Al/CuO nanothermite with different storage time. *Def. Technol.* **2020**. [CrossRef]

25. Abdallah, I.; Zapata, J.; Lahiner, G.; Warot-Fonrose, B.; Cure, J.; Chabal, Y.J.; Esteve, A.; Rossi, C. Structure and Chemical Characterization at the Atomic Level of Reactions in Al/CuO Multilayers. *ACS Appl. Energy Mater.* **2018**, *1*, 1762–1770. [CrossRef]

26. Lahiner, G.; Zapata, J.; Cure, J.; Richard, N.; Djafari-Rouhani, M.; Estève, A.; Rossi, C. A redox reaction model for self-heating and aging prediction of Al/CuO multilayers. *Combust. Theory Model.* **2019**, *23*, 700–715. [CrossRef]

27. Deal, B.E.; Grove, A.S. General Relationship for the Thermal Oxidation of Silicon. *J. Appl. Phys.* **1965**, *36*, 3770–3778. [CrossRef]

28. Fischer, S.; Grubelich, M. Theoretical energy release of thermites, intermetallics, and combustible metals. In Proceedings of the Twenty-Fourth International Pyrotechnics Seminar, Monterey, CA, USA, 27–31 July 1998; pp. 231–286.

29. Nabatame, T.; Yasuda, T.; Nishizawa, M.; Ikeda, M.; Horikawa, T.; Toriumi, A. Comparative studies on oxygen diffusion coefficients for amorphous and gamma-Al$_2$O$_3$ films using O-18 isotope. *Jpn. J. Appl. Phys.* **2003**, *42*, 7205–7208. [CrossRef]

30. Julien, B.; Cure, J.; Salvagnac, L.; Josse, C.; Esteve, A.; Rossi, C. Integration of Gold Nanoparticles to Modulate the Ignitability of Nanothermite Films. *ACS Appl. Nano Mater.* **2020**, *3*, 2562–2572. [CrossRef]

31. Zapata, J.; Nicollet, A.; Julien, B.; Lahiner, G.; Esteve, A.; Rossi, C. Self-propagating combustion of sputter-deposited Al/CuO nanolaminates. *Combust. Flame* **2019**, *205*, 389–396. [CrossRef]

32. Lahiner, G.; Nicollet, A.; Zapata, J.; Marín, L.; Richard, N.; Rouhani, M.D.; Rossi, C.; Estève, A. A diffusion–reaction scheme for modeling ignition and self-propagating reactions in Al/CuO multilayered thin films. *J. Appl. Phys.* **2017**, *122*, 155105. [CrossRef]

33. Wang, H.; Julien, B.; Kline, D.; Alibay, Z.; Rehwoldt, M.C.; Rossi, C.; Zachariah, M.R. Probing the Reaction Zone of Nanolaminates at ~µs Time and ~µm Spatial Resolution. *J. Phys. Chem. C* **2020**, *124*, 13679–13687. [CrossRef]

Publisher's Note: MDPI stays neutral with regard to jurisdictional claims in published maps and institutional affiliations.

Article

Controllable Electrically Guided Nano-Al/MoO₃ Energetic-Film Formation on a Semiconductor Bridge with High Reactivity and Combustion Performance

Xiaogang Guo [1,2], Qi Sun [3,*], Taotao Liang [4] and A. S. Giwa [5]

[1] Chongqing Key Laboratory of Inorganic Special Functional Materials, College of Chemistry and Chemical Engineering, Yangtze Normal University, Chongqing 408100, China; guoxiaogang@yznu.edu.cn

[2] Material Corrosion and Protection Key Laboratory of Sichuan Province, College of Chemistry and Environmental Engineering, Institute of Functional Materials, Sichuan University of Science and Engineering, Zigong 643000, China

[3] College of Life Sciences, Chongqing Normal University, Chongqing 401331, China

[4] Faculty of Materials and Energy, Southwest University, Chongqing 400715, China; liangtaotao@email.swu.edu.cn

[5] State Key Joint Laboratory of Environment Simulation and Pollution Control, School of Environment, Tsinghua University, Beijing 100084, China; giwasegun@live.com

* Correspondence: sunqi2017@cqnu.edu.cn

Received: 17 April 2020; Accepted: 14 May 2020; Published: 18 May 2020

Abstract: Film-forming techniques and the control of heat release in micro-energetic chips or devices create challenges and bottlenecks for the utilization of energy. In this study, promising nano-Al/MoO₃ metastable intermolecular composite (MIC) chips with an uniform distribution of particles were firstly designed via a convenient and high-efficiency electrophoretic deposition (EPD) technique at room temperature and under ambient pressure conditions. The mixture of isopropanol, polyethyleneimine, and benzoic acid proved to be an optimized dispersing agent for EPD. The kinetics of EPD for oxidants (Al) and reductants (MoO₃) were systematically investigated, which contributed to adjusting the equivalence ratio of targeted energetic chips after changing the EPD dynamic behaviors of Al and MoO₃ in suspension. In addition, the obtained nano-Al/MoO₃ MIC energetic chips showed excellent heat-release performance with a high heat release of ca. 3340 J/g, and were successfully ignited with a dazzling flame recorded by a high-speed camera. Moreover, the fabrication method here is fully compatible with a micro-electromechanical system (MEMS), which suggests promising potential in designing and developing other MIC energetic chips or devices for micro-ignition/propulsion applications.

Keywords: nano-Al/MoO₃ MIC; stable suspension; electrophoretic deposition; kinetics; micro initiator

1. Introduction

In recent decades, increasing attention has been paid to energetic fuels with high energy density (e.g., metastable intermixed composites (MICs) or nanothermites). They can generate superior combustion performance, so they are widely used in high-efficiency propellants [1], welding auxiliary devices [2], pyrotechnics [3], and specialized igniters or energetic chips [4] for a variety of military purposes and civilian applications. Generally, MICs are regarded as excellent fuels, which generally consist of metal-fuel (e.g., Al, Mg) and oxidizers that include metal oxides (e.g., MoO₃ [5], Fe₂O₃ [6], polyvinylidene fluoride [7], NiO [8], CuO [9], and iodates [10]). Compared to traditional micro-MICs, nano-structured MICs have been gradually verified as a promising candidate for highly reactive energetic materials or composites due to their higher heat-release properties, greater contact area between fuels and oxidizers, and faster detonation velocity [11–13]. Therefore, there is an emerging research area of interest in designing novel nano-structured MICs via facile techniques.

The Al/MoO_3 MIC, as a desirable energetic system, has continuously aroused great interest, owing to its high heat of reaction (4698 J/g) and adiabatic flame temperature (3547 °C, higher than that of Al/Fe_2O_3, Al/CuO, etc.). Recently, different fabrication methods have been explored to prepare Al/MoO_3 MICs or nanothermites for developing their exothermic performances, including the thermal co-evaporation method [14], magnetron sputtering [15,16], sonic wave-assisted physical mixing [17], and arrested reactive milling laser irradiation [18]. For example, M.R. Zachariah et al. designed Al/MoO_3 MICs with different multilayer internal structures via the magnetron sputtering method on a semiconductor bridge as a promising micro-energy storage device, and analyzed the condensed state reaction process in the obtained nano-multilayered films [19]. E.L. Dreizin et al. reported low-temperature exothermic reactions in fully-dense Al/MoO_3 nanocomposite powders fabricated by the arrested reactive milling technique [20]. In addition, Al/MoO_3 MICs fabricated by the traditional mechanical mixing technique were shown to fuel a dramatic combustion exothermic process with a high burning rate of 100 ± 4 m/s and a high pressurization rate of 35 kPa/μs [21]. Nevertheless, it is relatively difficult to simultaneously meet the requirements of being low cost and easy to operate, with high-film-forming efficiency, using most reported methods. It is worth noting that using a portable electrophoresis, electrophoretic deposition (EPD) has technically demonstrated advantages in controllability of the composition and deposition efficiency for the target products from the charged micro/nanoparticles [22,23], or polymer molecules [24,25] in a stable suspension. The fabrication of Al/CuO and Al/NiO energetic films with uniform distribution was demonstrated by the K. T. Sullivan group [26] and the D. X. Zhang group [27], respectively. In our previous research work, the EPD technique was successfully used to prepare an Al/Bi_2O_3 MIC system [28]. Moreover, for practical application, it is essential to combine MICs with micro-electromechanical system (MEMS) technology [29] (i.e., so-called "nanoenergetics-on-a-chip" technology), constructing miniature energy-demanding devices with wide applications. However, there are few reports of the controlled design of MIC (e.g., Al/MoO_3) chips via the EPD technique.

Thus, for these reasons, a novel nano-Al/MoO_3 MIC chip combined with Al/MoO_3 nanolaminates and a typical micro-semiconductor bridge was firstly designed via the facile EPD method using isopropanol, polyethyleneimine, and benzoic acid as an optimized dispersion system. As a type of binary energetic chip, the composition of the nano-Al/MoO_3 MIC can be affected mainly by the EPD dynamic behaviors of the fuel (Al) and oxidizer (MoO_3). Thus, further exploration of this aspect was analyzed and verified theoretically. Finally, the heat-release properties and ignition test of the product were investigated.

2. Experimental Section

2.1. Reagents and Materials

Polyethyleneimine, PEG-2000, benzoic acid, and nano-Al powders (99.9%) were purchased from Aladdin Inc. Corporation. (Shanghai, China). Isopropyl alcohol was purchased from Kelong Industrial Inc., (Chengdu, China). The other reagents (including hydrogen peroxide and ethanol) from Sinopharm Chemical Reagent Co., Ltd. (Shanghai, China) were used as analytical grade purity without further purification. Deionized water (18.2 Ω) was used in all experiments.

2.2. Controllable Design of Nano-Al/MoO₃ MIC

In the fabrication of the nano-Al/MoO_3 MIC, the EPD technique was exploited, and the corresponding detailed schematic diagram is displayed in Figure 1. Firstly, a classic one-step method was developed to prepare flake-like MoO_3 powders. To be specific, 0.25 M Mo powders were added into 200 mL deionized water with trace PEG2000 marked as mixture *A*, then H_2O_2 (30 wt%) was dripped into mixture *A* slowly, until the yellow molybdenum peroxide sol appeared. After ultrasonic treatment for 0.25 h, the obtained sol was moved into a hydrothermal reactor at 110 ± 2 °C for 4 h, and the MoO_3 powders were fabricated after repeated centrifugal cleaning at a rotation speed of 10,000 r/min and vacuum drying.

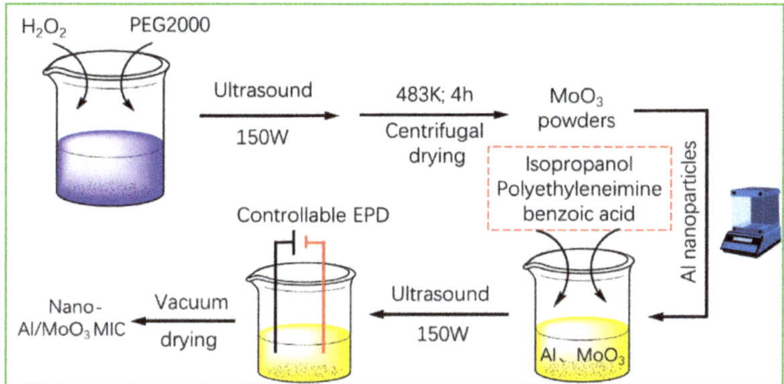

Figure 1. Schematic diagram of the facile fabrication of the nano-Al/MoO$_3$ metastable intermolecular composite (MIC) chip.

Then, the nano-Al and MoO$_3$ powders with different mass ratios were added into the optimized dispersant of isopropanol, polyethyleneimine, and benzoic acid with a volume ratio of 50:1:1 to obtain a stable suspension after ultrasonic treatment for 20 min at 25 °C. During EPD, a micro-ignition bridge was the working electrode, and the copper sheet with the same area was used as the counter electrode; the detailed size of electrodes is shown in Figure S1 (Supporting Information). The distance of the two electrodes ranged from 0.4 to 2.4 cm. In addition, the EPD process was conducted under different field strengths. The EPD time ranged from 0 to 16 min. After EPD, the working electrode was removed from the suspension, and dried in an oven at 80 °C for 1.5 h. The nano-Al/MoO$_3$ MIC chip was finally obtained after cooling to room temperature for the subsequent ignition experiments. The deposited efficiency (deposit weight per area (mg cm^{-2})) of the deposits was calculated by dividing the increased weight of the working electrode after EPD by the deposition area. Each experiment was repeated five times, and the average value of five parallel experiments was used as the final valid result.

2.3. Material Characterization

The morphology, element distribution, and crystalline structures of the nano-Al/MoO$_3$ MIC were measured with a field emission scanning electron microscope (FESEM, JSM-7800F, Tokyo, Japan) equipped with energy dispersive X-ray spectroscopy (EDX), and X-ray diffractometer (XRD-6000, Shimadzu, Tokyo, Japan) with a scanning rate of 5°/min, respectively. Atomic absorption spectroscopy (AAS, 180-80, Exter Analytical, Tokyo, Japan) was used to determine the mass or mole ratio of Al and MoO$_3$ in deposited energetic film. The heat-release (Q) of the energetic chip was analyzed by differential scanning calorimetry (DSC, STA449F3, NETZSCH, Berlin, Germany) measured in a temperature range from 25 to 1000 °C at a low heating rate of 10 K/min under a 99.999% argon flow. Ignition of the product was studied using home-made capacitor charge/discharge initiating equipment, and video recordings of the deflagration were recorded by a high-speed camera (Phantom v7.3, Vision Research, Inc., Wayne, NJ, USA) at an imaging speed of 10^4 f/s.

3. Results and Discussion

3.1. EPD Dynamic Studies

A successful EPD generally largely depends on various dispersing agents, and a large number of experimental studies have been carried out to compare the dispersion systems for EPD of specific sorts of particles [24,30,31]. After a large number of comparison attempts, the optimal dispersing agent of mixture of isopropanol, polyethyleneimine, and benzoic acid was used to fabricate the nano-Al/MoO$_3$ MIC by EPD at normal tempearture and pressure. For verifying the EPD controllability, dynamic behaviours of particles in optimized suspension were studied in detail. Shown in Figure 2a is the deposited efficiency (mg/cm^2) as a function of deposited time under applied electric field strengths ranging from 6 to 12 V/mm during EPD of Al/MoO$_3$ MIC films. It was obviously observed that the deposition efficiency increased with the EPD time when the field strength was fixed at 6 V/mm, and a similar trend was also seen in a higher field strength of 9 or 12 V/mm. The higher field strength provides a higher EPD efficiency at a certain deposited time (e.g., 10 min). Moreover, the deposited efficiency increased linearly with deposited time increasing from 0 to T_c (the critical time between linear and non-linear EPD dynamic in the critical region (black circle)) in Figure 2a,b, which is consistent with the research of the Zhang group [32]. In addition, T_c decreased with an increase of applied field strength; that is, T_c for 6 V/mm was larger than T_c for 9 and 12 V/mm, which was primarily due to the more severe precipitation and collision of particles under higher field strengths. Thus, the EPD process of Al/MoO$_3$ MIC can be more precisely controlled in the linear control region ($t < T_c$) in this study, which is due to the relatively complex relationship of deposited efficiency and EPD time in the nonlinear variation region for all field strengths. Furthermore, the effect of the distance of electrodes on the deposited efficiency of the nano-Al/MoO$_3$ MIC is analyzed in Figure 2c. When the solid loading concentration, EPD time, and applied field strength were set at 0.5 g/L, 8 min, and 12 V (blue line), respectively, the deposited efficiency increased gradually with the distance of electrodes rising to 1.2 from 0.4 cm. It then decreased slowly, as the distance of electrodes continued to increase. This result is perhaps caused by the more violent disturbance of particles under a smaller distance of electrodes, and the higher degree of the settlement of particles under a longer distance of electrodes that leads to a lower EPD efficiency. Similar change trends were observed at higher field strengths (Figure 2c), which provides a valuable reference for realizing controllable EPD of different particles.

In addition, the exothermicity of MICs is a key indicator that largely depends on the mass or mole ratio of fuel (e.g., Al) and oxidizer (e.g., MoO$_3$). Generally, in MIC energetic reactions, the equivalence ratio (Φ) is defined as the actual ratio of fuel to oxidizer divided by the stoichiometric ratio of fuel to oxidizer in an energetic reaction, that is $\Phi = $ (F/O)$_{actual}$/(F/O)$_{stoich}$ [26]. For the codeposition process of the Al and MoO$_3$ particles, the equivalence ratio in the starting suspension (Φ_s) was adjusted accurately in weighed samples, and the equivalence ratio in the deposited product (Φ_d) was determined by EDX and AAS techniques. Figure 2d displays the Φ_d of Al and MoO$_3$ particles in the Al/MoO$_3$ MIC chip as a function of Φ_s of nano-Al and MoO$_3$ particles in the starting suspension. Clearly, it can be seen that Φ_d increased linearly with Φ_s by EDX and AAS analysis, and the fitted equations were similar ($Y = 1.97X - 1.04$, $R^2 = 992$ for EDX analysis, and $Y = 2.02X - 0.96$, $R^2 = 998$ for AAS analysis). Thus, Φ_d in the nano-Al/MoO$_3$ MIC could be adjusted by changing Φ_s in suspension, which contributes to optimizing the proportion of components in product, and further developing the exothermic performance of the product.

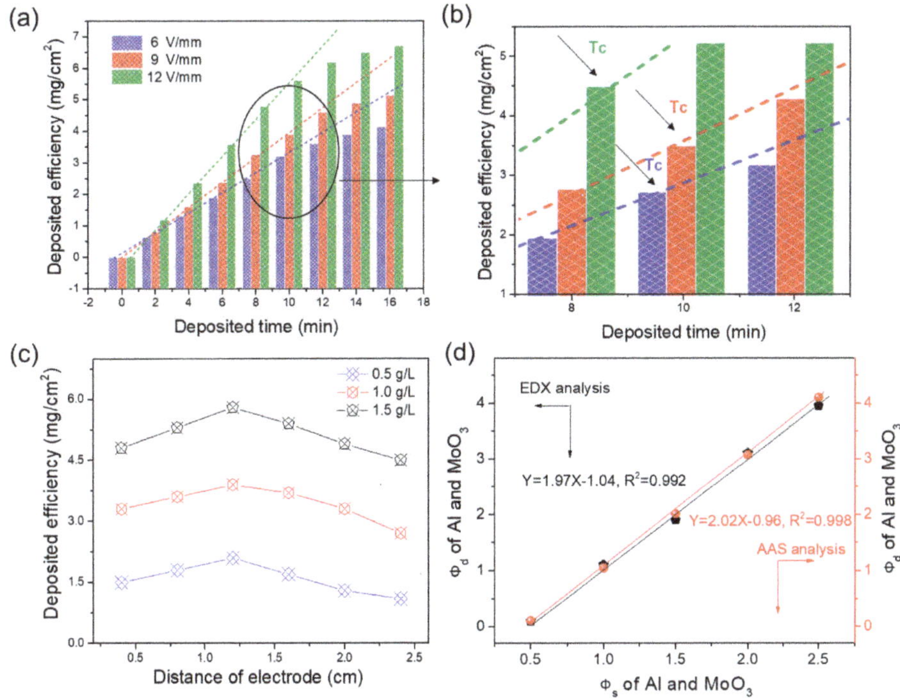

Figure 2. (**a**) Deposited efficiency (mg/cm^2) of nano-Al/MoO$_3$ MIC as functions of deposition time under different applied electric field strengths; (**b**) the local amplification image of the black circle in (**a**); (**c**) the relationship of deposited efficiency and the distance of two electrodes under different loading concentrations; and (**d**) Φ_d of Al and MoO$_3$ in the Al/MoO$_3$ MIC chip as a function of Φ_s of nano-Al and MoO$_3$ particles in the starting suspension.

3.2. Characteristics of Nano-Al/MoO$_3$ MIC

XRD analysis was used to investigate the crystal structures of the nano-Al/MoO$_3$ MIC in Figure 3. It can be clearly seen that two groups of distinct diffraction peaks are marked in good agreement with that of the standard spectra for pure Al (JCPDS card No. 04-0787; Fm-3m (225)) and MoO$_3$ (JCPDS card No. 35-0609; Pb nm (62)) on the product, demonstrating the successful co-EPD of the Al and MoO$_3$ particles. In addition, the fact that there are no other clear peaks of Al$_2$O$_3$ and Mo in Figure 3 indicates the high purity of the product, and that no thermite reactions took place during the EPD process.

The as-obtained nano-Al/MoO$_3$ MIC films via EPD are displayed in Figure 4. Regions of large-scale local sags, crests, and separations are not seen optically in the target product surface (Figure 4a), which exhibits significant coating characteristics and uniformity. Clearly, in the FESEM image of product in Figure 4b, the nano-Al/MoO$_3$ MIC appears to be uniformly distributed, without rare unmixed zones. The higher-resolution images in Figure 4c indicate that the nano-Al particles were scattered or distributed randomly in flake-like MoO$_3$, which significantly helps to enlarge the contact areas of Al and MoO$_3$, and shorten the mass transportation length (MSL) during the thermite exothermic reaction of nano-Al/MoO$_3$ MIC. Moreover, there were numerous gaps among the particles (Figure 4b,c), contributing to providing a large number of heat-release channels or multiple spatial streams, and further improving the exothermic performance of the product [27]. Moreover, the elemental compositions in the nano-Al/MoO$_3$ MIC were analyzed by the EDX technique, as shown in Figure 4d, where the EDX spectrum indicates that all expected elements of Al, Mo, and O existed in the energetic film surface, consistent with the results of the XRD analysis. It is worth noting that the mole ratio of Al, Mo, and O was close to 2:1:3 (0.336:0.16:0.50) (seen in Figure 4d and Table S1 in Supporting Information), and the corresponding reaction mole ratio of Al and MoO_3 was close to 2:1, which contributed to realizing a sufficient aluminothermic reaction ($2Al + MoO_3 \rightarrow Al_2O_3 + Mo + H_{Heat-release}$, $\Delta H = 4698$ J/g) [17,33]. In addition, we conducted a comparative study of FESEM mapping and the corresponding results are similar in three random regions (Table S1), where the mole ratio of aluminum to nickel is close to 1:1, which indicates the uniform distribution of the product. Moreover, the percentage errors of the mole ratio of elements are approximate 5% in six random regions, according to both EDX and AAS analysis, further demonstrating the homogeneously mixed nano-Al/MoO_3 MIC obtained by EPD.

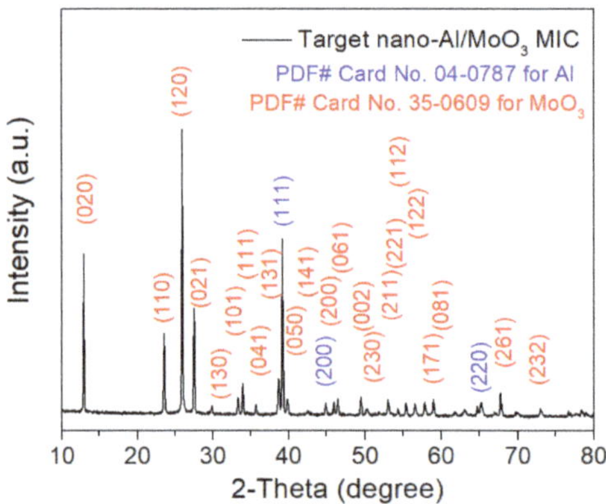

Figure 3. Typical X-ray diffractometer (XRD) pattern of the as-obtained nano-Al/MoO_3 MIC.

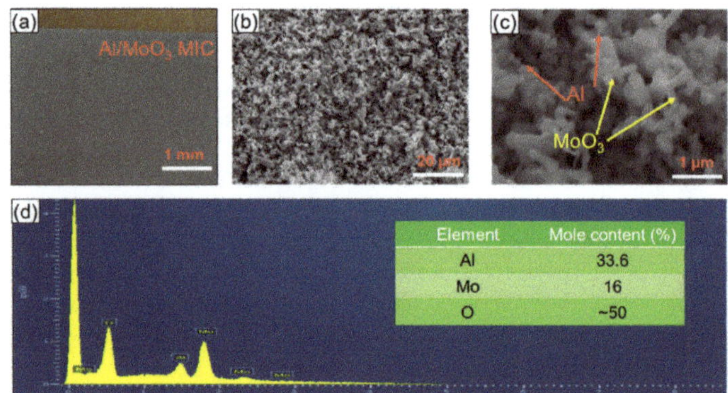

Figure 4. (**a**) Optical and (**b**,**c**) typical field emission scanning electron microscope (FESEM) images of the nano-Al/MoO_3 MIC films prepared using electrophoretic deposition (EPD); (**d**) energy dispersive X-ray spectroscopy (EDX) spectrum of the product with an inserted table showing the mole content (%) of all elements.

3.3. Thermal Studies

Exploration of exothermic performance is essential to energetic materials, including MICs [34–37], and is shown in Figure 5 in detail. Figure 5a displays the DSC data measured from the nano-Al/MoO_3 MIC with the Φ_d of ~1.0 and a low heating rate of 10 K/min. In addition to an unobservable exothermic peak at ~400 °C, probably due to the reaction between Al nano-particles with much smaller-sized MoO_3 particles, there are several observable exothermic peaks in Figure 5a, where the exothermic peak (green rectangle area and yellow rectangle area) is mainly because of the reaction between Al particles with smaller-sized MoO_3 particles [38], and the latter two exothermic peaks at 703.4 °C and

735.9 °C (blue rectangle area) are from the reaction of Al and bigger-sized MoO_3 particles, which is consistent with the results from the Zhu group [15]. In addition, there was also an endothermic peak (Figure 5a, primrose yellow rectangle area) at ca. 660 °C, caused by the melt of metal-Al. After a fitting calculation, the value of the heat-release of the nano-Al/MoO_3 MIC was as high as ~3340 J/g, which was >70% of the theoretical value, indicating the relatively sufficient thermite reaction. Furthermore, the effect of the deposited time on the heat-release of the product is analyzed in Figure 5b. There was a similar trend for different field strengths from 6 to 12 V/mm, that is, the output of heat was almost stable as the deposited time increased, showing the great controllability of EPD dynamic behaviors of Al or MoO_3 particles in suspension. The heat-release values as functions of Φ_d and Φ_s of Al and MoO_3 are clearly shown in the 3D histogram (Figure 5c). The heat-release values increased first and then gradually decreased with Φ_d of Al and MoO_3, and were highly associated with Φ_s of Al and MoO_3 in the starting suspension. The peak value of the heat release of the nano-Al/MoO_3 MIC can be obtained when Φ_d of Al and MoO_3 was close to 1.0.

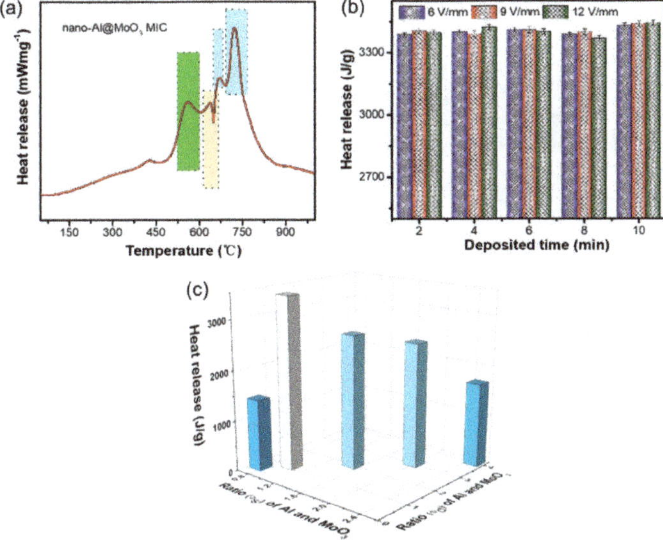

Figure 5. (a) Differential scanning calorimetry (DSC) curve of the obtained nano-Al/MoO_3 MIC; (b) the relationship of heat release and deposited time under different applied electric field strengths; and (c) 3D histogram of heat release as a function of Φ_d and Φ_s of Al and MoO_3 particles in the product.

The thermite reaction deflagration processes of an electric explosion for the nano-Al/MoO_3 MIC chip were realized a self-regulating capacitor charge/discharge initiating device, and recorded synchronously by a high-speed camera. The detonation schematic diagram is displayed in Figure 6a. When the ignition circuit was switched on, the energetic target chip was quickly detonated with a dazzling blaze. The corresponding flame propagation images of the nano-Al/MoO_3 MIC chip are observed in Figure 6b, where the interval time between adjoining images is 0.1 ms. The flame duration time of the nano-Al/MoO_3 MIC chips was >1 ms, and loud sounds during the ignition test indicated that the thermite reactions were so intense that energy was released quickly [39–42]. In addition, the observed intense deflagration was in accordance with the DSC results, which provided a facile route to nano-MIC energetic chips for MEMS application. In addition, the heat-release performance of MIC chips can be optimized by building a theoretical bridge between the equivalence ratio of oxidant and reductant in starting suspension (Φ_s) and target energetic films or initiators (Φ_d).

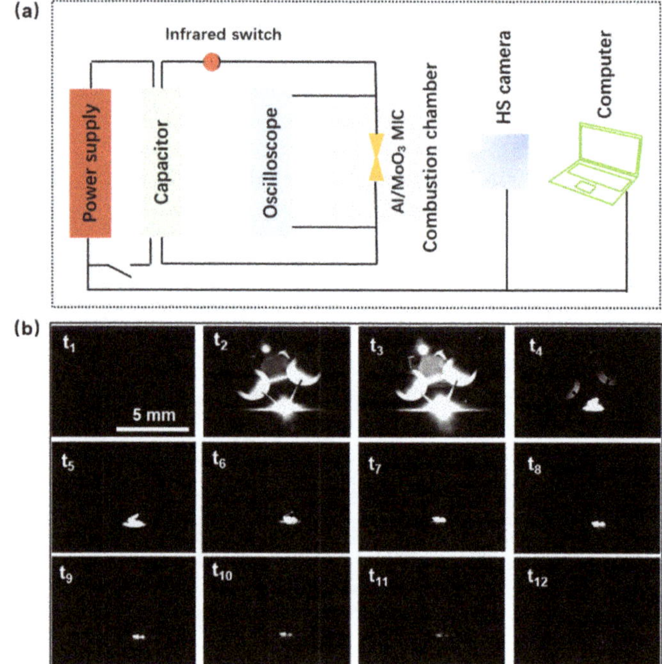

Figure 6. (**a**) Schematic of the ignition system for the micro nano-Al/MoO$_3$ MIC chip initiator, and ignition process recorded by a high-speed camera; (**b**) series of still images taken from a typical ignition deflagration study of the nano-Al/MoO$_3$ MIC fabricated by EPD process with $\Phi_d = 1.0$; the time interval between images is 0.1 ms ($t_n - t_{n-1} = 0.1$ ms, $n \geq 2$).

4. Conclusions

In this study, a novel Al/MoO$_3$ MIC chip initiator was firstly fabricated by a high-efficiency EPD technique in an optimized mixture dispersant of isopropanol, polyethyleneimine, and benzoic acid at normal temperature and pressure. The microstructures and chemical compositions of the product were demonstrated by FESEM, EDX, and XRD. The deposited energetic films exhibited even mixing between the oxidizer (Al) and reductant (MoO$_3$), contributing to enhancing their exothermic performance. The EPD dynamic behaviors of nano-Al and MoO$_3$ particles were studied, which can act as a theoretical bridge for connecting the Φ_s in starting suspension and Φ_d in energetic chips. DSC results showed the apparent exothermic peaks of nano-Al/MoO$_3$ MIC chips, due to the thermite reaction between Al and MoO$_3$, and the corresponding total heat-release was as high as ca. 3340 J/g when Φ_d of Al and MoO$_3$ was close to 1.0. In addition, the Al/MoO$_3$ MIC chip initiator can be successfully ignited with a typical capacitor charge/discharge ignition device, exhibiting outstanding detonation performance with a short burst time and a dazzling flame. In short, the design of the Al/MoO$_3$ MIC chip initiator in this study will provide a universal approach for fabricating other thermite energetic chips with wide civilian and military applications, especially in micro-initiation or micro-propulsion systems.

Supplementary Materials: The following are available online at http://www.mdpi.com/2079-4991/10/5/955/s1, Figure S1: The size specification of the working and counter electrodes used for EPD dynamic research and ignition test. All yellow rectangle zones are parts of electrodes without touching the optimized suspension., Table S1: Molar content results of different elements in products by EDX and AAS analysis in three random regions.

Author Contributions: X.G. and T.L. conceived and designed the experiments; T.L. and Q.S. conducted the experiments; X.G. and Q.S. analyzed the data; X.G. and T.L. contributed reagents/materials/analysis tools; X.G. and T.L. wrote the paper; X.G., Q.S., A.S.G. and T.L. revised the paper. All authors have read and agreed to the published version of the manuscript.

Funding: This research was funded by the National Natural Science Foundation of China (21805014), the scientific and technological research program of Chongqing Municipal Education commission (KJQN201801424), Natural Science Foundation of Chongqing (No cstc2019jcyj-msxmX0675), Yangtze Normal University (No. 2018QNRC10) and the opening project of material corrosion and protection key laboratory of Sichuan province (2018CL19).

Acknowledgments: The authors acknowledge the financial support from National Natural Science Foundation of China (research core funding No. 21805014), Chongqing Municipal Education commission (research core funding No. KJQN201801424), Natural Science Foundation of Chongqing (research core funding No. cstc2019jcyj-msxmX0675), Yangtze Normal University (young scientist research funding No. 2018QNRC10) and the opening project of material corrosion and protection key laboratory of Sichuan province (research core funding No. 2018CL19).

Conflicts of Interest: The authors declare no conflict of interest.

References

1. Zhang, C.; Sun, C.; Hu, B.; Yu, C.; Lu, M. Synthesis and characterization of the pentazolate anion cyclo-N_5^- in $(N_5)_6(H_3O)_3(NH_4)_4Cl$. *Science* **2017**, *355*, 374–376. [CrossRef]

2. Sullivan, K.T.; Zhu, C.; Duoss, E.B.; Gash, A.E.; Kolesky, D.B.; Kuntz, J.D.; Lewis, J.A.; Spadaccini, C.M. 3D Printing: Controlling Material Reactivity Using Architecture (Adv. Mater. 10/2016). *Adv. Mater.* **2016**, *28*, 1901. [CrossRef]

3. He, W.; Liu, P.-J.; He, G.-Q.; Gozin, M.; Yana, Q.-L. Highly Reactive Metastable Intermixed Composites (MICs): Preparation and Characterization. *Adv. Mater.* **2018**, *30*, 1706293. [CrossRef] [PubMed]

4. Cao, X.; Deng, P.; Hu, S.; Ren, L.; Li, X.; Xiao, P.; Liu, Y. Fabrication and Characterization of Nanoenergetic Hollow Spherical Hexanitrostibene (HNS) Derivatives. *Nanomater.* **2018**, *8*, 336. [CrossRef] [PubMed]

5. Liu, J.; Shao, S.; Fang, G.; Meng, B.; Xie, Z.; Wang, L. High-Efficiency Inverted Polymer Solar Cells with Transparent and Work-Function Tunable MoO_3-Al Composite Film as Cathode Buffer Layer. *Adv. Mater.* **2012**, *24*, 2774–2779. [CrossRef] [PubMed]

6. Grapes, M.; Reeves, R.V.; Fezzaa, K.; Sun, T.; Densmore, J.M.; Sullivan, K.T. In situ observations of reacting Al/Fe_2O_3 thermite: Relating dynamic particle size to macroscopic burn time. *Combust. Flame* **2019**, *201*, 252–263. [CrossRef]

7. Wang, J.; Qu, Y.; Gong, F.; Shen, J.; Zhang, L. A promising strategy to obtain high energy output and combustion properties by self-activation of nano-Al. *Combust. Flame* **2019**, *204*, 220–226. [CrossRef]

8. Wang, N.; Hu, Y.; Ke, X.; Xiao, L.; Zhou, X.; Peng, S.; Hao, G.; Jiang, W. Enhanced-absorption template method for preparation of double-shell NiO hollow nanospheres with controllable particle size for nanothermite application. *Chem. Eng. J.* **2020**, *379*, 122330. [CrossRef]

9. Dai, J.; Wang, F.; Ru, C.; Xu, J.; Wang, C.; Zhang, W.; Ye, Y.; Shen, R. Ammonium Perchlorate as an Effective Additive for Enhancing the Combustion and Propulsion Performance of Al/CuO Nanothermites. *J. Phys. Chem. C* **2018**, *122*, 10240–10247. [CrossRef]

10. Jian, G.; Feng, J.; Jacob, R.J.; Egan, G.C.; Zachariah, M.R. Super-reactive nanoenergetic gas generators based on periodate salts. *Angew. Chem. Int. Ed.* **2013**, *52*, 9743–9746. [CrossRef]

11. Ramachandran, R.; Vuppuluri, V.S.; Fleck, T.J.; Rhoads, J.F.; Gunduz, I.E.; Son, S.F. Influence of Stoichiometry on the Thrust and Heat Deposition of On-Chip Nanothermites. *Propellants Explos. Pyrotech.* **2018**, *43*, 258–266. [CrossRef]

12. Hea, W.; Lib, Z.-H.; Chena, S.; Yangc, G.; Yang, Z.; Liu, P.; Yana, Q.-L. Energetic metastable n-Al@PVDF/EMOF composite nanofibers with improved combustion performances. *Chem. Eng. J.* **2020**, *383*, 123146. [CrossRef]

13. Liu, P.; Liu, P.; Wang, M. Ignition and combustion of nano-sized aluminum particles: A reactive molecular dynamics study. *Combust. Flame* **2019**, *201*, 276–289. [CrossRef]

14. Suhaimi, B.F.; Palale, S.; Teh, L.K.; Mathews, N.; Mhaisalkar, S. Energy band and optical modeling of charge transport mechanism and photo-distribution of MoO_3/Al-doped MoO_3 in organic tandem cells. *Funct. Mater. Lett.* **2020**, *13*, 2051003. [CrossRef]

15. Xu, J.; Tai, Y.; Ru, C.; Dai, J.; Ye, Y.; Shen, R.; Zhu, P. Tuning the Ignition Performance of a Microchip Initiator by Integrating Various Al/MoO₃ Reactive Multilayer Films on a Semiconductor Bridge. *ACS Appl. Mater. Interfaces* **2017**, *9*, 5580–5589. [CrossRef]

16. Xu, J.B.; Tai, Y.; Shen, Y.; Dai, J.; Xu, W.; Ye, Y.H.; Shen, R.Q.; Hua, Y. Characteristics of energetic semiconductor bridge initiator based ondifferent stoichiometric ratios of Al/MoO₃ reactive multilayer filmsunder capacitor discharge conditions. *Sensor. Actuator A* **2019**, *296*, 241–248. [CrossRef]

17. Weir, C.; Pantoya, M.L.; Daniels, M.A. The role of aluminum particle size in electrostatic ignition sensitivity of composite energetic materials. *Combust. Flame* **2013**, *160*, 2279–2281. [CrossRef]

18. Wolenski, C.; Wood, A.; Mathai, C.J.; He, X.; McFarland, J.A.; Gangopadhyay, K.; Gangopadhyay, S.; Maschmann, M.R. Nanoscale surface reactions by laser irradiation of Al nanoparticles on MoO₃ flakes. *Nanotechnol.* **2018**, *30*, 045703. [CrossRef]

19. Xu, J.; Shen, Y.; Wang, C.; Dai, J.; Tai, Y.; Ye, Y.; Shen, R.; Wang, H.; Zachariah, M.R. Controlling the energetic characteristics of micro energy storage device by in situ deposition Al/MoO₃ nanolaminates with varying internal structure. *Chem. Eng. J.* **2019**, *373*, 345–354. [CrossRef]

20. Williams, R.A.; Schoenitz, M.; Ermoline, A.; Dreizin, E.L. Low-temperature exothermic reactions in fully-dense Al/MoO₃ nanocomposite powders. *Thermochim. Acta* **2014**, *594*, 1–10. [CrossRef]

21. Glavier, L.; Taton, G.; Ducéré, J.-M.; Baijot, V.; Pinon, S.; Calais, T.; Esteve, A.; Rouhani, M.; Rossi, C. Nanoenergetics as pressure generator for nontoxic impact primers: Comparison of Al/Bi₂O₃, Al/CuO, Al/MoO₃ nanothermites and Al/PTFE. *Combust. Flame* **2015**, *162*, 1813–1820. [CrossRef]

22. Zhou, Y.; Qian, W.; Huang, W.; Liu, B.; Lin, H.; Dong, C. Carbon Nanotube-Graphene Hybrid Electrodes with Enhanced Thermo-Electrochemical Cell Properties. *Nanomater.* **2019**, *9*, 1450. [CrossRef] [PubMed]

23. Tokunaga, S.; Itoh, Y.; Yaguchi, Y.; Tanaka, H.; Araoka, F.; Takezoe, H.; Aida, T. Electrophoretic Deposition for Cholesteric Liquid-Crystalline Devices with Memory and Modulation of Reflection Colors. *Adv. Mater.* **2016**, *28*, 4077–4083. [CrossRef] [PubMed]

24. Sarkar, A.; Daniels-Race, T. Electrophoretic Deposition of Carbon Nanotubes on 3-Amino-Propyl-Triethoxysilane (APTES) Surface Functionalized Silicon Substrates. *Nanomaterials* **2013**, *3*, 272–288. [CrossRef] [PubMed]

25. Lin, F.; Boettcher, S.W. Adaptive semiconductor/electrocatalyst junctions in water-splitting photoanodes. *Nat. Mater.* **2013**, *13*, 81–86. [CrossRef] [PubMed]

26. Sullivan, K.T.; Worsley, M.A.; Kuntz, J.D.; Gash, A.E. Electrophoretic deposition of binary energetic composites. *Combust. Flame* **2012**, *159*, 2210–2218. [CrossRef]

27. Zhang, D.; Li, X. Fabrication and Kinetics Study of Nano-Al/NiO Thermite Film by Electrophoretic Deposition. *J. Phys. Chem. A* **2015**, *119*, 4688–4694. [CrossRef]

28. Guo, X.; Lai, C.; Jiang, X.; Mi, W.; Yin, Y.; Li, X.; Shu, Y. Remarkably facile fabrication of extremely superhydrophobic high-energy binary composite with ultralong lifespan. *Chem. Eng. J.* **2018**, *335*, 843–854. [CrossRef]

29. Jang, N.-S.; Ha, S.-H.; Jung, S.-H.; Kim, S.H.; Lee, H.M.; Kim, J.-M. Fully packaged paper heater systems with remote and selectiveignition capabilities for nanoscale energetic materials. *Sensor. Actuator A* **2019**, *287*, 121–130. [CrossRef]

30. Zhang, H.; Kinnear, C.; Mulvaney, P. Fabrication of single-nanocrystal srrays. *Adv. Mater.* **2019**, 1904551.

31. Guo, X.; Li, X.; Wei, Z.; Li, X.; Niu, L. Rapid fabrication and characterization of superhydrophobic tri-dimensional Ni/Al coatings. *Appl. Surf. Sci.* **2016**, *387*, 8–15. [CrossRef]

32. Zhang, Z.; Huang, Y.; Jiang, Z. Electrophoretic Deposition Forming of SiC-TZP Composites in a Nonaqueous Sol Media. *J. Am. Ceram. Soc.* **1994**, *77*, 1946–1949. [CrossRef]

33. Seo, H.-S.; Kim, J.-K.; Kim, J.-W.; Kim, H.-S.; Koo, K.-K. Thermal behavior of Al/MoO₃ xerogel nanocomposites. *J. Ind. Eng. Chem.* **2014**, *20*, 189–193. [CrossRef]

34. Yu, C.; Zheng, Z.; Zhang, W.; Hu, B.; Chen, Y.; Chen, J.; Ma, K.; Ye, J.; Zhu, J. Sustainable Electrosynthesis of Porous CuN₃ Films for Functional Energetic Chips. *ACS Sustain. Chem. Eng.* **2020**, *8*, 3969–3975. [CrossRef]

35. Hu, L.; Yin, P.; Zhao, G.; He, C.; Imler, G.H.; Parrish, D.A.; Gao, H.; Shreeve, J.M. Conjugated Energetic Salts Based on Fused Rings: Insensitive and Highly Dense Materials. *J. Am. Chem. Soc.* **2018**, *140*, 15001–15007. [CrossRef]

36. Van der Heijdena, A.E.D.M. Developments and challenges in the manufacturing, characterization and scale-up of energetic nanomaterials—A review. *Chem. Eng. J.* **2018**, *350*, 939–948. [CrossRef]

37. Yana, Q.-L.; Zhao, F.-Q.; Kuo, K.K.; Zhang, X.-H.; Zeman, S.; DeLuca, L.T. Catalytic effects of nano additives on decomposition and combustion of RDX-, HMX-, and AP-based energetic compositions. *Prog. Energy Combust. Sci.* **2016**, *57*, 75–136. [CrossRef]

38. Tai, Y.; Xu, J.; Wang, F.; Dai, J.; Zhang, W.; Ye, Y.; Shen, R. Experimental and modeling investigation on the self-propagating combustion behavior of Al-MoO$_3$ reactive multilayer films. *J. Appl. Phys.* **2018**, *123*, 235302. [CrossRef]

39. Wang, J.; Jiang, X.J.; Zhang, L.; Qiao, Z.Q.; Gao, B.; Yang, G.C.; Huang, H. Design andfabricationofenergetic superlattice like-PTFE/Al with superior performance and application in functional micro-initiator. *Nano Energy* **2015**, *12*, 597–605. [CrossRef]

40. Guo, X.; Li, X.; Li, H.; Zhang, D.; Lai, C.; Li, W. A Comprehensive Investigation on the Electrophoretic Deposition (EPD) of Nano-Al/Ni Energetic Composite Coatings for the Combustion Application. *Surf. Coatings Technol.* **2015**, *265*, 83–91. [CrossRef]

41. Kim, S.B.; Kim, K.J.; Cho, M.H.; Kim, J.H.; Kim, S.H.; Kim, K.-T. Micro- and Nanoscale Energetic Materials as Effective Heat Energy Sources for Enhanced Gas Generators. *ACS Appl. Mater. Interfaces* **2016**, *8*, 9405–9412. [CrossRef] [PubMed]

42. Ma, X.X.; Zhu, Y.; Cheng, S.X.; Zheng, H.X.; Liu, Y.S.; Qiao, Z.Q.; Yang, G.C.; Zhang, K.L. Energetic composites based on nano-Al and energetic coordination polymers (ECPs): The "father-son" effect of ECPs. *Chem. Eng. J.* **2020**, *392*, 123719. [CrossRef]

Article

Preparation and Characterization of Al/HTPB Composite for High Energetic Materials

Alexander Vorozhtsov [1], Marat Lerner [1,2], Nikolay Rodkevich [2], Sergei Sokolov [1,*], Elizaveta Perchatkina [1] and Christian Paravan [3]

[1] National Research Tomsk State University, Lenin Avenue, 36, Tomsk 634050, Russia; abv1953@mail.ru (A.V.); lerner@ispms.tsc.ru (M.L.); perchatkinae@mail.ru (E.P.)
[2] Institute of Physics Strength and Material Science SB RAS, Avenue Akademicheskii, 2/4, Tomsk 634055, Russia; ngradk@ispms.tsc.ru
[3] Department of Aerospace Science and Technology, Politecnico di Milano, 34 via LaMasa, I-20156 Milan, Italy; christian.paravan@polimi.it
* Correspondence: sokolovsd95@gmail.com; Tel.: +7-923-406-77-01

Received: 13 October 2020; Accepted: 6 November 2020; Published: 8 November 2020

Abstract: Nanosized Al (nAl) powders offer increased reactivity than the conventional micron-sized counterpart, thanks to their reduced size and increased specific surface area. While desirable from the combustion viewpoint, this high reactivity comes at the cost of difficult handling and implementation of the nanosized powders in preparations. The coating with hydroxyl-terminated polybutadiene (HTPB) is proposed to improve powder handling and ease of use of nAl and to limit its sensitivity to aging. The nAl/HTPB composite can be an intermediate product for the subsequent manufacturing of mixed high-energy materials while maintaining the qualities and advantages of nAl. In this work, experimental studies of the high-energy mixture nAl/HTPB are carried out. The investigated materials include two composites: nAl (90 wt.%) + HTPB (10 wt.%) and nAl (80 wt.%) + HTPB (20 wt.%). Thermogravimetric analysis (TGA) is performed from 30 to 1000 °C at slow heating rate (10 °C/min) in inert (Ar) and oxidizing (air) environment. The combustion characteristics of propellant formulations loaded with conventional and HTPB-coated nAl are analyzed and discussed. Results show the increased burning rate performance of nAl/HTPB-loaded propellants over the counterpart loaded with micron-sized Al.

Keywords: HTPB; aluminum nanopowders; solid propellants; burning rate; coated aluminum

---

1. Introduction

The common composition of mixed high-energy material (HEM) as heterogeneous solid propellants is based on three main ingredients: an oxidizing agent, an organic binder, and a high energy density fuel. Ammonium perchlorate (AP) or ammonium nitrate (AN) are typically used as oxidizer, low-molecular-weight polybutadiene (usually cured by isocyanates) is a common binder, while metal powders are considered as fuels. Aluminum powder is the most readily available, intensively studied and efficient metallic fuel in various HEM formulations. As the aluminum particle size decreases, the ignition temperature and combustion time decrease [1,2]. In solid propellants, the replacement of micron-sized Al powders (µAl) with nanosized aluminum powders (nAl) leads to an increase in the burning rate and a decrease in the size of the condensed combustion products (aggregates/agglomerates) in the near-surface zone [3]. An analysis of the effects of Al powder size reduction on the burning rate of solid propellants and on the condensed combustion products formation is reported in [4]. In this latter study, the burning rate of solid propellants loaded with Al particles with size of 30 µm and 170 nm was measured [4]. Under the investigated conditions, the burning rate was almost doubled when µAl powders were replaced by nAl [4].

Nanosized metal powders are characterized by very high specific surface area. Nanoparticles have different chemical and physical properties when compared to micron sized powders. Due to their increased reactivity, nanopowders are very attractive ingredients for HEM formulations [5–12].

Despite the obvious advantages for inclusion nAl powders in the HEM formulations (i.e., faster reaction rates), there are some disadvantages that limit their use. Nanosized Al is more sensitive than the micron-sized counterpart to the influence of oxidizing or corrosive environment due to the large specific surface area [13,14]. This increased sensitivity to the environment may lead to significant active aluminum content losses during nAl storage [14]. As a consequence, nAl could release less energy during the combustion, thus decreasing the combustion performance of HEMs loaded with it. To suppress/limit the nAl aging during storage, the nanoparticles surface (typically passivated by Al_2O_3) can be coated by a protective layer [15–17]. The encapsulation of particles with hydrocarbon/fluorohydrocarbon coatings has shown the promising results in the limitation of aging process of nAl powder [18]. Hydroxyl-terminated polybutadiene (HTPB) is a suitable material for nAl capping when solid propellant applications are targeted, since it is a common binder in composite formulations for solid rocket propulsion [19,20]. The HTPB offers very low glass transition temperature, relatively low viscosity, high combustion heat and, once cured, high mechanical properties of the final product even at high powder filling ratios.

In this paper the use of HTPB as a protective coating for nAl is discussed. The nAl powder coating process includes the use of acetylacetone together with HTPB. In the proposed strategy the acetylacetone is applied on the particle first. Thus, protective organic hydrophobic layer is created on the surface of the nanoparticles. Then, HTPB is sorbed on the hydrophobic surface of the nanoparticles increasing the compatibility of the nAl with binder and propellant components, preventing a decrease in the content of active aluminum due to aging, and can also simplify propellant production and casting [18]. For powders added to the propellant, the same polymer is used as a binder as for coating the particles. However, when mixing the nanopowders and polymer, the viscosity of the mixture increases drastically, slightly decreasing with continued mixing. Since the mixture has a high viscosity, its preparation requires a lot of time and energy. To overcome this problem, an important technological stage is the preliminary preparation of pastes based on metal nanopowders and polymer binders, followed by their inclusion in the HEM formulation.

The propellant burning rate is a key parameter for the design of a solid rocket motor and, in particular for the thrust level [21]. Together with the oxidizer decomposition and sublimation, and the metal combustion mechanism, HTPB degradation is one of the significant stages of the propellant burning [22]. Thus, the study of the effect of the propellant components on the process of destruction of HTPB is an urgent task in the field of propellant improvement.

The main aim of this work is the experimental study the characteristics of different nAl-based HTPB-containing pastes (nAl/HTPB). Various pastes were prepared, the difference between them being the nAl-to-HTPB ratio. The work focuses on two nAl/HTPB (10 wt.% and 20 wt.%) that are contrasted in terms of their characteristics and effects on the burning behavior of solid propellants loaded with them.

2. Materials and Methods

Aluminum nanopowder (nAl) used in present work was produced by electric explosion of wire (EEW) method [23,24].

Hydroxyl-terminated polybutadiene (HTPB R45), acetylacetone of analytical grade and mineral spirit (used as solvents) with boiling point in the range of 70–100 °C were used without any further purification.

Ammonium perchlorate (AP) used for the preparation of the propellant formulations was used in the form of fine and coarse powders, the fractional composition of which is shown in Table 1.

Table 1. Ammonium perchlorate (AP) particle size distribution.

Parameter	AP_{coarse}	AP_{fine}
$D_{0.1}$, µm	139 ± 1.39	1.70 ± 0.02
$D_{0.5}$, µm	238 ± 2.38	11.9 ± 0.12
$D_{0.9}$, µm	392 ± 3.92	60.9 ± 0.61
D_{32}, µm	202 ± 2.02	3.57 ± 0.036
D_{43}, µm	253 ± 2.53	23.8 ± 0.24

$D_{0.1}$—the diameter below which 10% of the particles lay; $D_{0.5}$—the diameter below which 50% of the particles lay; $D_{0.9}$—the diameter below which 90% of the particles lay; D_{32}—the surface-based mean diameter; D_{43}—the volume-based mean diameter.

Microstructure and morphology of the aluminum nanopowders were captured by a JEM-100 CXII transmission electron microscope (TEM, JEOL Ltd., Tokyo, Japan) for both pristine air-passivated and HTPB-coated powders. Active aluminum content (C_{Al}) was determined by volumetric method. The content was estimated by the evolution of the hydrogen released by the powder reaction with a 5 M NaOH solution [25].

Aluminum nanopowders specific surface area was determined by nitrogen adsorption/desorption by the Brunauer-Emmett-Teller (BET) method with a SORBTOMETR-M (Katakon, Russia) surface area analyzer. A specimen mass of 50 mg was used for the measurements. Specific surface area samples had been heated at 120 °C for 30 min before the surface area measurement was taken. The average particle diameter based on the specific surface area was calculated assuming a particle size distribution of uniform spheres, as:

$$d = \frac{6000}{\rho S} \tag{1}$$

where d—nanoparticle diameter, nm, ρ—particle density, g/cm^3, S—BET specific surface area, m^2/g [26].

The reactivity and stability of the powder at various temperatures and heating rates was studied with use of thermogravimetric analysis (TGA) and differential thermal analysis (DTA) Seiko Exstar 6000 (Seiko Instruments Inc., Chiba, Japan). The samples of 5–10 mg weight were heated from ambient temperature to 1273 K at heating rate 10 °C/min in air atmosphere. The TGA/differential scanning calorimetry (DSC) experiments in Ar were performed by a Netzsch F5 Jupiter STA analyzer (NETZSCH-Gerätebau GmbH, Selb, Germany) under the same conditions of the DTA scans.

To prepare the nAl/HTPB composition, for example with HTPB content of 10 wt.% (nAl-H10), 250 mL of solvent, 45.0 g of nAl powder, and 0.225 g of acetylacetone were used. The amount of acetylacetone was 0.5% of the weight of the bare nanopowder. The resulting mixture was stirred with a high shear homogenizer HG-15D (Daihan Scientific, Seoul, Korea) at 5000 rpm for 30 min. Then a HTPB-based solution (5.0 g HTPB in 50 mL of solvent) was added to the mixture and was then stirred for another 30 min. The solvent was removed from the mixture using an IKA 10 RV (IKA®-Werke GmbH & Co. KG, Staufen im Breisgau, Germany) rotary evaporator; the resulting product was dried at a pressure of 1.33 mbar for 16 h. The content of components in the samples are listed in Table 2.

Table 2. The composition of the initial and composite powders.

Composite	Nominal Powder Composition	Notes
nAl	nAl100 (Al, Al$_2$O$_3$)	Initial powder, air-passivated, 100 nm (nominal size)
nAl-H10	nAl100 (90 wt.%) + HTPB (10 wt.%)	Acetylacetone: 0.5 wt.% of nAl100 mass
nAl-H20	nAl100 (80 wt.%) + HTPB (20 wt.%)	Acetylacetone: 0.5 wt.% of nAl100 mass

Resonant acoustic mixing (LabRAM I apparatus, Resodyn Acoustic Mixers, Butte, MT, USA) was used to uniformly disperse the propellant ingredients [27].

The binder was prepared starting from HTPB R45 resin cured with isophorone diisocyanate (IPDI). The curing ratio of the binder ([-NCO]/[-OH]) was 1.04. Dioctyl adipate (DOA) was used as a plasticizer, while dibutyltin diacetate was added (in excess) to the formulation as curing catalyst. For a comparative assessment the propellant formulation was prepared with the initial nAl powder as presented in Table 3. The nAl mass fraction in the tested formulation is the same on a molar basis. For all the tested formulations, the coarse-to-fine AP ratio is 65:10.

Table 3. Base composition of the propellants tested.

Ingredients	Mass Fraction, wt.%
AP (coarse $D_{nominal} = 200$ μm)	65
AP (fine $D_{0.5} = 18$ μm)	10
HTPB	17
nAl	8

The propellant burning rate (r_b) was determined from tests carried out in a laboratory combustion chamber equipped with windows for combustion process video recording. The tested samples are cut into parallelepipeds (4 × 4 × 30 mm). The lateral surface of the samples is inhibited to provide a one-dimensional regression of the combustion surface. A simplified scheme of the experimental setup is given in Figure 1. Nitrogen is used for the combustion chamber pressurization, and to prevent combustion products smoke from inhibiting the visualization. Quasi-steady chamber pressure is granted by electrovalves controlled by a digital pressure regulator. The r_b value is determined using by a proprietary software. The latter enables the regressing surface tracking during the combustion, thus providing the quasi-steady burning rate. For a given propellant, experimental results are reduced according to the standard Vielle's law [28]:

$$r_b = ap^n \tag{2}$$

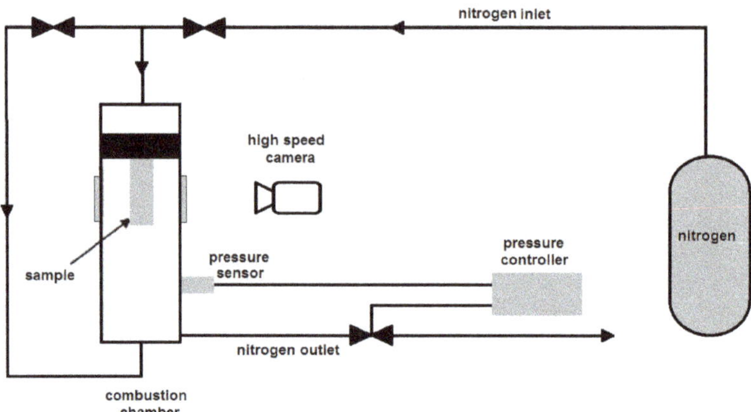

Figure 1. Schematic diagram of the lab-scale strand burner for r_b determination setup.

In the Equation (2), the r_b is typically expressed in mm/s, with p in bar.

Characterization of the nanopowders developed by Tomsk State University (TSU, Tomsk, Russia) is carried out in cooperation with the Space Propulsion Laboratory of the Politecnico di Milano (SPLAB-POLIMI, Milan, Italy). Investigation of the combustion of the propellants containing nanopowders is carried out by SPLAB-POLIMI.

3. Results and Discussion

In an earlier stage of the work, a number of nAl/HTPB composites were prepared. Manufactured materials featured an HTPB content in the range 10 to 50 wt.%. The composites with HTPB content of 10 wt.% and 20 wt.% are disperse solids, while for HTPB mass fractions >20%, the obtained materials are highly viscous masses of difficult handling. Thus, activities concentrated on nAl100-H10 and nAl100-H20.

3.1. nAl Characterization

A TEM image of the pristine, air-passivated aluminum nanopowder is presented in Figure 2. The shape of the visible nanoparticles is predominantly spherical. The native amorphous oxide layer consisting of bayerite α-Al(OH)$_3$ and γ-AlOOH boehmite [13] is on the surface of aluminum nanoparticles. Content of aluminum metal is usually about 90 wt.% [17,29]. Nanoparticles have a marked clustering tendency with formation of particles subject to cold cohesion reaching sizes up to hundreds of microns. The observed clusters are rather compact, yet the cold-cohesion particle-particle interactions are typically relatively weak, and clusters can be broken down (though reversibly) by mechanical stresses (e.g., ultrasound irradiation).

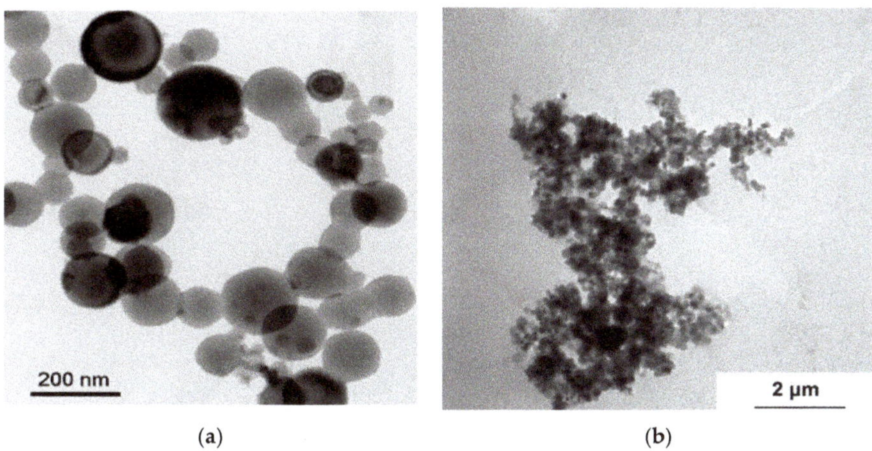

Figure 2. TEM images of the (a) pristine aluminum nanoparticles and (b) an aluminum nanoparticle cluster.

The content of active aluminum in the aluminum nanopowders used in the present work is (85.9 ± 0.8) wt.% as presented in Table 4. The BET—derived specific surface of the nanopowder is 12.6 ± 0.1 m^2/g, corresponding to an average diameter of the nanoparticles [Equation (1)] of ~175 nm.

Table 4. Active aluminum content in the tested nAl powders.

Powder	C_{Al}, wt.%	$C_{Al,\ Expected}$ [a], wt.%
nAl100	85.9 ± 0.8	-
nAl100-H10	74.7 ± 0.5	77.3
nAl100-H20	63.4 ± 2.8	68.7

[a] The expected aluminum content is estimated based on the nAl100.

Table 4 shows C_{Al} data for all the tested powders. For the nAl/HTPB, the actual C_{Al} is compared to the expected value resulting from the pristine powder (nAl100) and the nominal coating mass fraction

(10 wt.%, 20 wt.%). The difference between the actual and the expected values reported in the Table 4 is originated by the possible presence of (minor amounts of) residual solvent in the HTPB-containing pastes. From this point of view, this observation is supported by the slightly higher C_{Al} difference for nAl100-H20, which contains a higher HTPB mass fraction.

3.2. Characterization of the nAl/HTPB Pastes

An image of the nAl100-H10 is reported in Figure 3, where HTPB is seen to form a continuous organic layer, partially sorbed on the surface of the nanoparticles or particle clusters. Data reported in the Table 4 suggest the absence of powder-HTPB reactions modifying the powder composition: when preparing the composite, neither the formation of soluble aluminum compounds occurs when interacting with an organic substance, nor the sorption of the solvent by the particles of the composite. The HTPB deposition at the particle surface provides a protective layer shielding the nanoparticle surface from interaction with the environment, thus offering an increased aging resistance [21]. At the same time, the HTPB deposition promotes particle clustering (see Figure 3, where separated particles are encapsulated by the HTPB layer) [21], yet the impact of this effect should be evaluated considering the powder mixing and dispersion in the HEM matrix.

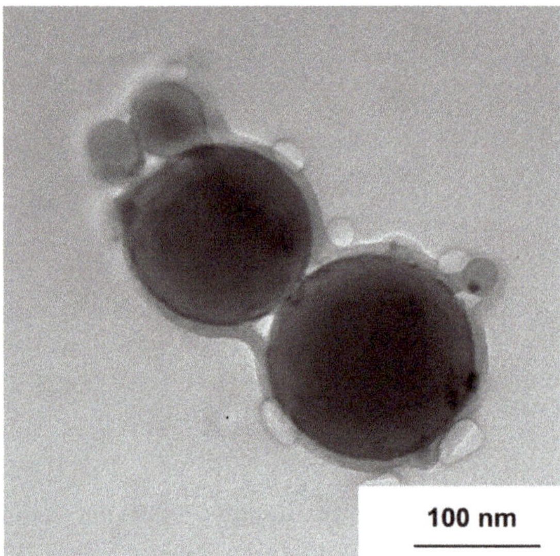

Figure 3. Aluminum nanoparticles coated with HTPB.

3.3. Thermogravimetric Analysis of the Pastes

3.3.1. Thermogravimetric Analysis of the Pastes in Ar

The TGA plots of HTPB and nAl-HTPB samples are shown in Figure 4. Pristine, uncured HTPB can be seen to undergo two-stage degradation in an argon atmosphere. The first stage of the destruction starts at about 225 °C and continues up to about 325 °C with a weight loss of ~11% at this stage. The second stage of the thermal degradation proceeds in the temperature range from 325 to 525 °C with almost complete mass loss. The rate of the HTPB degradation in this second stage, based on the TGA curve slope, is higher than that at the first stage.

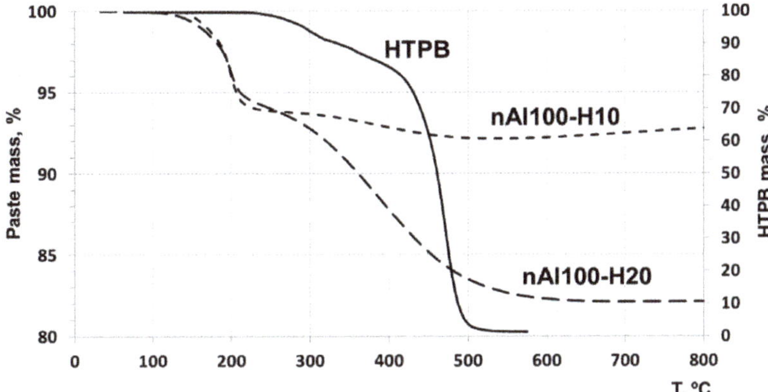

Figure 4. TGA of HTPB and nAl/HTPB pastes in Ar (10 °C/min).

The thermal degradation of HTPB in the nAl/HTPB pastes proceeds as a two-stage process too. However, the first stage of the destruction begins at a lower temperature (~100 °C) and ends at about 210 °C. Weight loss at this stage is 7–8%. The weight loss at this stage is independent of the paste composition. The destruction of the polymer can be assumed to occur on the nanoparticle surface and its surface oxide layer is a destruction catalyst, since the destruction rate in this case is higher than that in the lack of contact with alumina, despite the destruction proceeds at a lower temperature.

The thermal degradation of HTPB in the second stage proceeds at a lower rate compared with that at the first stage, although at a higher temperature. The thermal degradation end occurs in the range 500–550 °C, which suggests that polymer pyrolysis at this stage proceeds outside of its contact with alumina of the aluminum nanoparticle and the polymer degradation mechanism does not change, with a lower destruction rate due to a lower onset decomposition temperature.

3.3.2. Thermogravimetric Analysis of the Pastes in Air

TGA plots of HTPB and nAl/HTPB samples performed in air are shown in Figure 5.

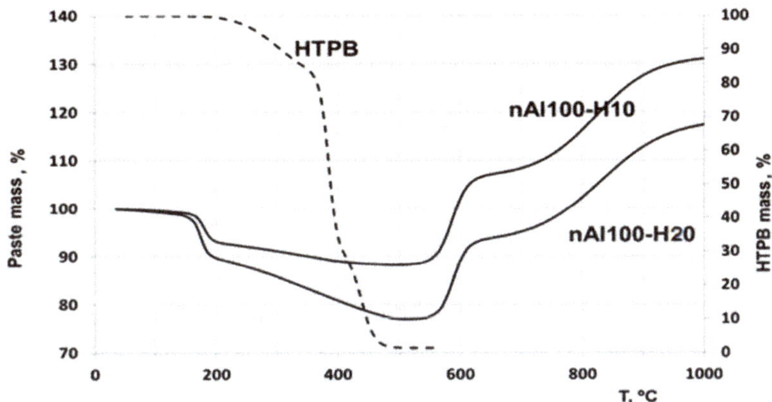

Figure 5. TGA of HTPB and nAl/HTPB pastes in air (10 °C/min).

Initial HTPB as can be seen from the plot undergoes two-stage mass-loss, as per the tests in Ar (Figure 4). The mass loss at the first stage of decomposition is insignificantly higher as compared to decomposition in Ar. The first stage of decomposition begins at about 200 °C and continues up to about 380 °C with a weight loss of ~7% at this stage. The second stage of reaction proceeds in the temperature range from 380 to 475 °C with almost complete weight loss. The rate of decomposition of HTPB at this stage, based on the TG curve slope is higher than that at the first stage.

The degradation/reaction of the HTPB applied on the surface of the nAl particles proceeds as in the Ar case in two-stages, however, the first stage of the decomposition begins at a lower temperature about 100 °C and ends at about 200 °C. Weight loss at this stage depends on the composition of the paste, weight loss being larger when increasing HTPB content. Above 200 °C the second stage of the destruction begins, with the completion to be about 500 °C. Aluminum nanoparticle oxidation begins at the temperature above 500 °C. In the temperature range 550–650 °C, oxidation of aluminum nanoparticles is observed, accompanied by their partial sintering [30]. Subsequent oxidation of aluminum is observed at temperature above 650 °C and continues at temperatures above 1000 °C.

3.4. Burning Rate

The propellant burning rate generally follows Vieille's law as shown in Equation (2). Data fittings and burning rate values obtained from propellant tests are shown in Figure 6 and Table 5.

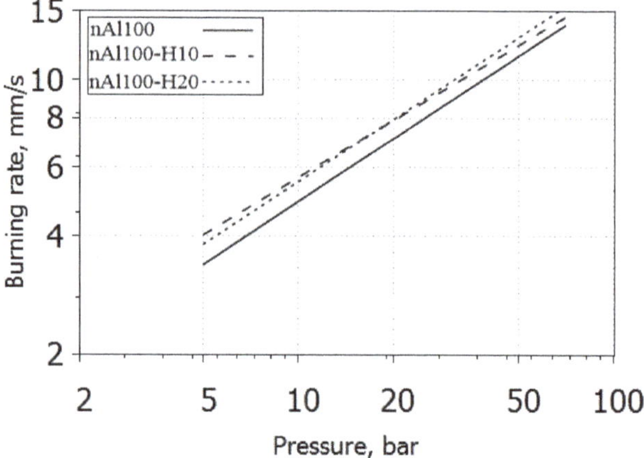

Figure 6. Burning rate of the propellants with aluminum nanopowders.

Table 5. Burning rate of the formulations tested.

Formulations	a_r, mm/(s bar ^{n_r})	n_r	R^2	r_b (40 bar), mm/s
AP_nAl100	1.444 ± 0.045	0.531 ± 0.012	0.997	10.2 ± 0.2
AP_nAl100-H10	1.856 ± 0.027	0.483 ± 0.005	0.999	11.0 ± 0.1
AP_nAl100-H20	1.636 ± 0.080	0.527 ± 0.018	0.992	11.3 ± 0.5

The data show the introduction of nAl/HTPB paste to provide an increase in the propellant burning rate. Compared to nAl100, the introduction of paste into the propellant formulation results in an increase in the rate coefficient *a* and a decrease in the value of the pressure exponent *n*. The burning rate increase suggest a better dispersion of the metal particles of the nAl/HTPB pastes in the propellant matrix. The value of pressure exponent for the propellants with introduced paste is somewhat lower compared with that for propellant with initial aluminum nanopowder. However, considering the interval of confidence on the ballistic exponent, its variations are relatively small (if any). This suggest the absence of changes in the combustion mechanism of the tested formulations.

4. Discussion

nAl/HTPB composites were prepared by wet mixing method. The preparation includes two stages. The first stage is the disruption of nAl particle clusters when stirring with acetylacetone and formation of the organic chemisorbed layer on the nanoparticle surface. This layer increases the hydrophobicity and compatibility with the polymer of the nanoparticle surface and facilitates the sorption of the HTPB molecules [31]. The second stage is the application of HTPB on the nanoparticle surface. TEM images show the homogeneous continuous layer of the polymer capping the nanoparticle/nanoparticles. The capping prevents direct contacts of the nanoparticles with each other (for separated particles), and hinders the penetration of oxidant molecules to the nanoparticle surface, which increases their stability during storage and processing. This effect provides the complete protection of the nanoparticles against environmental influences but retains their high reactivity which is demonstrated by nanoparticles slow heating rate oxidation and combustion tests [18], see Figure 5 and Table 5.

TGA analysis shows the lack of a chemical bond between the nanoparticle and HTPB molecules. Thermal degradation of HTPB during slow heating rate heating proceeds as intramolecular chemical process with minimal participation of oxygen. Only part of the HTPB molecules in contact with the nanoparticle surface is degraded at a lower temperature by the action of alumina layer as catalyst. So the HTPB destruction proceeds and aluminum oxidation sequential reactions occurring independently of each other. The thermal degradation of HTPB in the nAl/HTPB pastes at the second stage proceeds in the same temperature range as that of HTPB [31]. The degradation proceeding at this stage can be assumed to be noncatalytic process. The degradation process at this stage is suggested to proceed outside of the contact with alumina of the aluminum nanoparticle and the HTPB degradation mechanism is the same both for nAl/HTPB and HTPB.

The propellants with formulation comprising capped aluminum nanoparticles exhibit higher burning rate than the counterpart loaded with nAl100. Yet, the fundamentals of the burning process remain unchanged. Quantitative parameters of the burning law allow to consider the burning process to be faster and more stable when capped aluminum nanoparticles were added.

5. Conclusions

Different nAl/HTPB composites were prepared by a wet method. Of the prepared materials (HTPB mass fraction in the range 10 to 50 wt.%), two were selected for detailed analyses spanning from pre-burning characterizations to combustion analyses. The selection was made based on the availability of a disperse phase system with powder-like characteristics, and limited volability. In particular, the selected composites featured the following compositions (i) nAl100 (90 wt.%) + HTPB (10 wt.%) and (ii) nAl100 (80 wt.%) + HTPB (20 wt.%). In both materials acetylacetone served as compatibilizing agent for effective wetting of the nAl surface (Al_2O_3 from air passivation, with possibly, hydrated compounds) with HTPB. The nAl/HTPB composite can be an intermediate product for the subsequent manufacture of mixed high-energy materials while maintaining the quality and advantages of nAl. Capping of the nanoparticle surface with HTPB protects nanoparticles against environmental influence (i.e., material corruption due to ageing) and provides easier handling and manufacturing while granting good combustion performance as tesitifed by DTA-TGA scans and burning tests performed in a lab-scale burner. The slow heating rate, non-isothermal oxidation of the powders showed the effect of the nAl on the HTPB degradation process proceeding at lower temperature and with higher reaction rate than what is observed for the uncured polymer alone. On the other hand, no evidences of HTPB-aluminum interactions affecting the powder reactivity and oxidation mechanism were noticed. Thermal degradation effects were separately investigated in Ar and air to decouple the polymer pyrolysis effects from the composite oxidation mechanism.

The burning characteristics of the propellant formulations containing tested composites were analyzed. The inclusion of the composite in the propellant formulation provides faster burning rate with increasing burning stability at low pressure.

Author Contributions: Writing—original draft, N.R., S.S., E.P.; supervision, A.V.; writing—review and editing, A.V., N.R., S.S., C.P.; Methodology, M.L., C.P. All authors have read and agreed to the published version of the manuscript.

Funding: This work was carried out with financial support from the Ministry of Science and Higher Education of the Russian Federation (State assignment No. 0721-2020-0028) and (Agreement with Joint Institute for High Temperatures RAS No 075-15-2020-785).

Acknowledgments: This work was carried out with financial support from the Ministry of Science and Higher Education of the Russian Federation (State assignment No. 0721-2020-0028) and (Agreement with Joint Institute for High Temperatures RAS No 075-15-2020-785). The synthesis and microscopic characterization of the aluminum nanoparticles and Al/HTPB composites were performed according to the Government research assignment for ISPMS SB RAS, project No. III.23.1.1. The experimental skills and the work of Alberto Verga (SPLAB-POLIMI) were highly appreciated during the research activity.

Conflicts of Interest: The authors declare no conflict of interest.

References

1. Yetter, R.A.; Risha, G.A.; Son, S.F. Metal particle combustion and nanotechnology. *Proc. Combust. Inst.* **2009**, *32*, 1819–1838. [CrossRef]
2. Sundaram, D.S.; Yang, V.; Zarko, V.E. Combustion of nano aluminum particles (Review). *Combust. Explos. Shock Waves* **2015**, *51*, 173–196. [CrossRef]
3. Sossi, A.; Duranti, E.; Manzoni, M.; Paravan, C.; DeLuca, L.T.; Vorozhtsov, A.B.; Lerner, M.; Rodkevich, N.; Gromov, A.; Savin, N. Combustion of HTPB-based solid fuels loaded with coated nanoaluminum. *Combust. Sci. Technol.* **2013**, *185*, 17–36. [CrossRef]
4. Meda, L.; Marra, G.; Galfetti, L.; Inchingalo, S.; Severini, F.; DeLuca, L.T. Nanocomposites for rocket solid propellants. *Compos. Sci. Technol.* **2005**, *65*, 769–773. [CrossRef]
5. Abraham, A.; Nie, H.; Schoenitz, M.; Vorozhtsov, A.B.; Lerner, M.; Pervikov, A.; Dreizin, E.L. Bimetal Al–Ni nano-powders for energetic formulations. *Combust. Flame.* **2016**, *173*, 179–186. [CrossRef]
6. Noor, F.; Vorozhtsov, A.; Lerner, M.; Wen, D. Exothermic characteristics of aluminum-based nanomaterials. *Powder Technol.* **2015**, *282*, 19–24. [CrossRef]
7. Sakovich, G.V.; Arkhipov, V.A.; Vorozhtsov, A.B.; Bondarchuk, S.S.; Pevchenko, B.V. Investigation of combustion of HEM with aluminum nanopowders. *Nanotechnol. Russ.* **2010**, *5*, 91–107. [CrossRef]
8. Dreizin, E.L. Metal-based reactive nanomaterials. *Prog. Energy Combust. Sci.* **2009**, *35*, 141–167. [CrossRef]
9. Young, G.; Wang, H.; Zachariah, M.R. Application of Nano-Aluminum/Nitrocellulose Mesoparticles in Composite Solid Rocket Propellants. *Propellants Explos. Pyrotech.* **2015**, *40*, 413–418. [CrossRef]
10. Pang, W.; DeLuca, L.; Huixang, X.; Xuezhong, F.; Fengqu, Z.; Fangli, L.; Wuxi, X.; Yonghong, L. Effects of nano-metric aluminum powder on the properties of composite solid propellants. *Int. J. Energetic Mater. Chem. Propul.* **2015**, *14*, 265–282. [CrossRef]
11. Ivanov, Y.F.; Osmonoliev, M.N.; Sedoi, V.S.; Arkhipov, V.A.; Bondarchuk, S.S.; Vorozhtsov, A.B.; Kuznetsov, V.T. Productions of ultra-fine powders and their use in high energetic compositions. *Propellants Explos. Pyrotech.* **2003**, *28*, 319–333. [CrossRef]
12. Vorozhtsov, A.B.; Rodkevich, N.G.; Lerner, M.I.; Zhukov, A.S.; Bondarchuk, S.S.; Dyachenko, N.N. Metal nanoparticles in high-energetic materials practice. *Int. J. Energetic Mater. Chem. Propul.* **2017**, *16*, 231–241. [CrossRef]
13. Vorozhtsov, A.B.; Lerner, M.; Rodkevich, N.; Nie, H.; Abraham, A.; Schoenitz, M.; Dreizin, E.L. Oxidation of nano-sized aluminum powders. *Thermochim. Acta* **2016**, *636*, 48–56. [CrossRef]
14. Paravan, C.; Verga, A.; Maggi, F.; Galfetti, L. Accelerated aging of micron- and nano-sized aluminum powders: Metal content, composition and non-isothermal reactivity. *Acta Astron.* **2019**, *158*, 397–406. [CrossRef]
15. Ju, Z.Y.; An, J.L.; Guo, C.Y.; Li, T.R.; Jia, Z.Y.; Wu, R.F. The oxidation reaction and sensitivity of aluminum nanopowders coated by hydroxyl-terminated polybutadiene. *J. Energetic Mater.* **2020**, 1–14. [CrossRef]
16. Guo, L.; Song, W.; Hu, M.; Xie, C.; Chen, X. Preparation and reactivity of aluminum nanopowders coated by hydroxyl-terminated polybutadiene (HTPB). *Appl. Surf. Sci.* **2008**, *254*, 2413–2417. [CrossRef]
17. Lerner, M.I.; Glazkova, E.A.; Vorozhtsov, A.B.; Rodkevich, N.G.; Volkov, S.A.; Ivanov, A.N. Passivation of aluminum nanopowders for use in energetic materials. *Russ. J. Phys. Chem. B* **2015**, *9*, 56–61. [CrossRef]

18. Vorozhtsov, A.; Lerner, M.; Rodkevich, N.; Teplov, G.; Sokolov, S.; Perchatkina, E. Deagglomeration and Encapsulation of Metal and Bimetal Nanoparticles for Energetic Applications. In *Innovative Energetic Materials: Properties, Combustion Performance and Application*; Springer: Singapore, 2020; pp. 457–491.

19. Lu, Y.C.; Kuo, K.K. Thermal decomposition study of hydroxyl-terminated polybutadiene (HTPB) solid fuel. *Thermochim. Acta* **1996**, *275*, 181–191. [CrossRef]

20. Vorozhtsov, A.B.; DeLuca, L.T.; Reina, A.; Lerner, M.I.; Rodkevich, N.G. Effects of HTPB-coating on nano-sized aluminum in solid rocket propellant performance. *Sci. Technol. Energ. Mater.* **2015**, *76*, 105–109.

21. Pang, W.; DeLuca, L.T.; Wang, K.; Fu, X.; Li, J.; Xu, H.; Li, H. Performance of Composite Solid Propellant Containing Nanosized Metal Particles. In *Nanomaterials in Rocket Propulsion Systems*; Elsevier: Amsterdam, The Netherlands, 2019; pp. 263–298.

22. Dennis, C.; Bojko, B. On the combustion of heterogeneous AP/HTPB composite propellants: A review. *Fuel* **2019**, *254*, 115646. [CrossRef]

23. Lerner, M.I.; Glazkova, E.A.; Lozhkomoev, A.S.; Svarovskaya, N.V.; Bakina, O.V.; Pervikov, A.V.; Psakhie, S.G. Synthesis of Al nanoparticles and Al/AlN composite nanoparticles by electrical explosion of aluminum wires in argon and nitrogen. *Powder Technol.* **2016**, *295*, 307–314. [CrossRef]

24. Lerner, M.; Vorozhtsov, A.; Guseinov, S.; Storozhenko, P. Metal Nanopowders Production. In *Metal Nanopowders: Production, Characterization, and Energetic Applications*; Wiley: Hoboken, NJ, USA, 2014.

25. Chen, L.; Song, W.; Lv, J.; Chen, X.; Xie, C. Research on the methods to determine metallic aluminum content in aluminum nanoparticles. *Mater. Chem. Phys.* **2010**, *120*, 670–675. [CrossRef]

26. Kuśnieruk, S.; Wojnarowicz, J.; Chodara, A.; Chudoba, T.; Gierlotka, S.; Lojkowski, W. Influence of hydrothermal synthesis parameters on the properties of hydroxyapatite nanoparticles. *Beilstein J. Nanotechnol.* **2016**, *7*, 1586–1601.

27. Zare, A.; Harriman, T.A.; Lucca, D.A.; Roncalli, S.; Kosowski, B.M.; Paravan, C.; DeLuca, L.T. Mapping of aluminum particle dispersion in solid rocket fuel formulations. In *Chemical Rocket Propulsion*; Springer: Cham, Switzerland, 2017; pp. 673–688.

28. Maggi, F.; Dossi, S.; Paravan, C.; Galfetti, L.; Rota, R.; Cianfanelli, S.; Marra, G. Iron oxide as solid propellant catalyst: A detailed characterization. *Acta Astronaut.* **2019**, *158*, 416–424. [CrossRef]

29. Vorozhtsov, A.B.; Zhukov, A.S.; Ziatdinov, M.K.; Bondarchuk, S.S.; Lerner, M.I.; Rodkevich, N.G. Novel micro- and nanofuels: Production, characterization, and applications for high-energy materials. In *Chemical Rocket Propulsion*; Springer: Cham, Switzerland, 2017; pp. 235–251.

30. Lozhkomoev, A.S.; Rodkevich, N.G.; Vorozhtsov, A.B.; Lerner, M.I. Oxidation and oxidation products of encapsulated aluminum nanopowders. *J. Nanopart. Res.* **2020**, *22*, 1–13. [CrossRef]

31. Vorozhtsov, A.B.; Rodkevich, N.G.; Bondarchuk, I.S.; Lerner, M.I.; Zhukov, A.S.; Glazkova, E.A.; Bondarchuk, S.S. Thermokinetic investigation of the aluminum nanoparticles oxidation. *Int. J. Energetic Mater. Chem. Propul.* **2017**, *16*, 309–320. [CrossRef]

Article

Gaseous Products Evolution Analyses for Catalytic Decomposition of AP by Graphene-Based Additives

Shuwen Chen [1], Ting An [2], Yi Gao [3], Jie-Yao Lyu [1], De-Yun Tang [1], Xue-Xue Zhang [1], Fengqi Zhao [2],* and Qi-Long Yan [1],*

[1] Science and Technology on Combustion, Internal Flow and Thermostructure Laboratory, Northwestern Polytechnical University, Xi'an 710072, China; shuwenchen@nwpu.edu.cn (S.C.); jieyaolv@mail.nwpu.edu.cn (J.-Y.L.); tangdy@mail.nwpu.edu.cn (D.-Y.T.); xuexuezhang@mail.nwpu.edu.cn (X.-X.Z.)
[2] Science and Technology on Combustion and Explosion Laboratory, Xi'an Modern Chemistry Research Institute, Xi'an 710065, China; anting715@163.com
[3] School of Astronautics, Northwestern Polytechnical University, Xi'an 710072, China; gy0704@163.com
*   Correspondence: zhaofqi@163.com (F.Z.); qilongyan@nwpu.edu.cn (Q.-L.Y.)

Received: 13 March 2019; Accepted: 17 May 2019; Published: 24 May 2019

Abstract: A quantitative evaluation method has been developed to study the effects of nanoadditives on thermal decomposition mechanisms of energetic compounds using the conventional thermogravimetry coupled with mass spectrometry (TG/MS) technique. The decomposition of ammonium perchlorate (AP) under the effect of several energetic catalysts has been investigated as a demonstration. In particular, these catalysts are transition metal (Cu^{2+}, Co^{2+} and Ni^{2+}) complexes of triaminoguanidine (TAG), using graphene oxide (GO) as dopant. They have been well-compared in terms of their catalytic effects on the concentration of the released gaseous products of AP. These detailed quantitative analyses of the gaseous products of AP provide a proof that the proton transfer between $\cdot O$ and O_2 determines the catalytic decomposition pathways, which largely depend on the type of reactive centers of the catalysts. This quantitative method could be applied to evaluate the catalytic effects of any other additives on the thermal decomposition of various energetic compounds.

Keywords: thermolysis; energetic materials; GO-based catalysts; quantitative analyses; decomposition mechanisms

---

1. Introduction

Energetic materials (EMs) are widely used as propellants, explosives, and pyrotechnics. The conventional EMs, such as 1,3,5-trinitro-1,3,5-triazinane (RDX), 1,3,5,7-tetranitro-1,3,5,7-tetrazocane (HMX), and ammonium perchlorate (AP), are still playing dominant roles in the formulations. The nano-sized additives are usually used to improve their performances and more innovative additives have been designed and used during past decades [1–3]. The decomposition and combustion of EMs are the key parameters that have to be investigated before their applications, which are strongly connected with their compatibility, safety, and performance. The decomposition has been found to be the initial stage of combustion, and it should be well-evaluated at laboratory scale. AP is a well-known oxidizer in solid composite propellants. In order to get the burning behavior of propellants, researchers usually focus on their thermal property first [4,5]. The thermal behavior of AP has been widely studied during the past several decades [6,7]. Nanoadditives have high concentrations of dislocations and large surface areas, and therefore they normally show significant catalytic effects on the decomposition of AP [8–11]. However, the nanoadditives used today in formulations are mostly inert metal oxides or metal oxide composites. To some extent, they may reduce the energy content of the propellants.

In order to overcome the limitations mentioned above, one has to develop energetic metal complexes or metal organic frameworks (MOFs) with great thermal stability and compatibility as energetic catalysts. It has been recently shown that graphene oxide (GO) doped transition metal complexes are promising energetic catalysts [12,13]. It has been reported that the intrinsic exothermicity of GO (1600 J g^{-1}) is comparable to several hazardous chemicals and explosives [14]. In addition, GO could be considered as a stabilizing agent [15]. In our recent study, GO-doped transition metal (nickel, cobalt and copper) complexes of triaminoguanidine (TAG) were prepared, and their effects on the thermal properties and decomposition mechanisms of RDX studied [16]. These materials could not only catalyze the decomposition of RDX, but also improve the thermal stability of RDX due to their enhanced thermal conductivity. The decomposition behavior of GO complexes modified AP have been briefly studied based on conventional thermogravimetry coupled with mass spectrometry (DSC/TG) analysis [17]. The results showed that the hybrid catalysts can enhance the initial decomposition temperature and change the thermolysis mechanism by introducing different types of metal ions. But the detailed effect of nanoadditives, especially for GO-based energetic catalysts, on thermolysis products and reaction pathways of AP are not well investigated [18]. According to the literature, two mechanisms have been proposed for the thermal decomposition of AP: (a) electron transfer mechanism [19] and (b) proton transfer mechanism [20], depending on the temperature. A large amount of data has shown that the GO-TAG based catalysts have significant effects on heat releases and thermal stability of EMs, but the inherent chemical mechanisms are still not well-known.

Therefore, this paper intends to present a comprehensive quantitative analysis to clarify how nanoadditives affect the decomposition mechanisms of EMs. The catalytic effects of GO-based energetic additives on thermal decomposition chemical pathways and gaseous products of AP have been investigated as a typical demonstration. It is a novel method to get the catalyst mechanism by measuring the amount of gaseous decomposition products. Thermal analysis techniques including thermogravimetry coupled with mass spectrometry (TG/MS) were employed to evaluate the decomposition mechanisms [21–23]. This technique was used to evaluate the bond breaking and gaseous products formation during thermal decomposition of EMs. Thus, this work aims at studying the thermal decomposition of AP, in presence of various GO-TAG based energetic catalysts, on the basis of TG/MS technique.

2. Experimental Procedure

2.1. Sample Preparation

G-T-M and TAG-M composites were synthesized followed by the method reported in the previously published paper ([16], the preparation was summarized in the supporting materials). In order to get AP contained complex, a saturated solution of AP containing 160 mg of AP and 20 mL of acetone was prepared first. Then 40 mg TAG-M or G-T-M was added to this solution (M means Nickel, Cobalt or Copper ion), and stirred for 3 h at room temperature. The final G-T-M/AP and TAG-M/AP complexes were obtained after freeze drying. Warning—the solvent-free TAG-Cu complex would undergo self-ignition and fast deflagration to detonation transition reaction easily in the presence of oxygen.

G-T-M indicates GO-doped transition metals complexes of TAG. TAG-M is transition metals modified TAG. G-T-M/AP and TAG-M/AP correspond to G-T-M and TAG-M doping AP, respectively. The components of G-T-M/AP and TAG-M/AP have been summarized in Table 1.

Table 1. The compositions of ammonium perchlorate (AP)-based mixtures (in weight percent).

	AP	**GO**	**TAG**	**Metal**
G-T-Co/AP	80%	0.4%	5.0%	14.6% Cobalt
G-T-Cu$_2$/AP	80%	0.4%	5.7%	13.9% Copper
G-T-Ni/AP	80%	0.5%	6.6%	12.9% Nickel
TAG-Co/AP	80%	-	6.3%	13.7% Cobalt
TAG-Ni/AP	80%	-	7.1%	12.9% Nickel

2.2. Experimental Techniques

The gaseous products from thermal decomposition of AP and AP-based mixtures using TAG-M and G-T-M as energetic nanoadditives were detected using the TG-DSC/MS technique. This experiment was realized on a simultaneous thermoanalyzer STA 449 F3 coupled with a quadrupole mass spectrometer QMS 403 C Aëolos (Netzsch Group, Selb, Germany). An alumina pan with a pin-hole cover was used as a sample pan). The sample masses for these measurements were about 3 mg, with a heating rate of 10 °C/min in a temperature range of 40–500 °C, using an argon atmosphere (gas flow: 50 mL/min).

3. Results and Discussion

3.1. Possible Catalytic Decomposition Pathways of AP

According to the literature [24–26], there are two possible decomposition processes for AP; the low and high temperature stages. Since AP contains plenty of nuclei with tiny pores [27], they are likely to provide highly active sites on their surface at the low temperature decomposition stage, while the high-temperature decomposition stage takes place at the surfaces of the nanocrystals, and involves adsorption and desorption of ammonia as well as perchloric acid. The decomposition of AP generally follows two steps [28,29]; a solid-gas multiphase reaction of the first decomposition step at 300–330 °C, and a gas phase reaction of the second decomposition step at 450–480 °C. The possible decomposition reactions and transformations are as follows [27]:

$$NH_4ClO_4 \rightarrow NH_3 + HClO_4$$

$$HClO_4 \rightarrow 2O_2 + HCl$$

$$O_2 \rightarrow O_2^-$$

$$4NH_4^+ + O_2^- \rightarrow 2H_2O + 4NH_3$$

$$7ClO_4^- \rightarrow 2ClO_3 + ClO_2 + 2ClO + 9O_2 + Cl_2$$

$$4NH_3 + 5O_2 \rightarrow N_2O + NO + NO_2 + 6H_2O$$

As mentioned above, the TG/MS technique has been used to detect the products of thermal decomposition of AP and TAG-based catalysts coated AP. The major gas products have been found to be O_2, NO, NH_3, H_2O, ·O, which would be used to clarify the possible catalytic decomposition mechanisms.

As shown in Figure 1a, there are two obvious gas evolution peaks at around 308–323 °C and 438–445 °C for pure AP, which are attributed to the low-temperature decomposition and high-temperature decomposition, respectively. Compared with pure AP, G-T-Cu$_2$/AP shows only one gas release peak between 328 °C and 342 °C. The G-T-Cu$_2$/AP decomposes with a single but wide releasing peak, so G-T-Cu$_2$ has a catalyst effect on AP decomposition. In the presence of TAG-Ni, the decomposition peak of the first step has the same temperature range as that of pure AP, and the second peak temperature has decreased by about 40 °C. The addition of TAG-Ni has a larger effect on the releasing amount of NO and ·O. For the mixture of G-T-Ni/AP, as shown in Figure 1d, there is only one decomposition peak similar to that of G-T-Cu$_2$/AP, but with a higher peak temperature in the range of 364–383 °C. The results suggest that G-T-Ni can improve the thermal stability of AP compared with G-T-Cu$_2$.

A sharp decomposition peak can be observed for TAG-Co/AP with the range of 297–301 °C, so the presence of TAG-Co has a large catalytic effect on AP decomposition by decreasing the amount of released gases and decomposition temperature. All the gas products are releasing at almost the same temperature. Regarding G-T-Co/AP, the decomposition peak is smoother and the peak temperature increases by 10–20 °C with the addition of GO. It has been reported that G-T-Co can slightly increase the thermal stability of AP [17]. The results indicate that TAG-M complexes in the presence of GO can enhance thermal stability for AP, resulting in slower reaction rates at a higher temperature range.

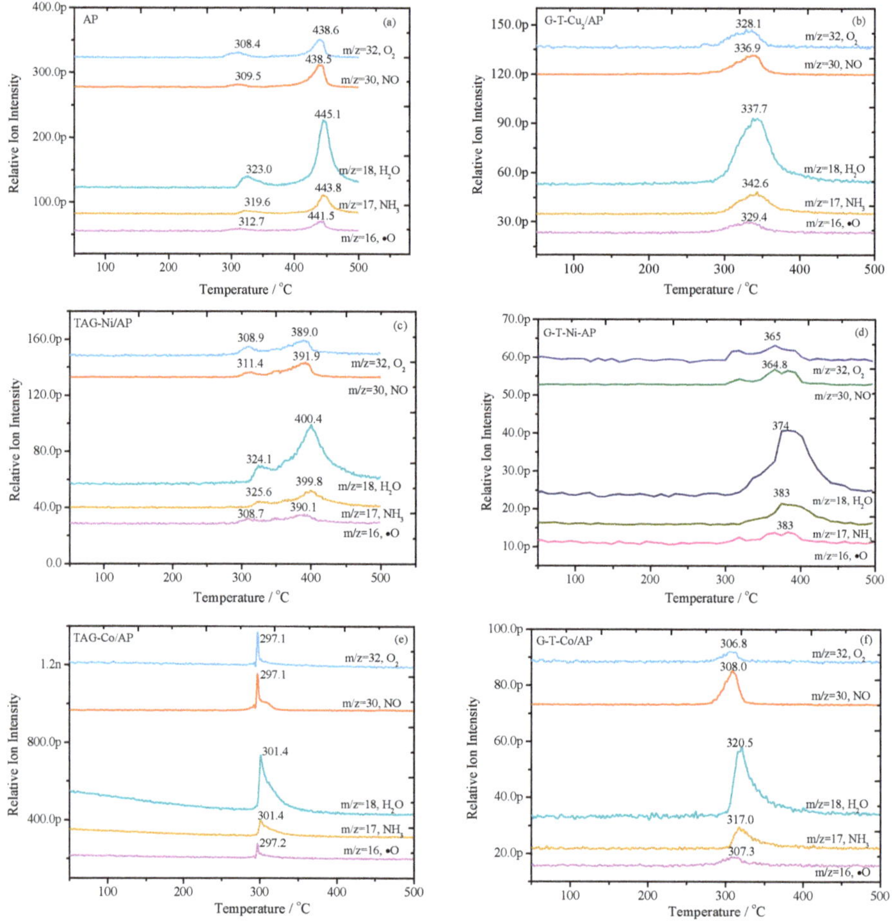

Figure 1. Temperature dependency of the ion flows from non-isothermal decomposition for pure AP and coated AP under a heating rate of 10 °C min^{-1} using thermogravimetry coupled with mass spectrometry (TG/MS) technique: (**a**) pristine AP; (**b**) G-T-Cu$_2$/AP; (**c**) TAG-Ni/AP; (**d**) G-T-Ni/AP; (**e**) TAG-Co/AP; (**f**) G-T-Co/AP; TAG—triaminoguanidine.

3.2. The Dependence of Gases' Evolution Processes on Temperature for Catalytic Decomposition of AP

The curves in Figure 1 were derived and summarized as a function of gaseous product types, as shown in Figure 2.

For the decomposition of pure AP (Figure 2f), the sequence of its gas releasing process is as follows: NO, ·O, O$_2$, NH$_3$ and H$_2$O. The releasing rate of NO and ·O increases rapidly at around 300 °C, and then slows down when the temperature reaches 350 °C. In the meantime, the other three kinds of gases start to release. The releasing rates of all the gaseous products reach their maximum values at the temperature of 420 °C, and start to decrease when the temperature is over 450 °C. The two releasing stages correspond to two decomposition peaks in Figure 1a. Compared with pure AP, the addition of TAG-M complexes could affect the gaseous releasing processes due to the change of chemical reaction pathways, which will be discussed in more detail in a later section.

There is only a trace amount of ·O (*m/z* = 16) radical being formed for all involved samples. The release of ·O is even earlier than that of pure AP in the presence of G-T-Co, G-T-Cu$_2$ and TAG-Co complexes, but TAG-Ni and G-T-Ni complexes postponed the reactions that would release ·O. The mass loss of AP in these cases should be largely caused by the transformation of $O_2 \rightarrow O_2^-$ that forms ·O radical, which has been demonstrated by the curve of $m/z = 16$ in Figure 1. Five kinds of TAG-based complexes have similar catalytic effects on reaction temperature, which lead to the formation of NH_3 ($m/z = 17$) and H_2O ($m/z = 18$), where the releasing rate of gas is increased. In comparison, G-T-Ni postpones the temperature at which NO ($m/z = 30$) starts to release. It also improves the thermal stability of AP, whereas the other four types of G-T-M complexes decrease the initial gas releasing temperature of AP. The initial decomposition temperature of AP is decreased, and the reaction rate is increased in formation of O_2 ($m/z = 32$), by using TAG-M complexes as additives.

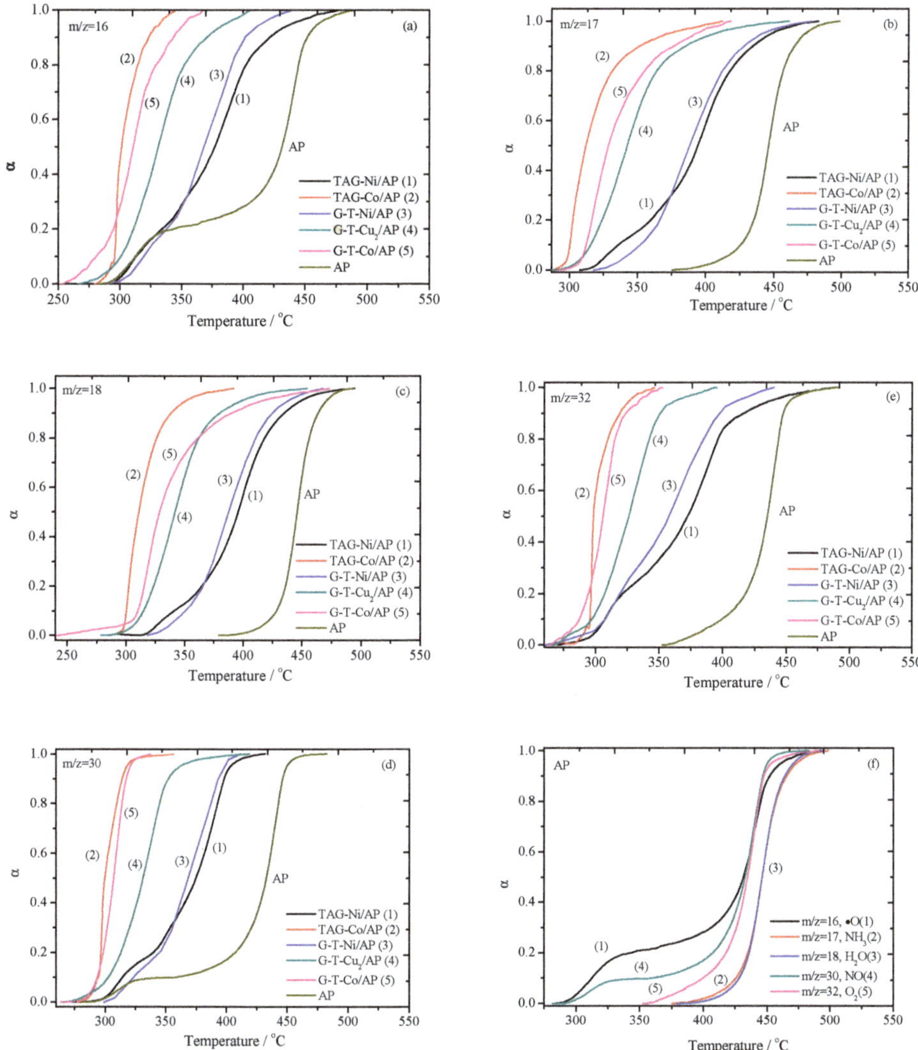

Figure 2. The dependence of accumulation (α is conversion rate) of the typical gaseous products on the temperature for AP-based mixtures: (**a**) $m/z = 16$, ·O; (**b**) $m/z = 17$, NH_3; (**c**) $m/z = 18$, H_2O; (**d**) $m/z = O_2$; (**e**) $m/z = 30$, NO; (**f**) summary of products evolution processes of pristine AP.

In the presence of TAG-Ni, all the gas release processes occur in two steps, which is similar to the case of pure AP. G-T-Ni/AP has a lower initial temperature for releasing O_2 and $\cdot O$ than that of TAG-Ni/AP. However, the initial releasing temperature for NO, NH_3 and H_2O of G-T-Ni/AP is higher than that of TAG-Ni/AP. If comparing G-T-Co/AP with TAG-Co/AP, the initial temperatures for the production of NO, O_2, $\cdot O$ and H_2O are much lower, where extra NH_3 was generated. The gas releasing rate of TAG-Co/AP is also higher than that of G-T-Co/AP throughout their decomposition. The G-T-Cu$_2$ has a significant catalytic effect on AP decomposition so that the stabilization of GO is excluded. It is clear that the initial temperature of G-T-Cu$_2$ has been decreased in comparison to pure AP. In summary, the addition of TAG-M complexes to AP may lead to significant changes in the chemical decomposition pathways.

3.3. Quantitative Analyses of Gaseous Products' Changes of AP in Presence of These Nanocatalysts

To study the catalytic effect of TAG-based complexes on the decomposition mechanisms of AP, the ion flow curves (Figure 1) have been integrated and analyzed. In order to make a quantitative comparison, the characteristic parameters of these curves are calculated and summarized in Table 2.

Table 2. A summary of TG/MS parameters of AP coated by different GO (graphene-oxide)-based catalysts.

	T_o	T_e	ΔT	n	n_m
$m/z = 16$, $\cdot O$					
TAG-Ni/AP	289.9	478.0	188.1	4.541×10^{-10}	9.69%
TAG-Co/AP	279.4	346.7	67.3	5.803×10^{-10}	5.81%
G-T-Ni/AP	289	440	151	1.969×10^{-10}	8.92%
G-T-Cu$_2$/AP	266.2	406.7	140.5	3.653×10^{-10}	8.93%
G-T-Co/AP	251.5	369.2	117.7	1.330×10^{-10}	7.79%
AP	282.7	491.2	208.5	5.734×10^{-10}	9.87%
$m/z = 17$, NH_3					
TAG-Ni/AP	307.8	483.1	175.3	7.052×10^{-10}	15.04%
TAG-Co/AP	287.6	412.5	124.9	16.89×10^{-10}	16.90%
G-T-Ni/AP	318	478	160	3.607×10^{-10}	16.34%
G-T-Cu$_2$/AP	287.2	461.5	174.3	6.389×10^{-10}	15.62%
G-T-Co/AP	291.2	418.4	127.2	2.595×10^{-10}	15.19%
AP	375.6	498.4	122.8	7.706×10^{-10}	13.27%
$m/z = 18$, H_2O					
TAG-Ni/AP	292.5	494.6	202.1	22.849×10^{-10}	48.73%
TAG-Co/AP	282.3	391.8	109.5	53.426×10^{-10}	53.47%
G-T-Ni/AP	318	468	150	11.052×10^{-10}	50.07%
G-T-Cu$_2$/AP	278.7	454.5	175.8	20.80×10^{-10}	50.85%
G-T-Co/AP	241.3	473.3	232	9.531×10^{-10}	55.79%
AP	379.2	493.5	114.3	26.328×10^{-10}	45.34%
$m/z = 30$, NO					
TAG-Ni/AP	273.7	431.9	158.2	5.153×10^{-10}	10.99%
TAG-Co/AP	277.7	356.5	78.8	14.65×10^{-10}	14.66%
G-T-Ni/AP	299	412	113	2.138×10^{-10}	9.69%
G-T-Cu$_2$/AP	265.3	419.0	153.7	5.010×10^{-10}	12.25%
G-T-Co/AP	263.7	337.6	73.9	2.552×10^{-10}	14.94%
AP	281.5	482.8	201.3	10.604×10^{-10}	18.26%
$m/z = 32$, O_2					
TAG-Ni/AP	270.8	488.9	218.1	7.291×10^{-10}	15.55%
TAG-Co/AP	266.4	346.3	79.9	9.146×10^{-10}	9.15%
G-T-Ni/AP	261	440	179	3.305×10^{-10}	14.97%
G-T-Cu$_2$/AP	265.0	395.5	130.5	5.055×10^{-10}	12.36%
G-T-Co/AP	262.9	352.6	89.7	1.075×10^{-10}	6.29%
AP	352.8	491.4	138.6	7.701×10^{-10}	13.26%

Notes: T_o, onset temperature of decomposition, in °C; T_e, the end temperature of decomposition, in °C; ΔT, $T_e - T_o$, in °C; n, the amount of each released gas; n_m, proportion of each product in total gaseous products.

Many investigations have shown that AP decomposition proceeds via the electron transfer from the cation NH_4^+ to anion ClO_4^-, and the catalytic processes with nanocrystalline additives involve the electron transfer between AP and nanoadditives. Boldyrev [28] concluded that the AP decomposition process is followed by a proton transfer from the cation NH_4^+ to anion ClO_4^-. We may assume that the addition of TAG-M complex has catalytic activity in AP decomposition. The amount of gaseous product is different depending on the type of catalysts. For TAG-M doped by GO, the nanocrystalline of TAG-M grows uniformly on the graphene nanosheets, which can increase the contact surface area between AP with the active electron transfer centers, thus resulting in accelerated thermolysis reaction processes. The GO-doped TAG-M complexes with different metal centers could either postpone or accelerate the initial decomposition of AP. According to the quantitative change of gases by TG-DSC/MS technique, the reaction pathways have been greatly changed by these catalysts, depending on the type of metal ions.

For releasing ·O, TAG-Ni and G-T-Ni have little influence on the initial temperature of decomposition, but G-T-Ni/AP can help increase the reaction rate compared with TAG-Ni/AP and AP. The addition of TAG-Co, G-T-Co or G-T-Cu$_2$ increases the initialization temperature and shortens the reaction time. For the reaction to produce NH_3, TAG-M complex decreased the initial reaction temperature of NH_3 by 57.6–88.4 °C while increasing the reaction time. For the reaction to produce NO, G-T-Ni/AP postpones the initial reaction temperature, while the other nano- complexes decrease the initialization temperature. All of these TAG-M complexes have increased the reaction rate.

TAG-Co has shortened the reaction that produces H_2O, whereas the other four types of complexes have prolonged the reaction time by 53.8% to 100%. The initial decomposition temperature of AP under the effects of all TAG-M complexes have been decreased by 61 °C to 138 °C. For the releasing of O_2, TAG-Co based complex has shortened the reaction time by 35–40%, but Ni-based and Cu-based composites prolong the reaction time by 29–57%. Cu-based component reduces the initial decomposition time and slightly decreases the reaction time by 5.8%.

The amount of each gas product is quantified by integrating the peak area of ion intensity curves in Figure 1. The percentage of major gaseous products for each TAG-M complex was calculated and summarized in Figure 3. The amount of H_2O and NH_3 was increased with the addition of G-T-Cu$_2$, while the amount of the other three kinds of gases was reduced. This suggests that G-T-Cu benefits the reaction of NH_4^+ and ClO_4^-, but suppresses the reaction of oxygen conversion. For TAG-Ni/AP and G-T-Ni/AP, the reactions of producing NO and ·O were suppressed, but the reactions producing H_2O, NH_3 and O_2 were promoted. The amount of change of gas released for G-T-Ni/AP is more significant than that of TAG-Ni/AP. For TAG-Ni/AP, the molar ratio of ·O and H_2O changed by −1.8% and 7.5%, respectively. Additionally, TAG-Ni/AP shows two decomposition peaks the same as that of pure AP. The TAG-Co based additives show a similar catalyst effect to that of TAG-Cu$_2$ complex. The reaction of ·O is very sensitive to TAG-Co, which reduces the gas amount by 40%. The decomposition curve of TAG-Co/AP is the sharpest and its reaction time is the shortest. G-T-Co/AP shows the greatest influence on the reaction of producing O_2, with a decrease in amount of gas by 52%.

The addition of TAG-M complex has an obvious catalytic effect on the whole decomposition efficiency of AP, where an increase of gas production of H_2O and NH_3 could be found. For the molar ratio of ·O, TAG-Ni/AP presents the lowest reduction rate of gas production at 1.8%, compared to 41.1% for that of TAG-Co/AP. The amount of O_2 generation was reduced for TAG-Co/AP, G-T-Co/AP, and G-T-Cu$_2$/AP, while it increased in TAG-Ni/AP and G-T-Ni/AP. Co-based complex shows an obvious reduction in O_2 product, among which G-T-Co/AP presents a reduction rate as high as 52.5%. The producing of NO is the most sensitive to G-T-Ni, with a reduction rate of 46.9%.

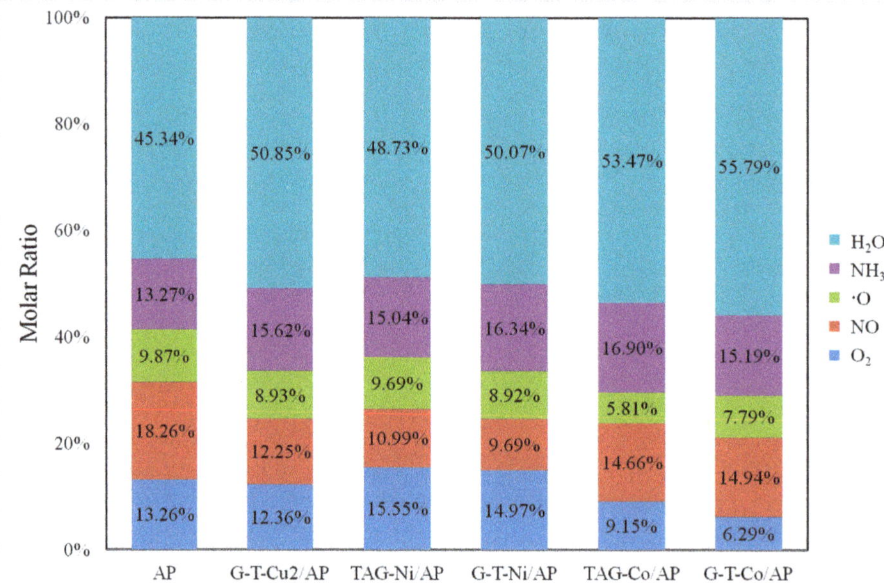

Figure 3. Comparison of molar ratio of gas products decomposed by AP coated with different TAG-M based catalysts.

Figure 4 shows that there is a pair of ions (NH_4^+ and ClO_4^-) in NH_4ClO_4 crystal lattice. The proton transfers from NH_4^+ to ClO_4^-, which leads to reaction *b* and the formation of NH_3 and $HClO_4$. Moreover, the graphene could be combined with metal oxides to improve the catalytic activity. For instance, the molar ratio change of G-T-Cu$_2$, TAG-Co, and G-T-Co show the same trend; the amount of H_2O and NH_3 is increasing, while the amount of ·O, NO and O_2 is decreasing. According to the amount of change, we can postulate that reactions *a*, *f*, and *e* were promoted (Figure 4).

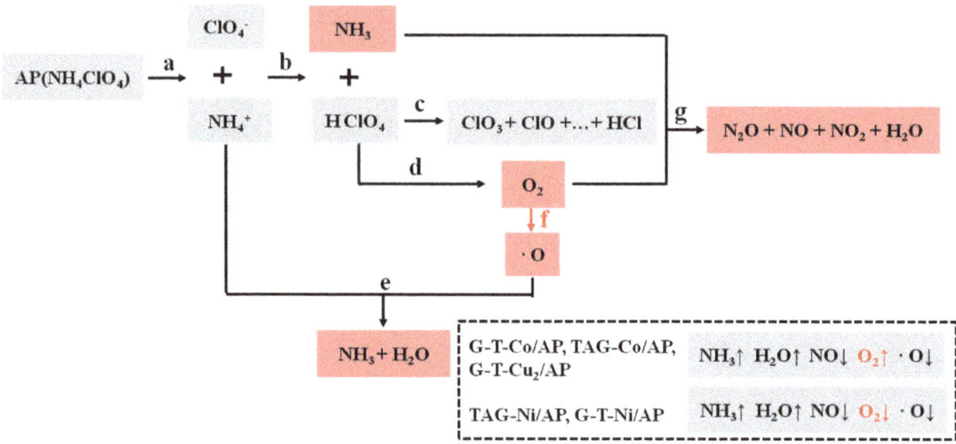

Figure 4. Thermal decomposition mechanisms of AP under the catalytic effects of TAG-M complexes (arrow up means increase and arrow down means decrease).

For TAG-Co catalyst dispersed in AP, the amount of $\cdot O$ has the minimum value, where one sharp decomposition peak was observed from the ion intensity curve. In the case of G-T-Co/AP, the amount of O_2 shows the maximum reduction, while graphene shows additional promotion to reaction f. In contrast, TAG-Ni catalyst promotes reaction d, but suppresses reaction g, while the amount of $\cdot O$ slightly decreases compared with that of pure AP. G-T-Ni shows catalytic effect on reaction d, but suppresses reactions f and g.

All TAG-M catalysts could promote the reactions of AP decomposition by increasing the amount of gas products. In addition G-T-Co, TAG-Co, and G-T-Cu$_2$ have stronger catalytic effects than nickel-based compounds, as they could make two-step gas releasing into one step. This conclusion is consistent with the thermal results from our previous research [17], which claimed that all TAG-M based composites could catalyze the first decomposition process of AP, and Co-based and Cu-based materials have stronger catalytic effects.

4. Conclusions

TG/MS was used to quantitatively study the effects of additives on decomposition mechanisms of energetic compounds, where catalytic decomposition of AP by GO-based catalysts has been selected as a typical example. It has been demonstrated that the detailed quantitative analyses of the gaseous products of AP would show the inherent mechanism changes under the effects of various additives. This method could be applied to analyze the decomposition mechanisms of any other energetic compounds.

The findings further support the literature on the catalytic decomposition kinetics of AP. The GO-based catalysts show improved catalytic efficiency due to their capability in increasing the conversion rates of NH_3 and H_2O. This can be explained by more O elements being transferred to react with NH_4^+, which enhances the initial decomposition heat, resulting in the combination of two decomposition peaks. The amount of O_2 decreases under the effect of Ni-based complexes, which promote the reaction in producing $\cdot O$. In particular, new findings suggest that TAG-Ni and G-T-Ni materials have higher stabilization effects on AP than the others. The Co-based and Cu-based composites could increase the release of O_2, so they have better catalytic effect on AP decomposition. The improved reactions between $\cdot O$ and the other radicals mean a better catalytic effect on the decomposition of AP.

Supplementary Materials: The following are available online at http://www.mdpi.com/2079-4991/9/5/801/s1, Figure S1: The G-T-M (M=Cu^{2+}, Ni^{2+} and Co^{2+}) coordination nanomaterials prepared by reaction of ammonized GO with corresponding metal nitrates: mononuclear coordination complexes could be formed based on triaminoguanidine ligand, Figure S2: The SEM photos of involved materials, where G-T-Cu, G-T-Co, G-T-Ni, TAG-Co and TAG-Ni are presented.

Author Contributions: Conceptualization, Q.-L.Y. and F.Z.; methodology, T.A. and X.-X.Z.; formal analysis, T.A. and Y.G.; investigation, Q.-L.Y. and S.C.; resources, J.-Y.L.; data curation, D.-Y.T.; writing—original draft preparation, S.C.; writing-review and editing, Q.-L.Y.; project administration, F.Z. and Q.-L.Y.; funding acquisition, Q.-L.Y.

Funding: This research was funded by the National Natural Science Foundation of China, grant number 51776176 and the Scientific Research Ordering Bureau of EDDM China with project number 61407200204.

References

1. Kappagantula, K.; Pantoya, M.L.; Hunt, E.M. Impact ignition of aluminum-teflon based energetic materials impregnated with nano-structured carbon additives. *J. Appl. Phys.* **2012**, *112*, 024902. [CrossRef]

2. Krause, H.H. New Energetic Materials. In *Energetic Materials: Particle Processing and Characterization*; Wiley-VCH Verlag GmbH & Co. KGaA: Weinheim, Germany, 2004; pp. 1–25.

3. Singh, G.; Kapoor, I.P.S.; Mannan, S.M.; Kaur, J. Studies on energetic compounds: Part 8: Thermolysis of Salts of HNO$_3$, and HClO$_4$. *J. Hazard. Mater.* **2000**, *79*, 1–18. [CrossRef]

4. Wang, Y.; Zhu, J.; Yang, X.; Lu, L.; Wang, X. Preparation of NiO nanoparticles and their catalytic activity in the thermal decomposition of ammonium perchlorate. *Thermochim. Acta* **2005**, *437*, 106–109.

5. Zhang, W.; Li, P.; Xu, H.; Sun, R.; Qing, P.; Zhang, Y. Thermal decomposition of ammonium perchlorate in the presence of Al(OH)$_3$·Cr(OH)$_3$ nanoparticles. *J. Hazard. Mater.* **2014**, *268*, 446–451. [CrossRef] [PubMed]

6. Xu, J.; Li, D.; Chen, Y.; Tan, L.; Kou, B.; Wan, F.; Jiang, W.; Li, F. Constructing sheet-on-sheet structured graphitic carbon nitride/reduced graphene oxide/layered MnO$_2$ ternary nanocomposite with outstanding catalytic properties on thermal decomposition of ammonium perchlorate. *Nanomaterials* **2017**, *7*, 450. [CrossRef] [PubMed]

7. Sergey, V.; Wight, C.A. Kinetics of Thermal Decomposition of Cubic Ammonium Perchlorate. *Chem. Mater.* **1999**, *11*, 3386–3393.

8. Liu, T.; Wang, L.; Yang, P.; Hu, B. Preparation of nanometer CuFe$_2$O$_4$, by auto-combustion and its catalytic activity on the thermal decomposition of ammonium perchlorate. *Mater. Lett.* **2008**, *62*, 4056–4058. [CrossRef]

9. Alizadeh-Gheshlaghi, E.; Shaabani, B.; Khodayari, A.; Azizian-Kalandaragh, Y.; Rahimi, R. Investigation of the catalytic activity of nano-sized CuO, Co$_3$O$_4$, and CuCo$_2$O$_4$, powders on thermal decomposition of ammonium perchlorate. *Powder Technol.* **2012**, *217*, 330–339. [CrossRef]

10. Chen, L.; Li, L.; Li, G. Synthesis of CuO nanorods and their catalytic activity in the thermal decomposition of ammonium perchlorate. *J. Alloys Compd.* **2008**, *464*, 532–536. [CrossRef]

11. Xu, H.; Wang, X.; Zhang, L. Selective preparation of nanorods and micro-octahedrons of Fe$_2$O$_3$, and their catalytic performances for thermal decomposition of ammonium perchlorate. *Powder Technol.* **2008**, *185*, 176–180. [CrossRef]

12. Yuan, Y.; Wei, J.; Wang, Y.; Shen, P.; Li, F.; Li, P.; Zhao, F.; Gao, H. Hydrothermal preparation of Fe$_2$O$_3$/graphene nanocomposite and its enhanced catalytic activity on the thermal decomposition of ammonium perchlorate. *Appl. Surf. Sci.* **2014**, *303*, 354–359. [CrossRef]

13. Memon, N.K.; Mcbain, A.W.; Son, S.F. Graphene Oxide/Ammonium Perchlorate Composite Material for Use in Solid Propellants. *J. Propuls. Power* **2016**, *32*, 682–686. [CrossRef]

14. Gao, X.; Jang, J.; Nagase, S. Hydrazine and Thermal Reduction of Graphene Oxide: Reaction Mechanisms, Product Structures, and Reaction Design. *J. Phys. Chem. C* **2010**, *114*, 832–842. [CrossRef]

15. Yan, Q.L.; Gozin, M.; Zhao, F.Q.; Cohen, A.; Pang, S.-P. Highly energetic compositions based on functionalized carbon nanomaterials. *Nanoscale* **2016**, *8*, 4799–4851. [CrossRef]

16. He, W.; Guo, J.H.; Cao, C.K.; Liu, X.-K.; Lv, J.-Y.; Chen, S.-Y.; Liu, P.-J.; Yan, Q.-L. Catalytic Reactivity of Graphene Oxide Stabilized Transition Metal Complexes of Triaminoguanidine on Thermolysis of RDX. *J. Phys. Chem. C* **2018**, *122*, 14714–14724. [CrossRef]

17. An, T.; He, W.; Chen, S.-W.; Zuo, B.-L.; Qi, X.F.; Zhao, F.Q.; Luo, Y.; Yan, Q.-L. Thermal Behavior and Thermolysis Mechanisms of AP under the Effects of GO-Doped Complexes of Triaminoguanidine. *J. Phys. Chem. C* **2018**, *122*, 14714–14724. [CrossRef]

18. Zhang, Y.; Liu, X.; Nie, J.; Yu, L.; Zhong, Y.; Huang, C. Improve the catalytic activity of α-Fe$_2$O$_3$, particles in decomposition of ammonium perchlorate by coating amorphous carbon on their surface. *J. Solid State Chem.* **2011**, *184*, 387–390. [CrossRef]

19. Jacobs, P.W.M.; Ng, W.L. Thermal decomposition of ammonium perchlorate single crystals. *J. Solid State Chem.* **1974**, *4*, 305. [CrossRef]

20. Jacobs, P.W.M.; Pearson, G.S. The thermal decomposition of ammonium perchlorate (i) introduction experimental analysis of gaseous products and thermal decomposition experiments. *Combust. Flame* **1969**, *13*, 419. [CrossRef]

21. Chaturvedi, S.; Dave, P.N. A review on the use of nanometals as catalysts for the thermal decomposition of ammonium perchlorate. *J. Saudi Chem. Soc.* **2013**, *17*, 135–149. [CrossRef]

22. Mallick, L.; Kumar, S.; Chowdhury, A. Thermal decomposition of ammonium perchlorate-A TGA-FTIR-MS study: Part I. *Thermochim. Acta* **2015**, *610*, 57–68. [CrossRef]

23. Sanoop, A.P.; Rajeev, R.; George, B.K. Synthesis and characterization of a novel copper chromite catalyst for the thermal decomposition of ammonium perchlorate. *Thermochim. Acta* **2015**, *606*, 34–40. [CrossRef]

24. Wang, X.B.; Li, J.Q.; Luo, Y.J. Preparation and Thermal Decomposition Behaviour of Ammonium Perchlorate/Graphene Aerogel Nanocomposites. *Chin. J. Explos. Propellants* **2012**, *35*, 76–80.

25. Wang, X.; Li, J.; Luo, Y.; Huang, M. A Novel Ammonium Perchlorate/Graphene Aerogel Nanostructured Energetic Composite: Preparation and Thermal Decomposition. *Sci. Adv. Mater.* **2014**, *6*, 530–537. [CrossRef]

26. Lan, Y.; Jin, M.; Luo, Y. Preparation and characterization of graphene aerogel/Fe$_2$O$_3$/ammonium perchlorate nanostructured energetic composite. *J. Sol-Gel Sci. Technol.* **2015**, *74*, 161–167. [CrossRef]
27. Liu, H.; Jiao, Q.; Zhao, Y.; Li, H.; Sun, C.; Li, X.; Wu, H. Cu/Fe hydrotalcite derived mixed oxides as new catalyst for thermal decomposition of ammonium perchlorate. *Mater. Lett.* **2010**, *64*, 1698–1700. [CrossRef]
28. Boldyrev, V.V. Thermal decomposition of ammonium perchlorate. *Thermochim. Acta* **2006**, *443*, 1–36. [CrossRef]
29. Rosser, W.A.; Inami, S.H.; Wise, H. Thermal decomposition of ammonium perchlorate. *Combust. Flame* **1968**, *12*, 427–435. [CrossRef]

 *nanomaterials*

Article

Transformation of Combustion Nanocatalysts inside Solid Rocket Motor under Various Pressures

Jun-Qiang Li [1,†], Linlin Liu [2,†], Xiaolong Fu [1], Deyun Tang [2], Yin Wang [2], Songqi Hu [2] and Qi-Long Yan [2,*]

[1] Xi'an Modern Chemistry Research Institute, Xi'an 710065, China; llijq@sohu.com (J.-Q.L.); fuxiaolong204@163.com (X.F.)
[2] Science and Technology on Combustion, Internal Flow and Thermo-structure Laboratory, Northwestern Polytechnical University, Xi'an 710072, China; lll@nwpu.edu.cn (L.L.); tangdy@mail.nwpu.edu.cn (D.T.); wongyin@mail.nwpu.edu.cn (Y.W.); pinecore@nwpu.edu.cn (S.H.)
*   Correspondence: qilongyan@nwpu.edu.cn
†   These authors contribute equally to this work.

Received: 31 January 2019; Accepted: 1 March 2019; Published: 6 March 2019

Abstract: In this paper, the dependences of the morphology, particle sizes, and compositions of the condensed combustion products (CCP) of modified double-base propellants (1,3,5-trimethylenetrinitramine (RDX) as oxidizer) on the chamber pressure (<35 MPa) and nickel inclusion have been evaluated under a practical rocket motor operation. It has been shown that higher pressure results in smaller average particle sizes of the CCPs. The CCPs of Ni-containing propellants have more diverse morphologies, including spherical particles, large layered structures, and small flakes coated on large particles depending on the pressure. The specific surface area (SSA) of CCPs is in the range of 2.49 to 3.24 $m^2\,g^{-1}$ for propellants without nickel are less dependent on the pressure, whereas it is 1.22 to 3.81 Ni-based propellants. The C, N, O, Al, Cu, Pb, and Si are the major elements presented on the surfaces of the CCP particles of both propellants. The compositions of CCPs from Ni-propellant are much more diverse than another one, but only three or four major phases have been found for both propellants under any pressure. The metallic copper is presented in CCPs for both propellants when the chamber pressure is low. The lead salt as the catalyst has been transformed in to Pb(OH)Cl as the most common products of lead-based catalysts with pressure lower than 15 MPa. When pressure is higher than 5 MPa, the nickel-based CCPs has been found to contain one of the following crystalline phases: $Pb_2Ni(NO_2)_6$, $(NH_4)_2Ni(SO_4)_2 \cdot 6H_2O$, $C_2H_2NiO_4 \cdot 2H_2O$, and NiO, depending on the pressure.

Keywords: solid propellants; condensed products; catalytic combustion; compositions; rocket motor

---

1. Introduction

In spite of great advancements in the field of new energetic materials (EMs), the 1,3,5-trimethylenetrinitramine (RDX) and ammonium perchlorate (AP) are still the most widely used ingredients as oxidizers in solid propellants [1]. In order to improve the combustion efficiency of solid propellants, one of the most effective ways is to add nano-sized catalysts. It has been shown that the catalysts could largely increase the burn rate and combustion efficiency of the solid propellants by changing the solid-state and gas-phase reaction mechanisms of between the oxidizers and fuels generated by the main ingredients such as binders, RDX, and AP. The effects of catalysts on the decomposition kinetics, reaction mechanisms, and burning rates, combustion characteristics have been widely investigated in the past decades. It has been shown that in case of nitramine-based propellants, the reaction of $CH_2O + NO_2 \rightarrow CO + NO + H_2O$ is proposed to be the most important one in their foam layer that determines the burn rate, where the evaporation and condensation of nitramines and nitric esters as the binder are responsible for the dominant mass transfer [2]. In the presence of extra oxidizer, such as potassium chlorate, the major gaseous decomposition products of RDX could be changed to CO, CO_2, HCN, NO_2, and H_2O due to instant consumption of CH_2O through the gas-phase transformation $CH_2O + O_2 \rightarrow HCO + HO_2$, resulting in higher reaction rate [3].

As one of the important groups of solid propellants, nitramine-containing modified double-base propellants are featured with low-emission (sometime smokeless), high mechanical strength, long shelf-life, and very low-pressure exponents. They are widely used in rocket motors of tactical missiles, with the higher burning rates (e.g., over 25 mm s^{-1} at 7 MPa) and low-pressure exponents (n < 0.2) [4]. In order to improve the energy content of this type of propellants, usually maximum 5wt% metal fuels (e.g., Al, Mg, B, and Ni) would be included [5]. The pressure exponent could be further decreased with extended pressure range by use of novel multi-functional catalysts [6]. The experiment phenomenon on combustion performance improvement has been widely reported by use of metal powder and novel catalysts as additives. However, the inherent mechanisms of these changes are still not so clear, even after much effort has been made on combustion mechanisms and the kinetic modeling of these processes [7]. In particular, some models have been developed to predict the comprehensive properties of the coarse condensed combustion products (CCPs) as a function of propellant formulation, burning conditions, agglomeration regularities, and geometric configuration of computational region [8,9]. Recent models on AP/HTPB propellants shows that under the condition of initial combustion pressure of 3.5 MPa and a pressure reduction rate of 1000 MPa/s, a narrow diffusion chemical reaction zone could be formed in the initial stage of depressurization. Further, the diffusion and premixed dual flame appears when the pressure drops to about 1.7 MPa [10]. For the same AP/HTPB system, three different reaction mechanisms have been proposed and simulated, which are based on the global chemistry using symbolic species for the three flames, a 12-species mechanism, and a 72-step reaction mechanism with 39-species, respectively [11]. Except for the numerical simulations, the chemical structure of the propellant flame could be predicted by ReaxFF reactive force field molecular dynamics and equilibrium thermodynamics simulations [12]. It has been found that the ReaxFF agrees considerably better with experimental results for minor species than the thermodynamic simulations.

It is widely accepted that the major combustion gaseous products from nitro-based propellants are H_2O, CO, CO_2, H_2, and N_2, whereas the HCN, NH_3, CH_4, nitrogen oxides, benzene, acrylonitrile, toluene, furan, aromatic amines, benzopyrene, and various polycyclic aromatic hydrocarbons are detected in minor concentrations. However, the dark zone of the propellants has much different chemicals as the intermediates with higher molecular weight than those of the luminous flame zone [13]. The gas phase reactions could become more complicated if new energetic ingredients are included. For instance, in the case of metal hydrides, the ZrH_2 was found to tune the decomposition behaviors of AP by enhancing the generation of NO in the high-temperature decomposition stage. The hydrogen released from ZrH_2 may promote the combustion reactions in gaseous phases and therefore induce the two-stage combustion behaviors of the corresponding propellants [14]. Furthermore, the catalysts play the key role in both condensed and gas phase reactions but with much more uncertainty; even the solid-state catalytic decomposition behavior could be well characterized. The commonly used and evaluated catalysts include nano metric metal oxides [15] and ferrocene derivatives [16]. It has been proposed that both iron oxide and copper chromite are primarily acting on the condensed phase, whereas the ferrocene first acts simply as a highly reactive fuel. The resulted ferric oxide from ferrocene in the condensed phase would further catalyze the gas phase reactions [17].

In order for a complete description of the combustion mechanisms, the CCPs have to be clarified with more details. It is much more difficult to detect the intermediates of CCP in the combustion processes of solid propellants, especially under the high temperature/pressure rocket motor operation conditions due to very limited diagnosis techniques. However, the CCPs could be finally analyzed after quenching. The CCPs usually refer to the solid products of aluminized [18] or boron-containing [19] propellants, which are usually in micron sizes after agglomeration, if using the nanosized Al [20]. It has been found that a particle-laden flame zone with a sensibly reduced particle size is disclosed in the case of nanometer Al. To better understand the metal particles burning process, the reactions of a suspension of solid particles in a rapidly-heated oxidizing gas have been investigated. It has been shown that there are two reaction-onset mechanisms, which leads to a nontrivial dependence of the total reaction time on the particle size and solid-fuel concentration within the suspension [21]. Except for the experimental setups, various new diagnosis techniques have been developed to monitor the combustion process of the solid propellants. For instance, the time-resolved synchrotron X-ray imaging to view the in-situ formed aluminum agglomerates at corresponding rocket chamber pressures [22]. This technique provides real time critical data for understanding the combustion behavior of aluminized solid propellants in real rocket motors. Secondly, the gas phase species of the propellants could be experimental determined by time-of-flight mass spectrometry (ToFMS) to verify theoretical results from the density functional theory (DFT) calculations [23].

Even so much abovementioned achievement has been made in term of diagnosis and modeling of the solid propellant combustion, there is still a huge challenge to clarify the detailed reaction processes in the flame zone. It is essential to evaluate the combustion mechanisms under the operation conditions of a real rocket motor, based on which the interior ballistic characteristics could be well predicted. Owing to the extremely large specific surface areas, nano-sized catalysts have significant catalytic effects in both condensed and gas phases during decomposition and subsequent combustion, via activation of the reactants and acceleration of their transition state formations. In order to determine the relationship between the compositions, the chemical structure of the CCPs of modified double-base (MDB) propellants with different additives under a wide range rocket motor flow conditions are investigated, using a recently assembled facility with the capacity to capture almost all of the CCPs. The results may also give some evidence on how the pressure change the chemical structure of the CCPs, which can help setup the experiments for the target structure from flame synthesis under high pressure.

2. Experimental Procedure

2.1. Preparation of the Propellant Samples and Nomenclature

The MDB propellant samples were prepared by a cast-curing technique at a temperature of 70 °C for three days. The detailed ingredients of the slurry for the first typical formulation are as follows: Nitrocellulose (24.0 wt%), Nitroglycerine (30 wt%), RDX (35.5 wt%), Al_2O_3 (2.0 wt%), Nickel (5.0 wt%), and Lead/Copper salicylates (3.5 wt%) as the catalysts. This sample is named as "LZ" in the following sections, where the ending number means the average maximum pressure in the rocket motor. For the second typical formulation are as follows: Nitrocellulose (24.0 wt%), Nitroglycerine (30.0 wt%), RDX (40.5 wt%), Al_2O_3 (2.0 wt%), and Lead/Copper salicylates (3.5 wt%). This sample is named as "JZ" in the following sections, where the ending number means the average maximum pressure in the rocket motor. The burn rate law for LZ and JZ propellants are $u = 8.36P^{0.422}$ (P = 1.0–10.0 MPa) and $u = 17.67P^{0.201}$ (P = 10–22 MPa), respectively.

2.2. Rocket Motor Assembling and Combustion Condensed Products Collection

Considering that the flame temperature of the MDB propellants is between 2500–3500 K, but the ingredients of the condensed products must be analyzed at ambient conditions (about 293 K), the CCPs have to be cooled down before all structure determinations. First, the reactions between the gaseous products with the condensed products need to be prevented during the cooling process. A facility has been designed in our lab to collect the CCPs and a schematic setup is shown in Figure 1 [24].

Air was first discharged from the collection tank using a vacuum pump and then the tank was filled with argon several times (no less than three times) to prevent reactions between the combustion products and oxygen from air. The propellant grain loaded in the solid rocket motor was ignited and the combustion products were ejected into the collection tank through a nozzle. After the combustion products were cooled down naturally and the CCPs were settled, which usually takes 2 h, the door of collection tank can be opened for the sampling process. The combustion chamber pressure can be changed by adjusting nozzle throat diameter and the chamber pressure was measured by the pressure transducer during ignition and combustion processes. The CCPs were dried at 80 °C under vacuum for 12 h before analysis.

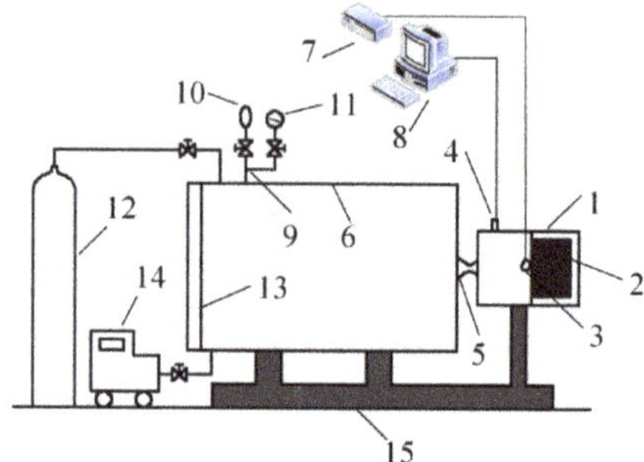

Figure 1. Schematic of the experimental facility: (1) solid rocket motor; (2) propellant grain; (3) igniter pad; (4) pressure transducer; (5) Laval nozzle; (6) collection tank; (7) ignition controller; (8) signal acquisition system; (9) gaseous product collecting piping/gas exhaust piping; (10) Teflon bag; (11) vacuum manometer; (12) hyperbaric argon cylinder; (13) door of collection tank; (14) vacuum pump; (15) steel holder and stabilizer; the shape of propellant grain, solid cylindrical coated with silicone rubber based heat insulation layer.

The combustion conditions of the propellants in this facility were almost the same as those in the solid rocket motor and the CCPs were widely dispersed in the stainless-steel collection tank with a large volume (about 2 m^3), decreasing the probability of chemical reactions among the condensed-phase products to a certain extent. Large wall heat transfer occurs after ejection due to the high thermal conductivity of stainless-steel, which is also beneficial with respect to cooling the products. In addition, the collection tank was filled with inert gas (argon) and this also played an important role in avoiding reactions among the combustion products by increasing the cooling rate and diluting the combustion reactants.

It should be noted that some gaseous products ejected from the solid rocket motor, such as H$_2$O may exist in the form of liquid or solid due to physical and chemical interactions during the cooling of the combustion products. The results of chemical analysis are reliable and can represent the real compositions of CCPs. Chamber pressure is important with respect to the compositions of the combustion products of the propellants. The experimental conditions, chamber pressures, and combustion time for the experiments of LZ and JZ formulations are shown in Table 1. In addition, the curves of chamber pressure vs. function time of two different MDB propellants are plotted in Figure 2.

Table 1. Experimental conditions and experimental results of LZ propellant grains.

Samples	D_t/mm	$P_{c,max}$/MPa	$P_{c,av}$/MPa	t_b/s
JZ-7	5.60	7.7	5.8	1.58
JZ-12	4.70	12.5	9.4	1.27
JZ-15	4.25	15.5	11.1	1.22
JZ-20	3.91	19.8	13.6	1.14
JZ-25	2.84	23.6	20.3	0.93
LZ-0	7.50	0.8	0.5	2.96
LZ-2	6.20	2.2	2.0	1.46
LZ-5	5.40	6.6	5.3	1.63
LZ-7	5.01	7.4	6.1	1.76
LZ-9	4.50	9.5	7.8	1.74
LZ-12	4.00	12.0	9.8	1.71
LZ-18	3.30	17.7	13.2	1.56
LZ-35	2.50	36.8	19.3	1.05

Notes: D_t is nozzle throat diameter. $P_{c,max}$ is maximum chamber pressure of rocket motor. $P_{c,av}$ is average chamber pressure of rocket motor. t_b is function time.

Figure 2 shows that the chamber pressure cannot maintain constant during combustion of the propellants, which is common in the case of small rocket motors with very short working times, and a proper grain selection can help in getting neutral burning only in the case of full-scale rocket motors [25]. In order to investigate the effect of pressure on the combustion products, the time-averaged pressure $P_{c,av}$ was used to characterize operation one. The pressure curve of the rocket motor could usually be divided into three parts, i.e., start-up phase, steady-state phase, and tail-off phase.

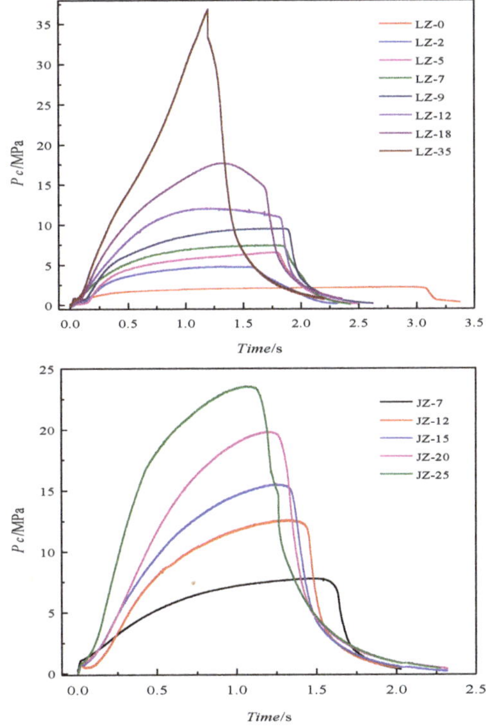

Figure 2. The pressure–time curves in the experimental rocket chamber for all firing measurements.

The steady-state phase clearly dominates the overall performance of the motor and the chamber pressure is usually calculated from this phase. Therefore, the starting point and end point of this phase must be determined. Figure 3 shows how to determine the function time of a solid rocket motor.

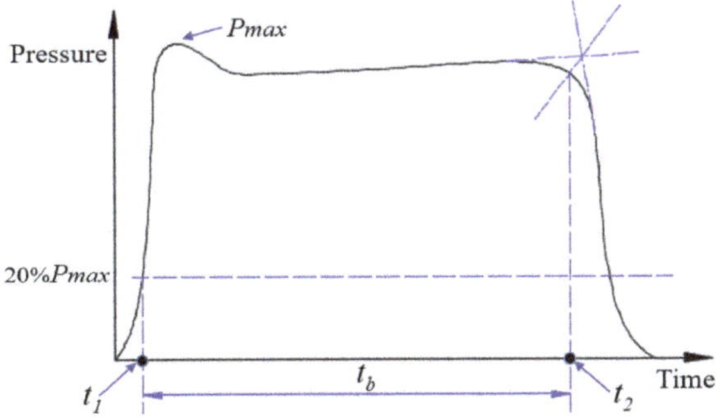

Figure 3. Function time diagram of solid rocket motor.

In these experiments, the starting point of the steady-state phase, t_1, is defined as the time corresponding to the point of 20% of the maximum pressure on P-t curves; t_2, corresponds to the transition point of the steady state to pressure drop. The time-averaged chamber pressure $P_{c,av}$ could be defined as follows:

$$P_{c,av} = \frac{\int_{t_1}^{t_2} P dt}{t_2 - t_1}$$

It is clear that with the increase of the chamber pressure, the function time becomes shorter due to increased burn rate. The curves become less flat, especially for the one that has peak pressure of 35 MPa, which is unexpectedly out of the calibrated scale (25 MPa) of the pressure sensor.

3. Results and Discussion

3.1. The Surface Morphologies of the Overall CCPs

As the first step, the obtained CCPs are divided into three parts based on the particle sizes. The residues with centimeter sizes were picked out by naked eyes (Figure S1), which belongs to the exfoliated or fragmented linear layer of the propellant charges due to pressure fluctuation and ablation. The particles in micron sizes are collected by sedimentation in ethanol, whereas the upper suspended part containing nano-sized particles are subject to the Transmission Electronic Microscopy (TEM) analyses in a later section. The morphologies of the micron-sized CCPs are investigated by the Scanning Electronic Microscopy (SEM) technique, which are shown in Figures 4 and 5.

As shown in Figure 4a–c, there are a large amount of spherical particles with diameters of 1–3 μm for JZ propellant, which does not contain Ni metal particles in comparison to LZ propellant. It is clear with the pressure increase, the average particle size decreases and also one could notice that once the pressure achieves 35 MPa, which is out of the range of the calibrated pressure sensor. The particles morphology seems return to be similar as the one obtained at lower pressure. It should be noted that the pressure of 35 MPa is the peak pressure that has very short duration as shown in Figure 2. In this case, the pressure attenuation is dramatically after the peak value, so that the CCPs formed are similar to those obtained under lower pressure. It is shown in Figure 4a_1–c_1 that there are a large amount of amorphous nano-particles deposited on those large spherical particles. The elemental analyses show that these materials in most cases are nanocrystalline inorganic residues including carbon soot particles.

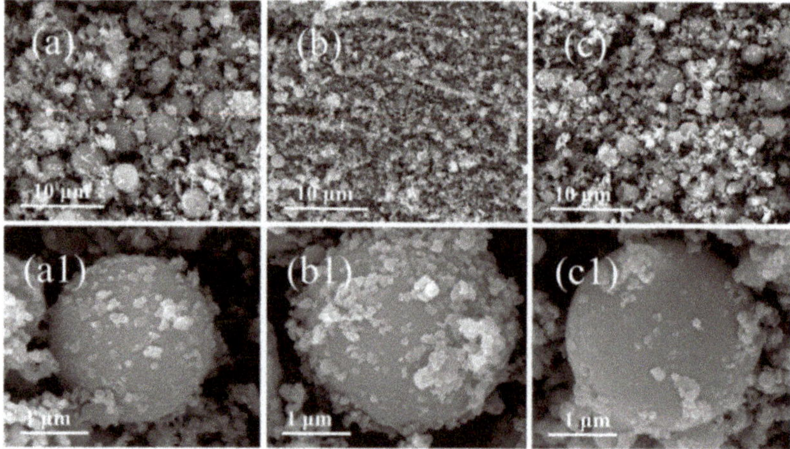

Figure 4. The SEM images of combustion products of JZ propellant grains: (**a,a₁**) JZ-15; (**b,b₁**) JZ-20; and (**c,c₁**) JZ-35.

Similarly, Figure 5 presents the surface structures of CCPs from LZ propellant at three different pressures as a comparison. With metallic Ni included, the LZ propellant has very different CCPs with a variety of morphologies, and the most typical ones are spherical particles, large layered structures, and small flakes coated on large particles. In general, the particle size of the spherical CCPs from LZ propellant is larger than those of JZ propellant. In order to clarify what elements are included in the CCPs, the EDS analyses have been done, with the spectra plotted in Figures S1 and S2, whereas the detailed elemental analyses are shown in Table S1, and the average atomic contents are summarized in Table 2.

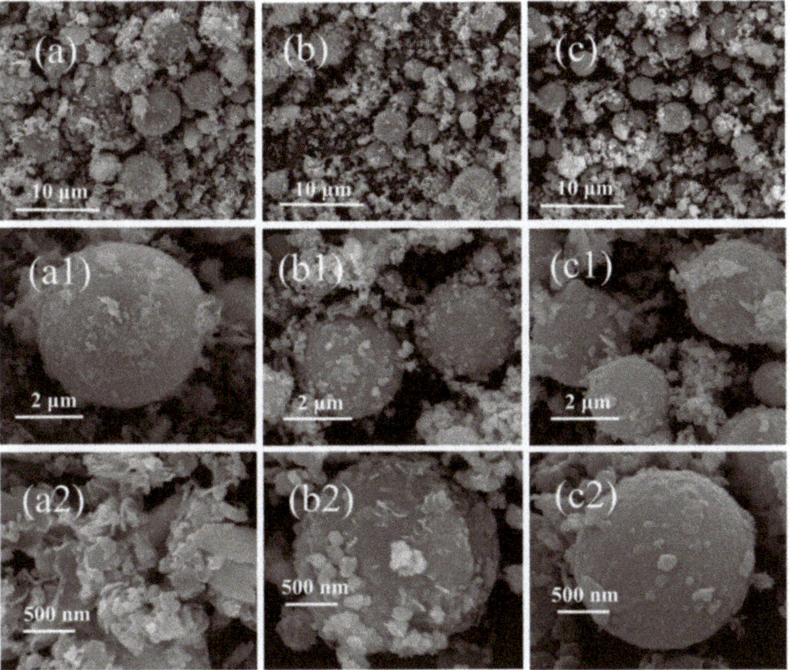

Figure 5. The SEM images of CCPs from LZ propellant (**a**,**a$_1$**,**a$_2$**) LZ-7; (**b**,**b$_1$**,**b$_2$**) LZ-18; (**c**,**c$_1$**,**c$_2$**) LZ-35.

Table 2. Element distribution results of JZ and LZ propellant grains combustion products.

Elements	Average Atomic Contents (%)					
	JZ-15	**JZ-20**	**JZ-25**	**LZ-7**	**LZ-18**	**LZ-35**
C K	30.27	31.09	44.43	40.13	42.33	26.08
N K	3.38	2.66	3.74	13.80	1.51	2.33
O K	43.74	41.84	34.92	31.27	34.60	44.02
Al K	9.55	10.18	7.60	4.16	7.50	11.80
Si K	3.21	3.06	2.31	1.51	3.52	3.41
P K	0.25	0.34	0.30	0.18	0.79	0.12
S K	1.10	1.26	0.91	-	-	-
Cl K	0.70	1.02	0.86	1.27	0.89	1.18
K K	0.68	0.62	0.8	0.99	0.60	0.70
Ca K	0.24	0.22	0.19	0.14	-	0.26
Fe K	0.99	0.93	0.42	0.66	0.86	0.53
Ni K	0.28	0.22	0.29	2.16	2.51	3.80
Cu K	4.33	5.00	1.96	1.41	1.74	2.52
Pb M	1.33	1.60	1.31	2.36	2.27	3.28

It is clear from Table 2 that, the C, N, O, Al, Cu, Pb, and Si are the major elements presented on the surfaces the CCP particles. For JZ propellants, with the increase of pressure, the content of carbon increases, and the O and Si decreases. The silicon element comes from the insulating material, which is based on a silicon rubber, indicating erosive burning to the insulation layer. However, higher pressure seems less erosive to the insulation layer for the JZ propellant free of metal fuel like nickel. It is opposite when the 5% of nickel is included as the metal fuel and flame stabilizer, where the erosive burning increases with the pressure, resulting in much higher Si content in the CCPs, e.g., it is 3.52% for LZ-18 and 3.06% for JZ-20, respectively. The organic copper and lead salts are mainly transformed into CCPs. However, when the pressure is very high for the JZ propellant, there is no interaction between catalysts with the metal fuel. The content of Cu would drop to very low as 1.96%, which is opposite of the LZ propellant. As shown in a later section, based phase analyses by X-ray Diffraction (XRD) show the higher pressure results in higher content of Cu and Pb in the CCPs, due to formation of some less volatile products such as minerals or alloys under the effect of nickel. The other minor elements such as Ca, K, and Cl are mainly from the complicated insulating materials. In terms of Al element, which comes from Al_2O_3 as a ballistic stabilizer, the content first increases and then decreases with the pressure for JZ propellant, whereas it largely increases with the pressure to even 11.8% when the peak pressure is 35 MPa for LZ propellant. The interaction between the Al_2O_3 with metallic nickel is significant and the higher pressure is favorable for such mutual reactions, resulting in more nonvolatile products, which is responsible for such an increase in Al element content.

3.2. The Particle Size Distributions of the Micron-sized CCPs

Based on the separation of the nano-sized and micron-sized CCP particles by sedimentation method in aqueous media, the micron-sized particles are subject to a size distribution analysis. The corresponding distribution curves are plotted in Figures S3 and S4, with the detailed parameters for these plots are summarized in Table S2. In order for better comparison of the average values of the distribution based on repeated experiments, the normalized curves are shown in Figures 6 and 7 for JZ and LZ propellants, respectively. It is clear from Figure 6 that the micron-sized CCPs from combustion of JZ propellants can be divided into three groups in terms of particle sizes: submicron ones, media ones (1–10 μm) and larger ones (>10 μm). Most of the particles have the diameters less than 80 μm for the JZ propellant burning at any pressure.

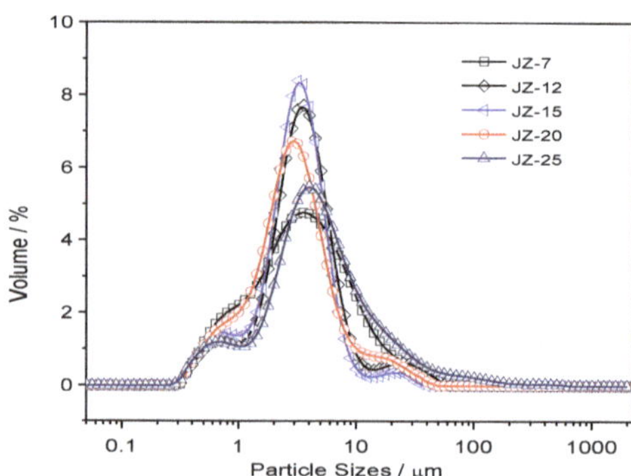

Figure 6. The particle size distributions of CCPs from combustion of JZ propellant grains under various pressures.

When the pressure is less than 12 MPa, the CCPs with submicron sizes are not well separated with the media sizes, as well as the larger sizes. It means that the size boundary is not so clear under lower pressure burning of JZ propellant. In comparison, when the pressure is higher than 15 MPa, the three peaks for three groups of particle sizes become more independent with more obvious boundaries.

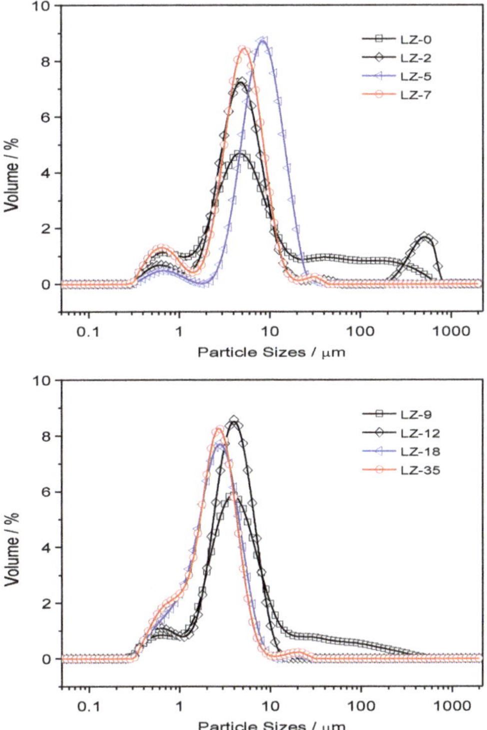

Figure 7. The particle size distributions of condensed combustion products (CCP) from combustion of LZ propellant grains under various pressures.

In terms of LZ propellants, the distributions are much different, which is highly dependent on the pressure. When the pressure is lower than 9 MPa, there are three major groups as the JZ propellant: micron and submicron (<5 μm), media sizes (5–15 μm), and large sizes (>15 μm). However, we can see that there is a strange narrow peak and large shoulder peak at the size over 200 μm at the pressure of ambient pressure and 2 MPa, which is probably due to presence of large agglomerated Nickel, which is too large to do the SEM analyses. It has been mentioned that the content of Al is largely dependent on the pressure and under lower pressure there is less interaction between Al and Ni in LZ propellant, resulting in lower burning efficiency and larger agglomeration.

When the pressure is over 18 MPa, the CCP from burning of LZ propellants, the amount of micron and submicron particles is greatly increased. Thus, the overall peak becomes narrower due to less agglomeration of the Nickel particles. For better quantitative comparison of the particle sizes, the detailed parameters are summarized in Table 3. It is clear that the specific surface area (SSA) of CCPs from the JZ propellant is less dependent on the pressure, which is in the range of 2.49 to 3.24 m^2 g^{-1}. It is 1.22 to 3.81 for the LZ propellant due to high pressure dependence of metallic nickel transformation and interaction of elemental Ni with Al_2O_3. The high pressure is favorable for more complete transformation and lower particle sizes [26]. The presence of less volatile products would decrease the size of CCP and, for instance, the formation of AlF_3 would greatly mitigate the agglomeration of Al combustion by using organic fluorine-contained additives [27].

Table 3. Particle size distributions of the CCPs for involved JZ and LZ propellants.

Samples	Obscuration	Residual	Concent.	Span	$D_{[4,3]}$ [a]	Uniformity	SSA	$D_{[3,2]}$ [b]	$d_{(0.1)}$ [c]	$d_{(0.5)}$ [d]	$d_{(0.9)}$ [e]
JZ-7	7.60	1.317	0.0027	3.154	4.981	1.050	3.06	1.959	0.794	3.179	10.819
JZ-12	8.90	1.144	0.0033	1.992	4.588	0.842	2.77	2.167	1.035	3.167	7.343
JZ-15	8.55	1.798	0.0029	1.685	3.483	0.606	3.08	1.945	0.875	2.876	5.721
JZ-20	9.31	1.125	0.003	2.512	3.959	0.905	3.24	1.855	0.834	2.661	7.519
JZ-25	8.46	0.623	0.0037	3.967	9.567	1.780	2.49	2.41	0.99	4.116	17.317
LZ-0	8.37	0.571	0.0044	18.859	33.566	5.830	2.13	2.818	1.084	5.298	101.003
LZ-2	10.39	0.865	0.0057	61.074	54.834	10.800	1.83	3.286	1.995	4.788	294.406
LZ-5	8.23	1.614	0.0069	1.392	8.51	0.445	1.22	4.929	3.761	7.714	14.498
LZ-7	8.92	1.985	0.0042	1.708	5.126	0.530	2.35	2.549	0.986	4.561	8.777
LZ-9	10.16	0.482	0.0047	7.373	15.425	3.230	2.29	2.619	1.308	4.06	31.245
LZ-12	9.05	0.741	0.0035	1.541	3.758	0.464	2.64	2.27	1.18	3.481	6.544
LZ-18	9.28	1.561	0.0027	1.638	2.676	0.530	3.53	1.7	0.839	2.349	4.685
LZ-35	10.00	1.736	0.0028	1.562	2.496	0.533	3.81	1.574	0.75	2.191	4.171

Notes: a, the weighted average of the particle size to the surface area; b, the weighted average of the particle size to the volume. c, particles whose diameter equals to or less than the value of $D_{(0.1)}$, and the sum of the volume fractions of which accounts for ten percent; it is the same for d and e; SSA, specific surface area, in m^2/g.

In this case, the polymorphic transition of the metal oxide is also possible, e.g., α-Al_2O_3 and θ-Al_2O_3 may be transformed into δ-Al_2O_3 and γ-Al_2O_3. In the presence of Al_2O_3 in LZ and JZ propellants as a stabilizer, it may react with carbon soot and many other organic residues to form various other condensed products instead of polymorphs of Al_2O_3. The cooling mode may also affect the particle size and phases content of the CCPs, but the comparative study under the same experimental conditions is fine [28].

3.3. The Interface Structure of the Nanosized CCPs

In order to observe the interface structures and the shape of the nano-sized CCPs, the TEM images of selected samples are collected and shown in Figures 5 and 6 for JZ and LZ propellants, respectively. The overall pictures of the nano-sized CCPs from combustion of JZ propellant under the pressure of 7, 12, and 15 MPa are shown in Figure 8a–c. The maximum sizes of the CCPs increases with the pressure and then decreases with the shape changes. The particle shapes become less uniform when the pressure increases. Generally, the spherical particles with diameters in the range of 200–400 nm could be formed under the pressure of 7 MPa. There are some needles and nano-flakes can be observed when the pressure is over 12 MPa, indicating new crystal phases. Large irregular aggregates could be found when the pressure achieves 25 MPa, where the spherical particle has a diameter of about 200–300 nm.

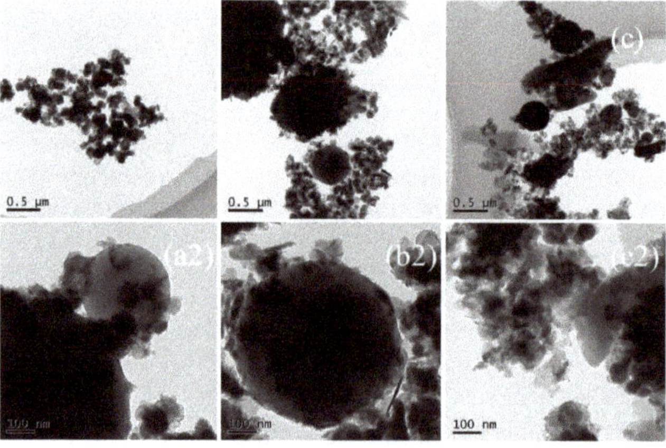

Figure 8. The TEM images of nano-sized CCPs of JZ propellant grains (a,a2) JZ-7; (b,b2) JZ-12; (c,c2) JZ-25.

In comparison, as shown in Figure 9a–c, the maximum sizes of the nano-sized CCPs increases with the pressure and the morphology changes as well. There are generally three types of particle shapes: stacked layers, spherical, and long needle-like. Under the ambient pressure, there are a large amount of small particles with diameter of less than 100 nm, which are deposited on stack-layered structures. When the pressure increases to 7 MPa, the layers become thicker and the needles appears with length over 500 nm and diameter less than 50 nm. With the pressure further increases, the length and diameter of the needle structure decreases (Figure 9c1). Generally, for LZ propellant, the spherical particles with diameters in the range of 50–100 nm under lower pressure, but it could be over 500 nm under the pressure of 35 MPa. Large irregular aggregates could be stick to the spherical particles at this high pressure. It has been proposed for metallized propellants that the irregular agglomerates usually larger than spherical agglomerates [29,30]. The formation of irregular agglomerates was proposed to occur by three steps: deformation of spherical metal drops, combination of particles with various shapes, and formation of irregular aggregations.

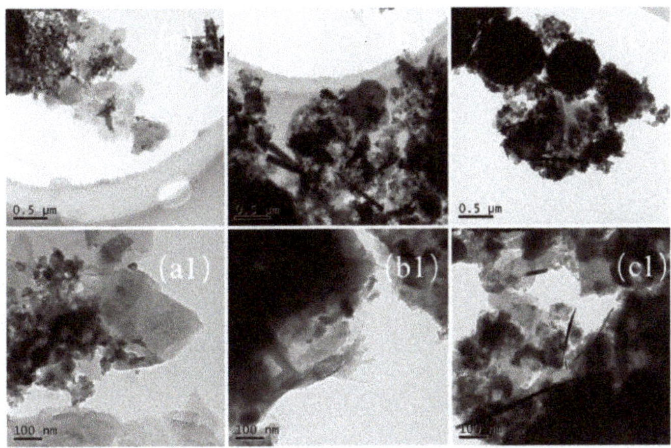

Figure 9. The TEM images of nano-sized CCPs of LZ propellant grains (**a,a₁**) LZ-0; (**b,b₁**) LZ-7; (**c,c₁**) LZ-35.

3.4. The Chemical Compositions of the CCPs

As the most important step, the determination of the crystal phases inside the CCPs has to be done, to show the inherent reaction mechanism in the gas phase under a real rocket motor flow condition. The powder X-ray (PXRD) spectra have been collected, with the crystal phases identified in Figures 10 and 11. The quantitative analyses of the content of the selected major phases are listed in Tables S4 and S5.

It has been shown from Figures 10 and 11 that the compositions of the crystal phases of CCPs are highly dependent on the pressure. In general, the CCPs from the LZ propellant are much more diverse than those from the JZ propellant. In the case of the JZ propellant under the attempted pressures, only three to four major phases have been identified under each pressure. The metallic copper is shown when the pressure is not higher than 20 MPa, the content slight decreases with the increase of pressure. It means the organic copper salts would transform to CuO first, which further participate in the reaction with carbon soot as an oxidizer. The presences of $AlCu_3$ at the pressure of 12 and 35 MPa and $AlCu_4$ at the pressure of 15 MPa indicate interaction of the resulted metallic Cu with the Al_2O_3. Only 13% of Al_2O_3 is found in CCP of JZ under pressure of 20 MPa, which means the reaction between Al_2O_3 with the other products in the gas phase is not complete at this pressure. It may be transformed to Al_2O at lower pressure (e.g., <7 MPa).

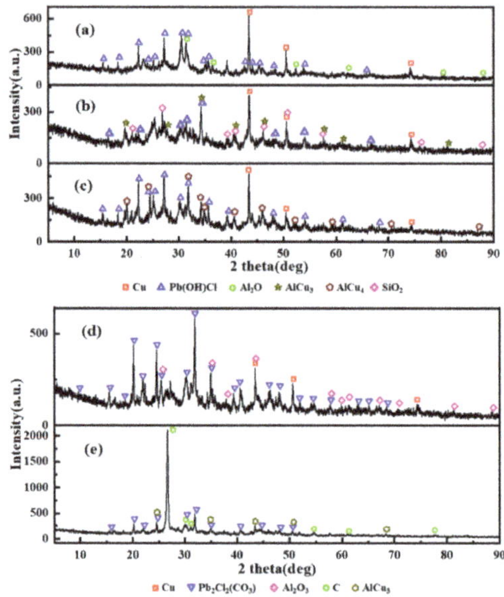

Figure 10. X-ray diffraction (XRD) results of JZ propellant grains combustion product; (**a–e**) represent pressure of 7, 12, 15, 20, and 25 MPa).

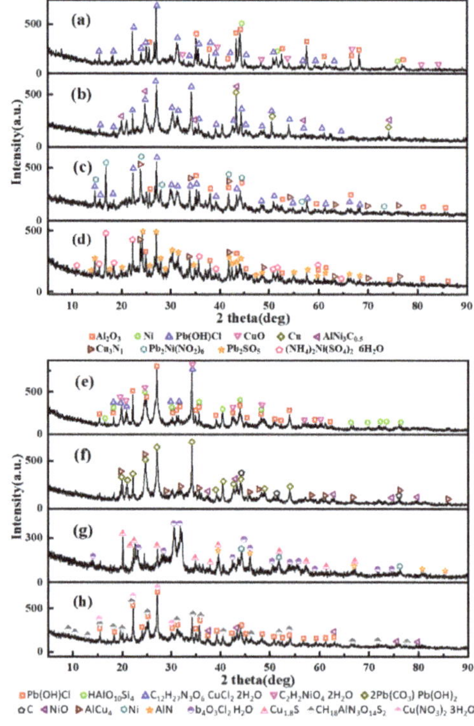

Figure 11. XRD results for condensed combustion products of LZ propellants; (**a–h**) represent pressure of 0, 2, 5, 7, 9, 12, 18, and 35 MPa.

The lead salt as the catalyst has been transformed in to Pb(OH)Cl at the pressure lower than 15 MPa or $Pb_2Cl_2(CO_3)$ when the pressure is higher than 20 MPa. The $Pb_2Cl_2(CO_3)$ is also so-called phosgenite as a naturally formed mineral, which are usually colorless and transparent. This mineral is rather sectile and consequently was earlier known as corneous lead. This finding may help in designing novel preparation method for this mineral by a flame synthesis process using lead-containing propellant as a precursor. Besides, a large amount of carbon soot was formed when the pressure achieves 35 MPa, due to extinguish of the CCP intermediates caused by fast pressure drop after the peak pressure (see in Figure 2). The unique product SiO_2 from JZ-12 is due to the contaminant of the exhausted insulating materials after oxidation.

In comparison, as shown in Figure 11 and Table S5, it is similar to LZ propellants, only three or four major crystalline components have been found under each pressure. The Pb(OH)Cl is the most common product of lead salts as the catalysts, whereas the copper salt could be transformed into various compounds depending on the pressure, e.g., it is CuO under ambient pressure. It is logical, then, that the gas phase reaction was not realized under the ambient pressure and the CuO as the condensed decomposition product of organic cooper salt did not change. The Cu_3N_1, $C_{12}H_{27}N_3O_6 \cdot CuC_{12} \cdot 2H_2O$, $Cu_{1.8}S$ and $Cu(NO_3)_2 \cdot 3H_2O$ are unique copper based compounds under the pressure of 7, 9, 18, and 35 MPa. The lead salt would transform to Pb_2SO_5, $2Pb(CO_3) \cdot Pb(OH)_2$ and $Pb_4O_3Cl_2 \cdot H_2O$ at the pressure of 7, 12, and 18 MPa. At the pressure of 5 MPa, the elemental lead even combines with nickel to form $Pb_2Ni(NO_2)_6$. It is interesting that at the same pressure of 12–15 MPa, no matter the LZ or JZ propellant, the copper aluminum alloy could be formed. It is $AlCu_4$ for the LZ propellant and $AlCu_3$ or $AlCu_4$ for the JZ propellant.

As mentioned earlier in this paper, the presence of metallic nickel in the formulation makes the compositions of CCPs largely dependent on the pressure. The diverse of the crystalline components of CCPs derives mostly from the nickel element. Under ambient pressure, the metallic nickel is presented due to incomplete combustion, and the other nickel-based compounds are not shown. With the increase of pressure (>5 MPa), the nickel-based CCPs are $Pb_2Ni(NO_2)_6$, $(NH_4)_2Ni(SO_4)_2 \cdot 6H_2O$, $C_2H_2NiO_4 \cdot 2H_2O$, and NiO. At the pressure of 9 MPa, the CCP contains Si-based compound $HAlO_{10}Si_4$, which is different from the JZ propellant (SiO_2). One could notice above that the high dependence of gas phase reactions on pressure, which results in a variety of CCPs in terms of different compositions and morphologies. In terms of Boron-based metal fuels, such as B-Mg and B-Al, the CCPs with ferrocene as the catalyst could be even more complicated. It has been shown that the condensed-phase products are mainly composed of B, C, B_4C (or $B_{12}C_2$), BN, Mg, MgO, $MgAl_2O_4$, Al, Al_2O_3, $AlCl_3$, $NH_4[Mg(H_2O)_6]Cl_3$, NH_4Cl, and Fe_3O_4 [24]. It was shown that the high pressure is not so benefit to high efficiency secondary combustion of boron in gas phase due to formation of less active boron carbide, graphite, and h-BN. As a comparison, the CCPs from aluminum boride and boron–aluminum blends are very similar, with the common ingredients of B, Al, B_2O_3, Al_2O_3, AlN, and Al_5O_6N, showing incomplete combustion due to low pressure [31]. In our case, the high pressure is favorable for secondary reaction of Ni-based compounds and the NiO is the major product under higher pressure, without metallic nickel detected.

4. Conclusions

In this paper, the dependences of the morphology, particle sizes, and compositions of the CCPs of modified double-base propellants on the chamber pressure and nickel inclusion have been evaluated under conditions for a real rocket motor operation. The following conclusions could be made:

(1) The peak chamber pressure has been successfully modulated by changing the size of the throat of the motor. The maximum pressure achieved herein is about 35 MPa, and the lowest one is the ambient pressure.

(2) With the pressure increase, the average particle size of the CCPs decreases. In presence of metallic Ni, the CCPs have a variety of morphologies depending on the pressure. The most typical ones are spherical particles, large layered structures, and small flakes coated on large particles. The SSA

of CCPs from the JZ propellant is less dependent on the pressure, which is in the range of 2.49 to $3.24\ m^2\ g^{-1}$, whereas it is 1.22 to 3.81 for the LZ propellant due to high pressure dependence of metallic nickel transformation and interaction of elemental Ni with Al_2O_3.

(3) The C, N, O, Al, Cu, Pb, and Si are the major elements presented on the surfaces the CCP particles. The micron-sized CCPs from combustion of both propellants can be divided into three groups in terms of particle sizes, but the LZ propellant containing metallic nickel has larger CCPs sizes.

(4) The compositions of CCPs from LZ propellant are much more diverse than those from JZ propellant, but only three to four major phases have been found under each pressure. The metallic copper is presented in CCPs for both propellants when the chamber pressure is low.

(5) The lead salt as the catalyst has been transformed in to Pb(OH)Cl as the most common products of lead-based catalysts at the pressure lower than 15 MPa or $Pb_2Cl_2(CO_3)$ when the pressure is higher than 20 MPa. The copper salt and nickel could be transformed into various compounds depending on the pressure and when pressure is greater than 5 MPa, the nickel-based CCPs contain one of the following phases: $Pb_2Ni(NO_2)_6$, $(NH_4)_2Ni(SO_4)_2 \cdot 6H_2O$, $C_2H_2NiO_4 \cdot 2H_2O$, and NiO.

Supplementary Materials: The following are available online at http://www.mdpi.com/2079-4991/9/3/381/s1, Figure S1: The EDS spectra of CCPs from JZ propellant grains, Figure S2: The EDS spectra of CCPs from LZ propellant grains; Figure S3: The particle size distribution curves for JZ propellant grains combustion under various pressures with repeated tests; Figure S4: The particle size distribution curves for LZ propellant grains combustion under various pressures with repeated tests; Table S1: Element distribution results of JZ and LZ propellant grains combustion products; Table S2: A summary of the CCPs particle distributions for JZ propellant grains. Table S3: Particle distribution of LZ propellant grains combustion products; Table S4: Chemical compositions of the CCPs of JZ propellants at various pressures; Table S5: Chemical compositions of the CCPs of LZ propellants at various pressures.

Author Contributions: Conceptualization, Q.-L.Y. and J.-Q.L.; methodology, L.L. and S.H.; formal analysis, D.T.; investigation, Q.-L.Y. and X.F.; resources, J.-Q.L.; data curation, Y.W.; writing—original draft preparation, Q.-L.Y.; writing—review and editing, Q.-L.Y.; project administration, J.-Q.L. and Q.-L.Y.; funding acquisition, J.-Q.L.

Funding: This research was funded by "National Natural Science Foundation of China, grant number 51776176" and "Fundamental Research Funds from Chinese H863 plan grant number 2018KC020167".

Acknowledgments: The basic research funding from Scientific Research Ordering Bureau of EDDM China with project number 61407200204 is also appreciated. Xiaojiang Li and Wuxi Xie prepared the propellant charges in MCRI. The graduate students Huan Liu, Jieyao Lyu and Peng Yang, who help a lot in the experiments of condensed products collections, are also greatly appreciated.

Conflicts of Interest: The authors declare no conflict of interest. The funders had no role in the design of the study; in the collection, analyses, or interpretation of data; in the writing of the manuscript, or in the decision to publish the results.

References

1. Yan, Q.-L.; Zhao, F.-Q.; Kuo, K.K.; Zhang, X.-H.; Zeman, S.; DeLuca, L.T. Catalytic Effects of Nano Additives on Decomposition and Combustion of RDX-, HMX-, and AP-Based Energetic Compositions. *Prog. Energy Combust. Sci.* **2016**, *57*, 75–136. [CrossRef]

2. Yan, Q.-L.; Li, X.-J.; Wang, Y.; Zhang, W.-H.; Zhao, F.-Q. Combustion mechanism of double-base propellant containing nitrogen heterocyclic nitramines (I): The effect of heat and mass transfer to the burning characteristics. *Combust. Flame* **2009**, *156*, 633–641. [CrossRef]

3. Dong, X.-F.; Yan, Q.-L.; Zhang, X.-H.; Cao, D.-L.; Xuan, C.-L. Effect of potassium chlorate on thermal decomposition of cyclotrimethylenetrinitramine (RDX). *J. Anal. Appl. Pyrol.* **2012**, *93*, 160–164. [CrossRef]

4. Yan, Q.-L.; Song, Z.-W.; Shi, X.-B.; Yang, Z.-Y.; Zhang, X.-H. Combustion mechanism of double-base propellant containing nitrogen heterocyclic nitramines (II): The temperature distribution of the flame and its chemical structure. *Acta Astronaut.* **2009**, *64*, 602–614. [CrossRef]

5. Wu, X.-G.; Yan, Q.-L.; Guo, X.; Qi, X.-F.; Li, X.-J.; Wang, K.-Q. Combustion efficiency and pyrochemical properties of micron-sized metal particles as the components of modified double-base propellant. *Acta Astronaut.* **2011**, *68*, 1098–1112. [CrossRef]

6.   Wang, Y.-L.; Yan, Q.-L.; An, T.; Chen, B.; Ji, Y.-P.; Wang, L.; Zhao, F.-Q. Unraveling the Effect of Anthraquinone Metal Salts as Wide-range Plateau Catalysts to Enhance the Combustion Properties of Solid Propellants. *Cent. Eur. J. Energy Mater.* **2018**, *15*, 376–390. [CrossRef]

7.   Jackson, T.L. Modeling of heterogeneous propellant combustion: A survey. *AIAA J.* **2012**, *50*, 993–1006. [CrossRef]

8.   Babuk, V.A.; Nizyaev, A.A. Modeling of evolution of the coarse fraction of condensed combustion products on a surface of burning aluminized propellant and within a combustion products flow. *Int. J. Energy Mater. Chem. Propuls.* **2017**, *16*, 23–38. [CrossRef]

9.   Babuk, V.A.; Budnyi, N.L.; Ivonenko, A.N.; Nizyaev, A.A. Simulation of Characteristics of Condensed Products in a Combustion Chamber. *Combust. Explos. Shock Waves* **2018**, *54*, 301–308. [CrossRef]

10.   Ye, Z.-W.; Yu, Y.-G. Numerical Simulation of Quenched Combustion Model for AP/HTPB Propellant under Transient Depressurization. *Propell. Explos. Pyrot.* **2017**, *42*, 1085–1094. [CrossRef]

11.   Gaduparthi, T.; Pandey, M.; Chakravarthy, S.R. Gas phase flame structure of solid propellant sandwiches with different reaction mechanisms. *Combust. Flame* **2016**, *164*, 10–21. [CrossRef]

12.   Jensen, T.L.; Moxnes, J.F.; Unneberg, E.; Dullum, O. Calculation of decomposition products from components of gunpowder by using ReaxFF reactive force field molecular dynamics and thermodynamic calculations of equilibrium composition. *Propell. Explos. Pyrot.* **2014**, *39*, 830–837. [CrossRef]

13.   Anderson, W.R.; Meagher, N.E.; Vanderhoff, J.A. Dark zones of solid propellant flames: Critically assessed datasets, quantitative model comparison, and detailed chemical analysis. *Combust. Flame* **2011**, *158*, 1228–1244. [CrossRef]

14.   Yang, Y.; Zhao, F.; Xu, H.; Pei, Q.; Jiang, H.; Yi, J.; Xuan, C.; Chen, S. Hydrogen-enhanced combustion of a composite propellant with ZrH_2 as the fuel. *Combust. Flame* **2018**, *18*, 67–76. [CrossRef]

15.   Pang, W.; DeLuca, L.T.; Fan, X.; Maggi, F.; Xu, H.; Xie, W.; Shi, X. Effects of Different Nano-Sized Metal Oxide Catalysts on the Properties of Composite Solid Propellants. *Combust. Sci. Technol.* **2016**, *188*, 315–328. [CrossRef]

16.   Ishitha, K.; Ramakrishna, P.A. Studies on the role of iron oxide and copper chromite in solid propellant combustion. *Combust. Flame* **2014**, *161*, 2717–2728. [CrossRef]

17.   Sinditskii, V.P.; Chernyi, A.N.; Marchenkov, D.A. Mechanisms of combustion catalysis by ferrocene derivatives. 1. Combustion of ammonium perchlorate and ferrocene. *Combust. Explos. Shock Waves* **2014**, *50*, 51–59. [CrossRef]

18.   Bai, Y.-J.; Xu, T.-W.; Li, Y. Parameters of combustor condensed phase particles form analysis for overload simulation test. *J. Solid Rocket Technol.* **2017**, *40*, 409–413, 419.

19.   Wu, Q.; Chen, L.-Q.; Wang, Y.-X.; Yang, Y.-X. Test method of combustion efficiency for condensed products of boron-based propellant in secondary combustion chamber of solid ducted rocket. *J. Solid Rocket Technol.* **2014**, *37*, 134–138.

20.   DeLuca, L.T.; Galfetti, L.; Colombo, G.; Maggi, F.; Bandera, A.; Babuk, V.A.; Sinditskii, V.P. Microstructure effects in aluminized solid rocket propellants. *J. Propuls. Power* **2010**, *26*, 724–733. [CrossRef]

21.   Soo, M.; Goroshin, S.; Bergthorson, J.M.; Frost, D.L. Reaction of a particle suspension in a rapidly-heated oxidizing gas. *Propell. Explos. Pyrot.* **2015**, *40*, 604–612. [CrossRef]

22.   Kalman, J.; Demko, A.R.; Varghese, B.; Matusik, K.E.; Kastengren, A.L. Synchrotron-based measurement of aluminum agglomerates at motor conditions. *Combust. Flame* **2018**, *196*, 144–146. [CrossRef]

23.   Patidar, L.; Thynell, S.T. Quantum mechanics investigation of initial reaction pathways and early ring-opening reactions in thermal decomposition of liquid-phase RDX. *Combust. Flame* **2017**, *178*, 7–20. [CrossRef]

24.   Liu, L.-L.; He, G.-Q.; Wang, Y.-H.; Hu, S.-Q. Chemical analysis of primary combustion products of boron-based fuel-rich propellants. *RSC Adv.* **2015**, *5*, 101416–101426. [CrossRef]

25.   Miller, W.H.; Barrington, D.K. A Review of Contemporary Solid Rocket Motor Performance Prediction Techniques. *J. Spacecr. Rockets* **1970**, *7*, 225–237.

26.   Arkhipov, V.A.; Zarko, V.E.; Zharova, I.K.; Zhukov, A.S.; Kozlov, E.A.; Aksenenko, D.D.; Kurbatov, A.V. Solid propellant combustion in a high-velocity cross-flow of gases. *Combust. Explos. Shock Waves* **2016**, *52*, 497–513. [CrossRef]

27.   Wang, W.-L.; Li, J.-M.; Yang, R.-J.; Liu, Z.; Li, S.-P. Influence of Organic Fluorine-contained Additives on Condensed Combustion Products of Aluminized Polyether Propellants. *Acta Armament.* **2017**, *38*, 704–710.

28. Hu, X.; Pang, A.-M.; Wang, Y.; Tang, Q.; Han, L.-P. Effects of collection methods on characteristics of condensed combustion products of NEPE propellant. *J. Solid Rocket Technol.* **2017**, *40*, 65–69.
29. Ao, W.; Liu, P.; Yang, W. Agglomerates, smoke oxide particles, and carbon inclusions in condensed combustion products of an aluminized GAP-based propellant. *Acta Astronaut.* **2016**, *129*, 147–153. [CrossRef]
30. Popenko, E.M.; Gromov, A.A.; Pautova, Y.I.; Chaplina, E.A.; Ritzhaupt-Kleissl, H.-J. SEM-EDX study of the crystal structure of the condensed combustion products of the aluminum nanopowder burned in air under the different pressures. *Appl. Surf. Sci.* **2011**, *257*, 3641–3644. [CrossRef]
31. Liang, D.; Xiao, R.; Liu, J.; Wang, Y. Ignition and heterogeneous combustion of aluminum boride and boron–aluminum blend. *Aerosp. Sci. Technol.* **2019**, *84*, 1081–1091. [CrossRef]

 *nanomaterials* 

Article

Thermal Decomposition Behavior and Thermal Safety of Nitrocellulose with Different Shape CuO and Al/CuO Nanothermites

Ergang Yao [1], Ningning Zhao [2],*, Zhao Qin [1], Haixia Ma [3], Haijian Li [1], Siyu Xu [1], Ting An [1], Jianhua Yi [1] and Fengqi Zhao [1],*

[1]   Science and Technology on Combustion and Explosion Laboratory, Xi'an Modern Chemistry Research Institute, Xi'an 710065, China; yaoerg@126.com (E.Y.); qzhao87@163.com (Z.Q.); h.j.Li@outlook.com (H.L.); xusy99@163.com (S.X.); anting715@163.com (T.A.); npecc_yjh2819@163.com (J.Y.)
[2]   School of Science, Xi'an University of Technology, Xi'an 710054, China
[3]   School of Chemical Engineering, Northwest University, Xi'an 710069, China; mahx@nwu.edu.cn
*   Correspondence: zhaonn@xaut.edu.cn (N.Z.); npecc@163.com (F.Z.)

Received: 15 March 2020; Accepted: 7 April 2020; Published: 11 April 2020

Abstract: Bamboo leaf-like CuO(b) and flaky-shaped CuO(f) were prepared by the hydrothermal method, and then combined with Al nanoparticles to form Al/CuO(b) and Al/CuO(f) by the ultrasonic dispersion method. The phase, composition, morphology, and structure of the composites were characterized by X-ray powder diffraction (XRD), transmission electron microscopy (TEM), scanning electron microscopy (SEM), and energy scattering spectrometer (EDS). The compatibility of CuO, Al/CuO and nitrocellulose (NC) was evaluated by differential scanning calorimetry (DSC). The effects of CuO and Al/CuO on the thermal decomposition of NC were also studied. The results show that the thermal decomposition reactions of CuO-NC composite, Al/CuO-NC composite, and NC follow the same kinetic mechanism of Avrami-Erofeev equation. In the cases of CuO and Al/CuO, they could promote the O-NO$_2$ bond cleavage and secondary autocatalytic reaction in condensed phase. The effects of these catalysts have some difference in modifying the thermolysis process of NC due to the microstructures of CuO and the addition of Al nanopowders. Furthermore, the presence of Al/CuO(f) can make the Al/CuO(f)-NC composite easier to ignite, whereas the composites have strong resistance to high temperature. Compatibility and thermal safety analysis showed that the Al/CuO had good compatibility with NC and it could be used safely. This contribution suggests that CuO and Al/CuO played key roles in accelerating the thermal decomposition of NC.

Keywords: nanoenergetic material; compatibility; nonisothermal reaction kinetics; thermal safety; catalytic action

1. Introduction

Nanosize metal oxides exhibit excellent electrical, optical, magnetic, and catalytic properties, because they have high specific surface area and surface energy, and active sites. Therefore, the preparation and application of nanosize metal oxides have attracted widespread attention. Recently, nanosize CuO has been extended studies for the potential applications in ion batteries [1,2], gas sensors [3], catalysts [4,5], magnetoelectric effects [6], etc. It has become the typical representative of nanosize transition metal oxides. In the field of explosives and solid propellants [7], nanosize CuO, as an important catalyst, has been applied for promoting combustion. There are a lot of reports regarding the catalytic effects of nanometal oxides to the main components of solid propellants [8]. CuO can decrease the decomposition peak temperature and activity energy of ammonium perchlorate (AP), and increase the decomposition reaction rate and releasing heat of AP [9,10]. When adding nanosize CuO to solid propellant formulations, it can increase the burning rate and decrease the pressure index [11].

The temperature of thermite reaction between CuO and Al can reach 2840 K, and its volume energy density is approximately three times higher than TNT [12]. The theoretical combustion heat of the Al/CuO thermite can reach to 3324.45 $kJ\cdot kg^{-1}$, which is higher than that of Al/ZnO (3256.33 $kJ\cdot kg^{-1}$), Al/CdO (2045.85 $kJ\cdot kg^{-1}$), and Al/Bi$_2$O$_3$ (1511.26 $kJ\cdot kg^{-1}$). When the Al/CuO nanothermites are used as a combustion catalyst, it can effectively improve the combustion performance of solid propellants due to the excellent characteristics of high energy density, high burning rate, high temperature of reaction product, and no need for oxygen during the combustion of Al/CuO nanothermites [13–15].

As a main energetic component, nitrocellulose (NC) with a nitrogen amount higher than 12% is widely used for gun and rocket propellants [16–19]. However, the applications of NC are extremely limited, due to the low burning temperature, the high impact sensitivity, the high friability, and the low density [20,21]. Additionally, the exothermic degradation of NC exists the potential hazard during the preparation, storage, and use [22,23]. It has been extensively studied experimentally and theoretically for revealing the pyrolysis mechanism and improving the energetic characteristics [24]. As an important research means, thermal analysis technology, such as differential scanning calorimetry (DSC) and thermogravimetric (TG), play an important role in obtaining the thermal decomposition performance of energetic materials. Additionally, the kinetic analysis that can obtain the kinetic parameters (pre-exponential factor, activation energy, and reaction model) by using DSC or TG data are also very useful for understanding the thermal decomposition reaction mechanism [25,26]. Accordingly, the non-isothermal kinetics approach of energetic materials based on the thermal analysis technique is employed to obtain the kinetic parameters of the thermal decomposition.

It is extremely importance that all the materials in the system are compatible due to the particularity of the energetic material itself. This means that they do not interact chemically and physically with each other in the energetic materials system. Poor compatibility might give rise to safety hazards in handing and deteriorated performance. Therefore, if we want to apply a novel substance or material to explosives, propellants, and pyrotechnics system, the first issue that must be considered is the compatibility of the new material with each other components. Various additives have been mixed with NC for either enhancing its stability or improving its pyrolysis properties. Current research shows that the additive, such as nanometal oxide, can accelerated the rupture of the O-NO$_2$ bonds [27] of NC and generated the NO$_2$ gas. A large amount of NO$_2$ can be adsorbed, which further enhances the secondary autocatalytic reaction of NC, due to the high specific surface area of Cr$_2$O$_3$ nanoparticles [28]. The DSC method is one of the most commonly used methods, not only for evaluating the chemical compatibility between components in the mixtures system at high temperatures, but also for investigating the thermal safety characteristic and the thermal decomposition behavior of the NC with catalyst [29–35].

In this study, the Bamboo leaf-like CuO(b) and flaky-shaped CuO(f) were prepared by the hydrothermal method. The differential scanning calorimetry (DSC) was used to evaluate the compatibility between CuO, Al/CuO, and NC. The thermal behavior and nonisothermal decomposition kinetics of the different CuO and Al/CuO to NC was investigated and the thermal decomposition kinetic mechanism function was explored. The DSC method evaluated the thermal safety characteristic of the NC composite system with CuO or Al/CuO as catalyst, which has the advantages of cheap, small quantity of sample required, and the capability of quickly select samples with better thermal decomposition performance.

2. Experimental

2.1. Materials

CuCl$_2\cdot$2H$_2$O, NaOH and anhydrous ethanol were purchased from Xilong Chemical Reagent Co. (Guangzhou, China), and they were all of analytical grade. Al nanopowders (chemical grade), approximately 50 nm in average diameter, was purchased from Jiaozuo Banlv Nano Material

Engineering Co. Ltd. (Jiaozuo, China). NC (12.6% N) was obtained from Xi'an Modern Chemistry Research Institute (Xi'an, China). All of the chemicals were used without further purification. The deionized water was used in entire experiment course.

2.2. Preparation of Bamboo Leaf-like CuO(b) and Al/CuO(b)

0.47 g of $CuCl_2 \cdot 2H_2O$ was dissolved in 20 mL deionized water, and then 10 mL NaOH solution ($1\ mol \cdot L^{-1}$) were added dropwise. After stirring for 30 min, the mixture was poured into a hydrothermal reaction vessel and reacted at 120 °C for 8 h. After the reaction is completed and natural cooling afterwards, the precipitate was separated and washed with anhydrous ethanol and deionized water several times. Subsequently, the precipitate was dried in an oven at 60 °C. Finally, the bamboo leaf-like CuO(b) was obtained.

After the CuO(b) were mixed up with Al nanopowders according to the mole rate of 1.33:1 for Al:CuO [36,37], the Al/CuO(b) nanothermite was prepared by the ultrasonic dispersion method.

2.3. Preparation of Flaky-shaped CuO(f) and Al/CuO(f)

0.47 g of $CuCl_2 \cdot 2H_2O$ was dissolved in 20 mL deionized water, and then 10 mL NaOH solution ($1\ mol \cdot L^{-1}$) was added dropwise. After stirring for 30 min, the mixture was poured into a hydrothermal reaction vessel and reacted at 180 °C for 16 h. After the reaction is completed and natural cooling afterwards, the precipitate was separated and then washed with anhydrous ethanol and deionized water for several times. Subsequently, the precipitate was dried in an oven at 60 °C. Finally, the flaky-shaped CuO(f) was obtained.

The Al/CuO(f) nanothermite was prepared by the ultrasonic dispersion method after the CuO(f) were mixed up with Al nanopowders according to the mole rate of 1.33:1 for Al:CuO [36,37].

2.4. Samples Characterization

An FEI Quanta 400 (FEI Co., Hillsboro, OR, USA) field-emission environment scanning electron microscope (SEM) characterized the morphology and structure of the sample. The acceleration voltage was 30 kV and current was 4 A. OXFORD INCAIE350 energy-dispersive X-ray spectroscopy (EDS) from OXFORD Instruments INC (Oxford, UK) was used to roughly examine the composition of the sample. The discharge voltage was 4–10 kV and the distance between the electrodes was exactly 1 mm. The morphology and size of the sample were investigated via transmission electron microscopy (TEM) and high-resolution TEM using HITACHI H-7650B (Tokyo, Japan). The operating voltage was 80 kV and resolution was 0.2 nm (lattice image). The purity and phase structure of the sample were confirmed by powder X-ray diffraction (XRD) analysis on a Rigaku D/MAX-3C (Tokyo, Japan) X-ray powder Diffractometer. The radiation source was Cu K_α (λ = 1.5418 Å) at 40 kV and 40 mA, and the range of 2θ was 10 ° to 80 °.

2.5. Measurement of Thermal Decomposition Properties and Compatibility

The thermal behavior and kinetic analysis of the samples were performed while using differential scanning calorimetry (DSC). The effects of CuO and Al/CuO on the thermal decomposition properties of nitrocellulose (NC) were investigated using DSC (Q2000, TA Co., New Castle, DE, USA) under an N_2 atmosphere at a flow rate of $50\ mL \cdot min^{-1}$. The test temperature range was room temperature to 300 °C, and the heating rate was 5.0, 10.0, 15.0, 20.0, 25.0, and $30.0\ °C \cdot min^{-1}$. The mass of each sample was about 0.22 mg and the sample cell is an aluminum crucible. The CuO-NC and Al/CuO-NC composites were obtained by physical mixing in an agate mortar at room temperature. The mass ratio of CuO: NC for the CuO-NC composites, and Al/CuO: NC for the Al/CuO-NC composites was 1:19.

The DSC thermal analysis method was used to assess the compatibility of different CuO and Al/CuO nanothermites with NC. The results come from the above-mentioned DSC test at the heating rate of $10.0\ °C \cdot min^{-1}$. The evaluation standard for compatibility of ingredients can refer to [29,30].

3. Results and Discussion

3.1. Morphology and Structure

Figure 1 shows the SEM images of CuO prepared at different reaction temperatures. When the reaction temperature is 80 °C, the obtained CuO is spindle shaped in the whole, as shown in Figure 1a. That is thick in the middle and thin in both ends. There are many stripe protrusions in the middle spindle shaped structure, and the ends of spindle shaped structure are curly. When the reaction temperature is 120 °C, the obtained CuO is bamboo leaf-shaped (Figure 1b). There are a few protrusions on the surface, and the leaves are thin. When the reaction temperature is 160 °C, the obtained CuO is in the boat form (Figure 1c). Its size becomes smaller than the structure that is shown in Figure 1a,b. When the reaction temperature is 160 °C, the size of obtained CuO is the smallest, and there are raised fragments on the surface (Figure 1d). These experimental results show that the size of the obtained CuO gradually decreases with the temperature increases and its morphology also obviously changed.

Figure 1. Scanning electron microscope (SEM) images of the as-prepared CuO with different reaction temperatures. (**a**) 80 °C; (**b**) 120 °C; (**c**) 160 °C; and, (**d**) 180 °C.

Figure 2 shows the SEM images of CuO prepared at 120 °C with different reaction times. When the reaction time is 2 h, the structure of obtained CuO is similar to the "shrimp head" with multiple curled "shrimp beard" at the end, as shown in Figure 2a. When the reaction time is 4 h, most of the obtained CuO structures are bamboo leaf-shaped (Figure 2b). There are a lot of protrusions on the surface, and the leaves are thin. When the reaction time is 8 h, the obtained CuO is also bamboo leaf-shaped (Figure 2c). There are a few protrusions at the middle region of surface, and the two ends are flaky-shaped. As the reaction time extend to 12 h (Figure 2d) or 16 h (Figure 2e), the morphology of the obtained CuO is found not significantly different from the CuO that is shown in Figure 2c. However, the agglomeration degree of CuO is obviously increased. This shows that the morphology of the obtained CuO is changed obviously with the extension of the reaction time. As the reaction time exceeds 8 h, the morphology is changed very little and it has significant aggregate.

Figure 2. SEM images of the as-prepared CuO at 120 °C with different reaction times. (**a**) 2 h; (**b**) 4 h; (**c**) 8 h; (**d**) 12 h; and, (**e**) 16 h.

TEM and SEM analysis were performed to probe the elemental composition and morphology of the bamboo leaf-like CuO(b) obtained by reacting at 120 °C for 8 h and corresponding Al/CuO nanothermites, the results are exhibited in Figure 3. It can be seen from the TEM image of bamboo leaf-like CuO(b) (Figure 3a) that the maximum leaf of CuO(b) is about 1.6 μm in length, 300 nm in width, and 57 nm in thickness. The leaf surface is not smooth. There are many folds and a few leaves have protrusions in the middle. The protrusion is about 25 nm wide and about 90 nm long. Figure 3b is a high-resolution transmission electron microscope (HRTEM) image of the CuO and its corresponding fast Fourier transform (FFT) pattern. It is found that the lattice fringes with a lattice distance of 0.27 nm appeared at this position, corresponding to the (110) plane. Figure 3c is the selected area electron diffraction (SAED) image of the CuO sample. From Figure 3c, one can see that there are some clear scattered points, which indicate that the CuO leaves are single crystal structure and well-crystallized. Figure 3d is the SEM image of Al/CuO nanothermites. The spherical Al nanopowders and bamboo leaf-like CuO stick to each other, as shown in Figure 3d. Additionally, there are some agglomerate for the Al nanopowders because of their small size.

We present their SEM images and energy-dispersive X-ray spectroscopy (EDS) patterns to analyze the morphology and composition of the above flaky-shaped CuO(f) and its corresponding Al/CuO(f) nanothermites samples, as shown in Figure 4a–d. It can be seen from Figure 4a that the shape of as-prepared CuO is flaky-shaped, the surface is not smooth, and it is covered with depressions and protrusions. Figure 4c is the SEM image of Al/CuO(f) nanothermites sample obtained by ultrasonic dispersion method with Al nanopowders and flaky-shaped CuO(f). Some spherical aluminum nanoparticles with smaller particle sizes are coated on the surface of some CuO fragments, as shown in Figure 4c. The self-agglomeration phenomenon of the Al nanopowders is weak due to the ultrasonic effect. The EDS and XRD results of flaky-shaped CuO(f) (Figures 4b and 5d) show that the sample obtained by the hydrothermal method is the pure CuO phase. Figure 4d is the EDS pattern of the Al/CuO(f) nanothermites. In combination with its XRD characterization results (Figure 5c), it can be obtained that the Al/CuO(f) sample is a mixture of Al and CuO. Because there is no other characteristic diffraction peaks, except the Al and CuO. Figure 4e is a TEM image for the flaky-shaped CuO(f). It can be seen that the fragments of flaky-shaped CuO(f) are irregular in shape and have wrinkles on the surface. The thickness is about 90 nm and the dispersion is good. From the enlarged view of the flaky-shaped

CuO(f) (Figure 4f), we can see that the folds on the fragments are clear. The largest fragment in size is approximately 0.75 μm wide and about 2.5 μm long, and the protrusions on the surface are about 0.13 μm wide. Figure 4g is a HRTEM image of flaky-shaped CuO(f) and its corresponding FFT pattern (insert). It is found that the lattice fringes with a lattice distance of 0.27 nm appeared at this position, corresponding to the (110) plane. Figure 4h is a SAED pattern of the flaky-shaped CuO(f) sample. The clear scattered points presented in Figure 4h indicate that the flaky-shaped CuO(f) is a single crystal structure with well-crystallized.

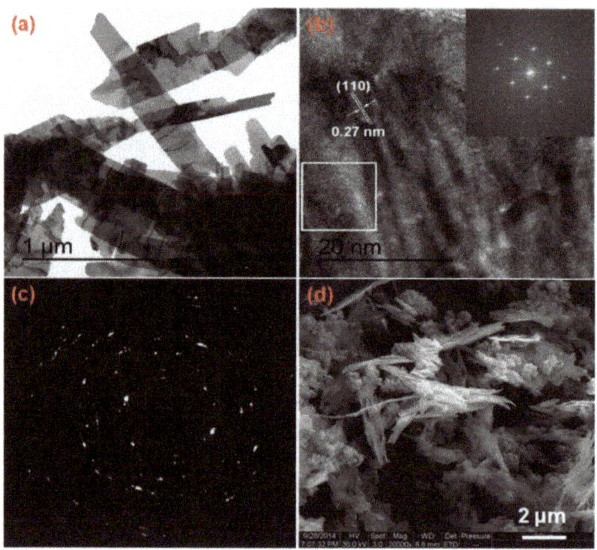

Figure 3. (**a**) Transmission electron microscopy (TEM) image of bamboo leaf-like CuO(b); (**b**) High-resolution transmission electron microscope (HRTEM) image of bamboo leaf-like CuO(b) and the corresponding fast Fourier transform(FFT) pattern (insert); (**c**) Selected area electron diffraction (SAED) pattern of bamboo leaf-like CuO(b); and, (**d**) SEM image of Al/CuO(b) nanothermites.

The XRD was used to analyze the phase microstructure, and the results are shown in Figure 5. From the diffraction pattern, the characteristic diffraction peaks of the bamboo leaf-like CuO(b) and flaky-shaped CuO(f) observed at the 2θ values of 32.95, 35.84, 39.05, 49.05, 53.79, 58.71, 61.76, 66.33, 68.19, 72.77, and 75.48 ° can be assigned as the (110), (−111), (111), (−202), (020), (202), (−311), (−113), (310), (311), and (−222) planes of CuO (JCPDS No. 65-2309). Additionally, the bamboo leaf-like CuO(b) and flaky-shaped CuO(f) are attributed to the monoclinic system, space group C2/c(15) with a = 4.662 Å, b = 3.416 Å, c = 13.7495 Å, and $\alpha = \beta = \gamma$ = 90 °. The XRD pattern also reveals that there are no unknown crystalline phase and impurities in the bamboo leaf-like CuO(b) and flaky-shaped CuO(f) samples. After the bamboo leaf-like CuO(b) and flaky-shaped CuO(f) were mixed with Al nanopowders, it also presented the characteristic diffraction peaks of Al in the Al/CuO(b) and Al/CuO(f) nanothermites. The characteristic diffraction peaks correspond to the (111), (200), (220), and (311) planes of face-centered cubic structure Al (JCPDS No. 65-2869), as shown in Figure 5a,c. Additionally, there are also no other unknown crystalline phase and impurities in Al/CuO(b) and Al/CuO(f) nanothermites. This means that there is no chemical reaction between Al nanopowders and CuO.

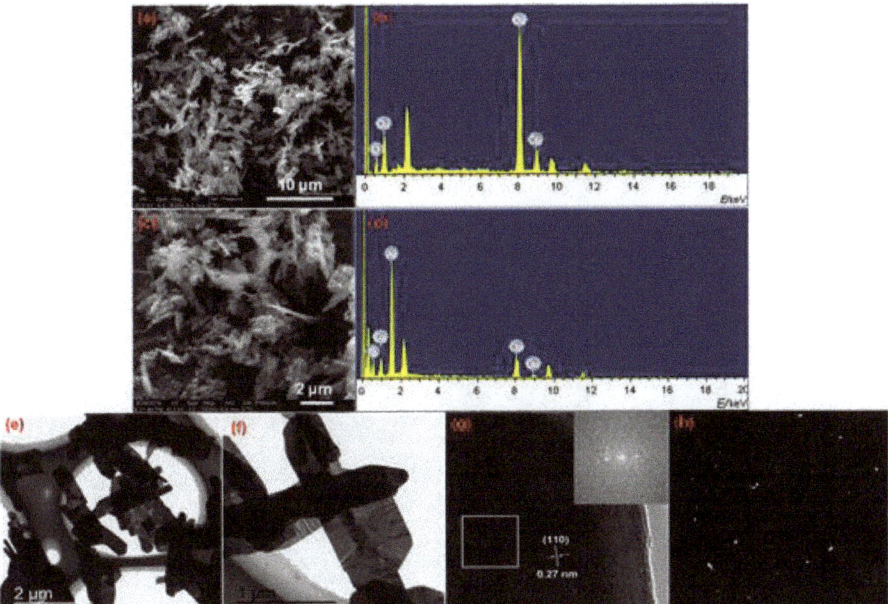

Figure 4. (**a**) SEM image of flaky-shaped CuO(f); (**b**) Energy-dispersive X-ray spectroscopy (EDS) pattern of flaky-shaped CuO(f); (**c**) SEM image of Al/CuO(f) nanothermites; (**d**) EDS pattern of Al/CuO(f) nanothermites; (**e,f**) are TEM images of flaky-shaped CuO(f); (**g**) HRTEM image of flaky-shaped CuO(f) and the corresponding FFT pattern (insert); and, (**h**) SAED pattern of flaky-shaped CuO(f).

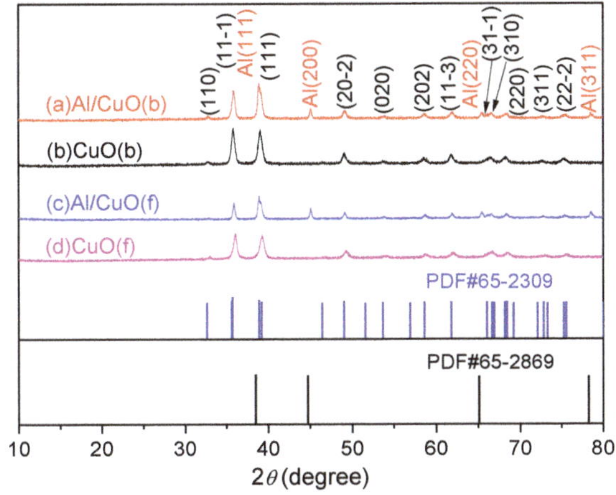

Figure 5. X-ray diffraction (XRD) patterns of different CuO and Al/CuO nanothermites. (**a**) Al/CuO(b); (**b**) CuO(b); (**c**) Al/CuO(f); and, (**d**) CuO(f).

3.2. Effect of CuO and Al/CuO on Thermal Decomposition of NC

The composite materials of CuO(b)-NC, CuO(f)-NC, Al/CuO(b)-NC, and Al/CuO(f)-NC were prepared by mixing NC with CuO or Al/CuO to analyze the effect of bamboo leaf-like CuO(b), flaky-shaped CuO(f) and their corresponding Al/CuO(b), Al/CuO(f) nanothermites on the thermal decomposition properties of NC. Figure 6 shows their SEM images. The CuO(b)-NC composite has a long rod shape and a rough surface. Around the rod shape structures, there are some CuO(b) particles. As can be seen from the magnified SEM image of CuO(b)-NC (Figure 6b), the surface of the NC is rough, and the bamboo leaf-shaped CuO(b) is adherent to the surface. The Al/CuO(b)-NC composite showed to be rod-shaped, and the bulk Al/CuO(b) grains are dispersed on the region near the NC short fibers, as shown in Figure 6c. It can be seen from Figure 6d that the Al/CuO(b) is adherent to the surface of the NC fibers, and there is some agglomeration on the spherical Al nanopowders and CuO fragments. The shape of CuO(f)-NC composite is rod-shaped or block-shaped with different length and size, as shown in Figure 6e. The magnified image of CuO(f)-NC (Figure 6f) shows that the surface of the NC is rough and a large number of CuO fragments adherent on the surface. The SEM images of Al/CuO(f)-NC is similar to the Al/CuO(b)-NC, and the structure is a short rod shape and the surface is rough, as shown in Figure 6g. There are also some agglomerations on the spherical Al nanopowders and CuO fragments (see Figure 6h).

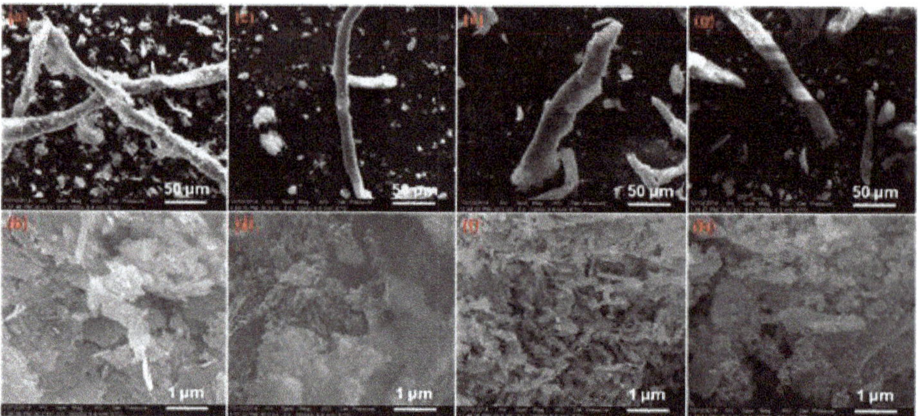

Figure 6. SEM images of different CuO-nitrocellulose (NC) and Al/CuO-NC composites. (**a,b**) CuO(b)-NC; (**c,d**) Al/CuO(b)-NC; (**e,f**) CuO(f)-NC; and, (**g,h**) Al/CuO(f)-NC.

3.2.1. Compatibility Analysis

Figure 7 is the DSC experimental result of NC, CuO(b)-NC, Al/CuO(b)-NC, CuO(f)-NC, and Al/CuO(f)-NC at a heating rate of 10 °C·min^{-1}. There is only one exothermic peak in the thermal decomposition process of the five materials, and their peak temperatures are 209.7 °C, 209.2 °C, 209.5 °C, 209.2 °C, and 209.3 °C. When compared with the NC, the thermal decomposition peak temperatures of CuO(b)-NC, CuO(f)-NC, Al/CuO(b)-NC, and Al/CuO(f)-NC are lower than that of NC at 0.5 °C, 0.1 °C, 0.5 °C, and 0.4 °C, respectively. These results indicate that there is no reaction at a low temperature between NC and other Al and CuO reactants, and the compatibility of the bamboo leaf-like CuO(b), flaky-shaped CuO(f), Al/CuO(b), Al/CuO(f) with NC is good. Therefore, the composites can be used as a component in the preparation of propellants and explosives.

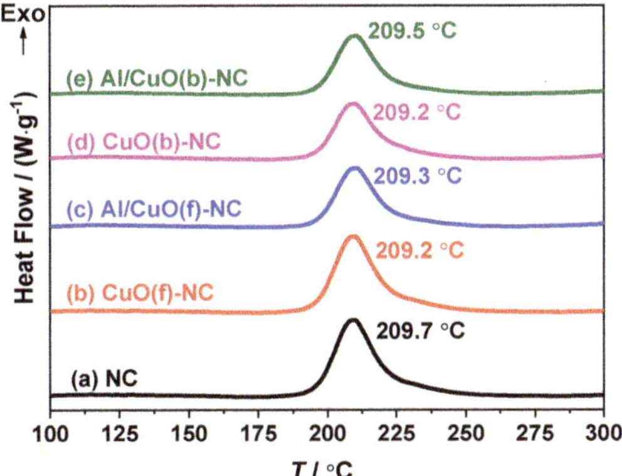

Figure 7. Differential scanning calorimetry (DSC) curves of NC, CuO(b)-NC, Al/CuO(b)-NC, CuO(f)-NC, and Al/CuO(f)-NC obtained at a heating rate of 10 °C·min^{-1}.

To obtain a better understanding of the effect of the thermite reaction between Al nanopowders and CuO to the thermal decomposition of NC, the thermal reaction characteristics of Al/CuO(b) and Al/CuO(f) were investigated by DSC. Figure 8 shows the DSC curves for Al/CuO(b) and Al/CuO(f) nanothermites at a heating rate of 10 °C·min^{-1}. One can see that the exothermic peak temperature of and Al/CuO(f) nanothermites are almost the same, but the releasing heat per unit mass of Al/CuO(b) (1153 J·g^1) is obviously higher than the Al/CuO(f) (681.5 J·g^1). The weak endothermic peaks were observed at about 654 °C from the DSC curves of Al/CuO(b) and Al/CuO(f) nanothermites. These are the melting peak of Al. The molten aluminum continues to react with CuO. The releasing heat per unit mass of the Al/CuO(b) after the melting of aluminum is approximately 101.0 J·g^1, and the Al/CuO(f) is very little. It can also see that the temperature of main exothermic peak for Al/CuO(b) and Al/CuO(f) nanothermites is far lower than the melting temperature of aluminum. The main exothermic peak temperatures of the thermite reaction are also much higher than the thermal decomposition of CuO-NC and Al/CuO-NC, as shown in Figure 7. This can further explain there being no reaction at low temperature between NC and other Al or Al/CuO.

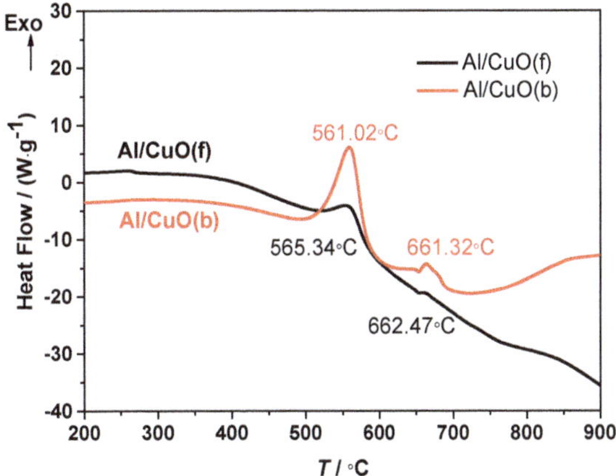

Figure 8. DSC curves of Al/CuO(f) and Al/CuO(b) at a heating rate of 10 °C·min^{-1}.

3.2.2. Non-isothermal Kinetic Analysis

In order to explore the reaction mechanism of the intense exothermic decomposition process of NC, CuO(b)-NC, CuO(f)-NC, Al/CuO(b)-NC, and Al/CuO(f)-NC, the thermal decomposition reaction kinetics was investigated by the non-isothermal DSC method. The kinetic parameters, apparent activation energy (E_a), pre-exponential factor (A), and kinetic model $f(\alpha)$ were obtained. From the non-isothermal DSC curves at different heating rates 5.0, 10.0, 15.0, 20.0, 25.0, and 30.0 °C·min^{-1}, one can obtained the values of the extent of conversion (α) to corresponding temperature (T) by integrating the peak area of DSC curves at different heating rates. The E_α can be obtained from the slope of the liner plot of lgβ_i versus T_i in the isoconversional Flynn-Wall-Ozawa's method (Equation (1)) [38]. Subsequently, the values of E_α to α can be obtained by repeating the procedure for a set of different α.

$$\lg\beta_i = \lg\left[\frac{AE_\alpha}{RG(\alpha)}\right] - 2.315 - 0.4567\frac{E_\alpha}{RT_{\alpha,i}} \quad i = 1,\ 2,\ 3,\ 4,\ 5,\ 6 \tag{1}$$

where, β is heating rate (K·min^{-1}); A is pre-exponential factor (s^{-1}); α is the extent of conversion; E_α is activation energy (J·mol^{-1}) at α; R is universal gas constant (8.314 J·mol^{-1}·K^{-1}); $G(\alpha)$ is the integral form of the reaction model; and, T is temperature (K). Some of the reaction models used in the non-isothermal kinetic analysis are listed in the literature [39].

The isoconversional methods (i.e. Flynn-Wall-Ozawa's method) do not need to know the reaction model, and the more accurate activation energy can also be obtained by using a set of non-isothermal curves under different heating rates [40,41]. Figure 9 shows the dependences of E_α to α for NC, different CuO-NC, and Al/CuO-NC. The isoconversional results of E_α to α are important in detecting and treating the multistep kinetics. The E_α-α curves of the decomposition process for the five samples have almost the same characteristic, as can be seen from Figure 9. The activation energies of NC, different CuO-NC, and Al/CuO-NC have little changes in the range of 0.100~0.750. Additionally, the ranges can be selected to calculate the non-isothermal reaction kinetics parameters.

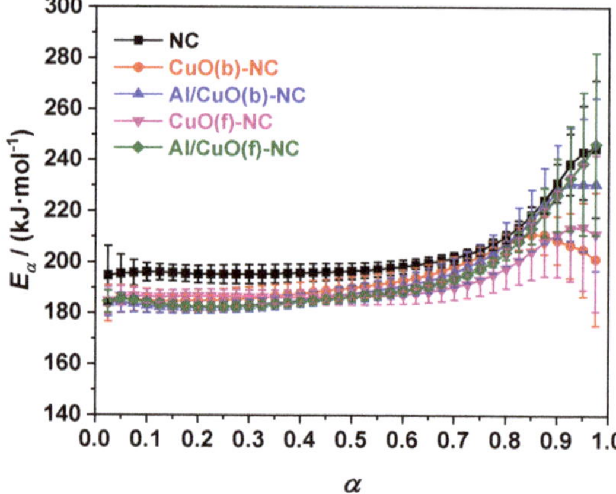

Figure 9. E_α vs α curves of NC, CuO(b)-NC, Al/CuO(b)-NC, CuO(f)-NC, and Al/CuO(f)-NC by the isoconversional Flynn-Wall-Ozawa's method.

In order to obtain the kinetic parameters, six integral methods (MacCallum-Tanner (Equation (2)), Šatava-Šesták (Equation (3)), Agrawal (Equation (4)), General integral (Equation (5)), Universal integral (Equation (6)), Ozawa (Equation (7)) and one differential method (Kissinger (Equation (8))) were employed. By substituting the α-T data and forty-one types of kinetic model functions into the integral and differential equations, the apparent activation energy (E_a), pre-exponential factor (A), and the most probable kinetic model $f(\alpha)$ were obtained by the logical choice method with the best linear correlation coefficient (r) [27]. Generally, the normal range of E_a and $\lg(A/\text{s}^{-1})$ for energetic materials is approximately 80 to 250 kJ·mol^{-1} and 7 to 30, respectively. The values of E_a, $\lg(A/\text{s}^{-1})$, and corresponding linear correlation coefficient (r) that was obtained by different methods at different heating rates are listed in Tables 1 and 2. The values of E_a and $\lg(A/\text{s}^{-1})$ obtained from each single non-isothermal DSC curve is in good agreement with the calculated values that were obtained by Kissinger's method and Ozawa's method. Therefore, one can conclude that the reaction mechanism of the intense exothermic decomposition process of NC, CuO(b)-NC, CuO(f)-NC, Al/CuO(b)-NC, and Al/CuO(f)-NC is classified as Avrami-Erofeev equation: $f(\alpha) = 3/2(1-\alpha)[-\ln(1-\alpha)]^{1/3}$ (differential form) and $G(\alpha) = [-\ln(1-\alpha)]^{2/3}$ (integral form).

$$\lg G(\alpha) = \lg\left(\frac{AE_a}{\beta R}\right) - 0.4828 E_a^{\,0.4357} - \frac{0.449 + 0.217 E_a}{0.001}\frac{1}{T} \tag{2}$$

$$\lg G(\alpha) = \lg\left(\frac{AE_a}{\beta R}\right) - 2.315 - 0.4567\frac{E_a}{RT} \tag{3}$$

$$\ln\left[\frac{G(\alpha)}{T^2}\right] = \ln\left[\frac{AR}{\beta E_a}\frac{1-2\left(\frac{RT}{E_a}\right)}{1-5\left(\frac{RT}{E_a}\right)^2}\right] - \frac{E_a}{RT} \tag{4}$$

$$\ln\left[\frac{G(\alpha)}{T^2}\right] = \ln\left[\frac{AR}{\beta E_a}\left(1 - \frac{2RT}{E_a}\right)\right] - \frac{E_a}{RT} \tag{5}$$

$$\ln\left[\frac{G(\alpha)}{T-T_0}\right] = \ln\left(\frac{A}{\beta}\right) - \frac{E_a}{RT} \tag{6}$$

$$\lg\beta = \lg\left[\frac{AE_{eO\ (\text{or pO})}}{RG(\alpha)}\right] - 2.315 - 0.4567\frac{E_{eO\ (\text{or pO})}}{RT_{e\ (\text{or p})}} \tag{7}$$

$$\ln\left(\frac{\beta}{T_p^2}\right) = \ln\left(\frac{AR}{E_K}\right) - \frac{E_K}{RT_p} \tag{8}$$

where, β is heating rate (K·min^{-1}); A is pre-exponential factor (s^{-1}); E_a is apparent activation energy (J·mol^{-1}); R is universal gas constant (8.314 J·mol^{-1}·K^{-1}); $G(\alpha)$ is the integral form of the reaction model; α is the extent of conversion; T is temperature (K); T_0 is the initial temperature (K) at which DSC curve deviates from the baseline of the non-isothermal DSC curve; T_e is the onset temperature (K); T_p is the peak temperature (K); E_{eO} (J·mol^{-1}); and, E_{pO} (J·mol^{-1}) are the apparent activation energy obtained from T_e and T_p by Ozawa's method, respectively; E_K is the apparent activation energy (J·mol^{-1}) obtained from T_p by Kissinger's method.

Table 1. Calculated values of kinetic parameters of decomposition reaction for NC, bamboo leaf-like CuO(b)-NC, and Al/CuO(b)-NC.

Method	$\beta/(^\circ C \cdot min^{-1})$	NC			CuO(b)-NC			Al/CuO(b)-NC		
		$E_a/(kJ \cdot mol^{-1})$	$lg(A/s^{-1})$	r	$E_a/(kJ \cdot mol^{-1})$	$lg(A/s^{-1})$	r	$E_a/(kJ \cdot mol^{-1})$	$lg(A/s^{-1})$	r
MacCallum-Tanner	5.0	208.0 ± 3.5	20.6 ± 0.4	0.9983	180.3 ± 5.3	17.5 ± 0.5	0.9927	176.2 ± 5.1	17.1 ± 0.4	0.9917
	10.0	205.4 ± 4.2	20.3 ± 0.4	0.9984	183.3 ± 5.8	17.9 ± 0.2	0.9935	184.3 ± 4.7	18.0 ± 0.3	0.9920
	15.0	209.3 ± 3.4	20.7 ± 0.4	0.9988	187.3 ± 5.7	18.3 ± 0.4	0.9930	184.2 ± 5.8	18.0 ± 0.4	0.9925
	20.0	209.3 ± 2.7	20.7 ± 0.6	0.9987	182.6 ± 4.8	17.8 ± 0.6	0.9958	184.6 ± 5.7	18.0 ± 0.2	0.9927
	25.0	211.8 ± 5.1	20.9 ± 0.4	0.9990	192.6 ± 4.8	18.9 ± 0.4	0.9934	181.7 ± 4.4	17.7 ± 0.4	0.9943
	30.0	210.1 ± 4.5	20.8 ± 0.4	0.9982	187.7 ± 4.1	18.4 ± 0.3	0.9962	182.3 ± 4.3	17.8 ± 0.3	0.9959
Šatava-Šesták	5.0	204.6 ± 3.9	20.2 ± 0.4	0.9983	178.4 ± 3.2	17.3 ± 0.6	0.9927	174.6 ± 3.7	16.9 ± 0.4	0.9917
	10.0	202.1 ± 5.2	19.9 ± 0.6	0.9984	181.2 ± 4.0	17.7 ± 0.5	0.9935	182.2 ± 6.0	17.8 ± 0.3	0.9920
	15.0	205.8 ± 5.4	20.4 ± 0.3	0.9988	185.0 ± 5.1	18.1 ± 0.3	0.9930	182.1 ± 6.7	17.8 ± 0.2	0.9925
	20.0	205.8 ± 4.2	20.3 ± 0.3	0.9987	180.6 ± 4.7	17.6 ± 0.5	0.9958	182.5 ± 6.3	17.8 ± 0.4	0.9927
	25.0	208.1 ± 3.6	20.6 ± 0.5	0.9990	190.0 ± 5.6	18.7 ± 0.3	0.9934	179.8 ± 5.3	17.5 ± 0.2	0.9943
	30.0	206.6 ± 5.6	20.4 ± 0.4	0.9982	185.4 ± 5.0	18.1 ± 0.3	0.9962	180.3 ± 4.9	17.6 ± 0.3	0.9959
Agrawal	5.0	207.2 ± 5.6	20.5 ± 0.5	0.9982	179.7 ± 4.0	17.5 ± 0.3	0.9920	175.7 ± 4.5	17.1 ± 0.3	0.9909
	10.0	204.5 ± 5.3	20.2 ± 0.6	0.9983	182.6 ± 4.8	17.8 ± 0.5	0.9929	183.6 ± 4.0	17.9 ± 0.3	0.9913
	15.0	208.4 ± 4.8	20.6 ± 0.4	0.9987	186.5 ± 5.8	18.3 ± 0.5	0.9924	183.4 ± 3.4	17.9 ± 0.2	0.9919
	20.0	208.3 ± 4.0	20.6 ± 0.6	0.9986	181.9 ± 3.5	17.8 ± 0.3	0.9954	183.8 ± 4.6	18.0 ± 0.3	0.9920
	25.0	210.7 ± 5.2	20.8 ± 0.6	0.9990	191.7 ± 4.4	18.8 ± 0.5	0.9928	180.9 ± 3.6	17.6 ± 0.2	0.9938
	30.0	209.0 ± 4.7	20.7 ± 0.6	0.9981	186.8 ± 6.7	18.3 ± 0.5	0.9958	181.5 ± 4.8	17.7 ± 0.4	0.9955
General integral	5.0	205.8 ± 6.2	19.0 ± 0.4	0.9985	178.3 ± 6.7	16.0 ± 0.6	0.9919	174.3 ± 5.3	15.6 ± 0.2	0.9908
	10.0	203.3 ± 6.08	18.7 ± 0.3	0.9983	181.3 ± 6.9	16.4 ± 0.3	0.9928	182.3 ± 5.9	16.5 ± 0.3	0.9912
	15.0	207.2 ± 4.9	19.1 ± 0.5	0.9987	185.3 ± 6.8	16.8 ± 0.5	0.9923	182.2 ± 6.1	16.5 ± 0.3	0.9918
	20.0	207.2 ± 5.3	19.1 ± 0.4	0.9986	180.7 ± 3.8	16.3 ± 0.5	0.9954	182.7 ± 5.7	16.5 ± 0.4	0.9920
	25.0	209.6 ± 6.2	19.3 ± 0.6	0.9989	190.6 ± 4.4	17.4 ± 0.3	0.9928	179.8 ± 4.3	16.2 ± 0.3	0.9937
	30.0	208.0 ± 4.7	19.2 ± 0.5	0.9981	185.8 ± 3.5	16.9 ± 0.4	0.9958	180.4 ± 5.9	16.3 ± 0.4	0.9955
Universal integral	5.0	207.2 ± 4.3	20.5 ± 0.6	0.9982	179.7 ± 6.9	17.5 ± 0.4	0.9920	175.7 ± 3.9	17.1 ± 0.2	0.9909
	10.0	204.5 ± 5.3	20.2 ± 0.3	0.9983	182.6 ± 6.5	17.8 ± 0.5	0.9929	183.6 ± 4.8	17.9 ± 0.2	0.9913
	15.0	208.4 ± 4.6	20.6 ± 0.5	0.9987	186.5 ± 5.1	18.3 ± 0.3	0.9924	183.4 ± 5.7	17.9 ± 0.4	0.9919
	20.0	208.3 ± 3.9	20.6 ± 0.4	0.9986	181.9 ± 6.1	17.8 ± 0.3	0.9954	183.8 ± 3.5	18.0 ± 0.4	0.9920
	25.0	210.7 ± 5.4	20.8 ± 0.6	0.9990	191.7 ± 6.1	18.8 ± 0.4	0.9928	180.9 ± 5.5	17.7 ± 0.2	0.9938
	30.0	209.0 ± 6.1	20.7 ± 0.4	0.9981	186.8 ± 5.6	18.3 ± 0.5	0.9958	181.5 ± 4.4	17.7 ± 0.4	0.9955
Mean		207.5 ± 4.8	20.2 ± 0.5		184.5 ± 5.2	17.8 ± 0.4		181.1 ± 5.0	17.4 ± 0.3	
Flynn-Wall-Ozawa		185.7 ± 1.9 (E_{eO}) 197.6 ± 6.4 (E_{pO})		0.9998 0.9979	178.7 ± 3.6 (E_{eO}) 187.0 ± 4.8 (E_{pO})		0.9919 0.9987	176.3 ± 5.3 (E_{eO}) 183.6 ± 2.6 (E_{pO})		0.9982 0.9996
Kissinger		199.7 ± 6.8 (E_K)	19.8 ± 0.7	0.9977	188.6 ± 5.0 (E_K)	18.6 ± 0.6	0.9986	185.0 ± 2.7 (E_K)	18.2 ± 0.3	0.9996
Mean(E_{eO}, E_{pO}, E_K)		194.3 ± 5.0			184.8 ± 4.5			181.6 ± 3.5		

Table 2. Calculated values of kinetic parameters of decomposition reaction for flaky-shaped CuO(f)-NC and Al/CuO(f)-NC.

Method	$\beta/(°C·min^{-1})$	CuO(f)-NC			Al/CuO(f)-NC		
		$E_a/(kJ·mol^{-1})$	$lg(A/s^{-1})$	r	$E_a/(kJ·mol^{-1})$	$lg(A/s^{-1})$	r
MacCallum-Tanner	5.0	190.4 ± 6.6	18.6 ± 0.4	0.9917	177.6 ± 5.7	17.2 ± 0.6	0.9889
	10.0	197.2 ± 4.0	19.3 ± 0.4	0.9921	177.7 ± 3.5	17.2 ± 0.6	0.9920
	15.0	192.0 ± 5.8	18.8 ± 0.5	0.9928	176.3 ± 4.7	17.1 ± 0.5	0.9936
	20.0	202.3 ± 6.8	19.9 ± 0.6	0.9932	187.8 ± 4.9	18.4 ± 0.4	0.9952
	25.0	189.1 ± 6.1	18.4 ± 0.5	0.9925	179.9 ± 3.9	17.5 ± 0.4	0.9930
	30.0	198.0 ± 3.7	19.4 ± 0.6	0.9941	190.6 ± 5.5	18.7 ± 0.4	0.9942
Šatava-Šesták	5.0	187.9 ± 6.7	18.4 ± 0.5	0.9917	175.9 ± 3.3	17.1 ± 0.5	0.9889
	10.0	194.4 ± 5.7	19.1 ± 0.4	0.9921	176.0 ± 4.5	17.1 ± 0.2	0.9920
	15.0	189.5 ± 5.3	18.5 ± 0.5	0.9928	174.7 ± 3.3	17.0 ± 0.4	0.9936
	20.0	199.2 ± 4.1	19.6 ± 0.5	0.9932	185.5 ± 6.1	18.2 ± 0.5	0.9952
	25.0	186.7 ± 6.7	18.2 ± 0.5	0.9925	178.0 ± 6.1	17.3 ± 0.5	0.9930
	30.0	195.1 ± 5.1	19.1 ± 0.5	0.9941	188.1 ± 3.8	18.4 ± 0.3	0.9942
Agrawal	5.0	189.7 ± 4.7	18.6 ± 0.5	0.9910	177.1 ± 6.6	17.2 ± 0.3	0.9879
	10.0	196.4 ± 3.7	19.3 ± 0.6	0.9915	177.1 ± 5.4	17.2 ± 0.5	0.9913
	15.0	191.2 ± 6.4	18.7 ± 0.5	0.9921	175.6 ± 5.3	17.1 ± 0.3	0.9929
	20.0	201.3 ± 5.5	19.8 ± 0.5	0.9926	187.0 ± 3.8	18.3 ± 0.3	0.9947
	25.0	188.2 ± 5.1	18.4 ± 0.5	0.9918	179.1 ± 3.7	17.4 ± 0.5	0.9923
	30.0	196.9 ± 3.2	19.3 ± 0.4	0.9936	189.7 ± 3.5	18.6 ± 0.3	0.9937
General integral	5.0	188.4 ± 5.3	17.1 ± 0.5	0.9909	175.6 ± 4.6	15.7 ± 0.3	0.9878
	10.0	195.1 ± 4.7	17.8 ± 0.6	0.9914	175.8 ± 4.2	15.8 ± 0.3	0.9912
	15.0	190.1 ± 6.1	17.3 ± 0.5	0.9921	174.4 ± 5.3	15.6 ± 0.4	0.9929
	20.0	200.3 ± 6.5	18.4 ± 0.5	0.9925	185.9 ± 5.8	16.9 ± 0.5	0.9947
	25.0	187.2 ± 3.8	16.9 ± 0.4	0.9918	178.0 ± 5.0	16.0 ± 0.6	0.9922
	30.0	196.0 ± 5.5	17.9 ± 0.6	0.9936	188.6 ± 3.2	17.2 ± 0.4	0.9936
Universal integral	5.0	201.3 ± 4.1	19.8 ± 0.6	0.9926	177.1 ± 5.5	17.2 ± 0.4	0.9879
	10.0	196.4 ± 3.4	19.3 ± 0.4	0.9915	177.1 ± 5.0	17.2 ± 0.4	0.9913
	15.0	190.1 ± 4.0	17.3 ± 0.4	0.9921	175.6 ± 6.5	17.1 ± 0.4	0.9929
	20.0	191.2 ± 6.3	18.7 ± 0.5	0.9921	187.0 ± 6.8	18.3 ± 0.3	0.9947
	25.0	188.2 ± 3.5	18.4 ± 0.6	0.9918	179.1 ± 6.7	17.4 ± 0.3	0.9923
	30.0	196.9 ± 4.1	19.3 ± 0.5	0.9936	189.7 ± 3.4	18.6 ± 0.3	0.9937
Mean		193.5 ± 5.1	18.7 ± 0.5		180.6 ± 4.9	17.3 ± 0.4	
Flynn-Wall-Ozawa		$177.8 \pm 5.1\ (E_{eO})$ $183.7 \pm 8.3\ (E_{pO})$		0.9983 0.9959	$171.5 \pm 6.8\ (E_{eO})$ $185.9 \pm 3.9\ (E_{pO})$		0.9968 0.9991
Kissinger		$185.1 \pm 8.7\ (E_K)$	18.2 ± 1.0	0.9956	$187.4 \pm 4.1\ (E_K)$	18.5 ± 0.4	0.9990
Mean(E_{eO}, E_{pO}, E_K)		182.2 ± 7.4			181.6 ± 4.9		

Substituting the E_a and A values (listed in Tables 1 and 2) of NC, CuO(b)-NC, Al/CuO(b)-NC, CuO(f)-NC, Al/CuO(f)-NC, and $f(\alpha)=3/2(1-\alpha)[-\ln(1-\alpha)]^{1/3}$ into Equation (9) [42]. The kinetic equations of the intense exothermic decomposition process of NC, CuO(b)-NC, CuO(f)-NC, Al/CuO(b)-NC, and Al/CuO(f)-NC can be described as Equations (10)–(14), respectively.

$$\frac{d\alpha}{dT} = \frac{A}{\beta} f(\alpha) \exp\left(-\frac{E}{RT}\right) \tag{9}$$

$$\frac{d\alpha}{dT} = \frac{10^{20.4}}{\beta}(1-\alpha)[-\ln(1-\alpha)]^{1/3} \exp(-2.5 \times 10^4/T) \tag{10}$$

$$\frac{d\alpha}{dT} = \frac{10^{18.0}}{\beta}(1-\alpha)[-\ln(1-\alpha)]^{1/3} \exp(-2.2 \times 10^4/T) \tag{11}$$

$$\frac{d\alpha}{dT} = \frac{10^{17.6}}{\beta}(1-\alpha)[-\ln(1-\alpha)]^{1/3} \exp(-2.2 \times 10^4/T) \tag{12}$$

$$\frac{d\alpha}{dT} = \frac{10^{18.9}}{\beta}(1-\alpha)[-\ln(1-\alpha)]^{1/3}\exp(-2.3\times10^4/T) \tag{13}$$

$$\frac{d\alpha}{dT} = \frac{10^{17.5}}{\beta}(1-\alpha)[-\ln(1-\alpha)]^{1/3}\exp(-2.2\times10^4/T) \tag{14}$$

3.2.3. Thermal Safety Analysis

The simple model derived earlier in the frame of Semenov's thermal explosion theory [43] was established for estimating the thermal ignition temperature (T_{be0}) and the critical temperatures of thermal explosion (T_{bp0}) of as-prepared NC-based complexes. All of the calculations are based on the differential scanning calorimetry (DSC) experiments and the safety parameters, such as the thermal conductivity, particle size, mechanical properties, and pressure dependence of melting point et al., are not involved [44,45].

The values of T_{e0} and T_{p0} can be calculated in accordance with the Equation (15) in order to obtain the self-accelerating decomposition temperature (T_{SADT}) of as-prepared samples. Then the value of T_{SADT} can be obtained by the Equation (16). The values of T_{e0} (T_{SADT}) and T_{p0} for NC, CuO(b)-NC, CuO(f)-NC, Al/CuO(b)-NC, and Al/CuO(f)-NC are listed in Table 3.

$$T_{ei(or\ pi)} = T_{e0(or\ p0)} + a\beta_i + b\beta_i{}^2 + c\beta_i{}^3 \tag{15}$$

where T_e is the onset temperature; T_p is the peak temperature; β is the heating rate; T_{e0} and T_{p0} are the onset and peak temperatures corresponding to $\beta\rightarrow0$, respectively; and, a, b, and c are the polynomial coefficients; i= 1, 2, $\cdots$, 6.

$$T_{SADT} = T_{e0} \tag{16}$$

Table 3. Calculated values of kinetic parameters of decomposition reaction for NC, CuO(b)-NC, Al/CuO(b)-NC, CuO(f)-NC, and Al/CuO(f)-NC.

Sample	E_a/ (kJ·mol^{-1})	lg(A/s^{-1})	T_{e0}/°C	T_{p0}/°C	T_{be0}/°C	T_{bp0}/°C	$\Delta S^{\neq}$/ (J·mol^{-1}·K^{-1})	$\Delta H^{\neq}$/ (kJ·mol^{-1})	$\Delta G^{\neq}$/ (kJ·mol^{-1})
NC	207.5	20.2	181.8	197.0	191.4	206.7	138.4	199.7	134.6
CuO(b)-NC	184.5	17.8	176.4	189.3	180.2	193.5	91.7	188.6	146.2
Al/CuO(b)-NC	181.1	17.4	169.3	184.4	178.6	194.3	84.5	185.0	146.3
CuO(f)-NC	193.5	18.7	178.6	190.0	189.3	206.3	109.1	185.0	133.9
Al/CuO(f)-NC	180.6	17.3	152.9	191.8	161.3	201.9	83.2	187.4	148.7

The T_{be0} can be obtained substituting the values of E_{eO} (listed in Tables 1 and 2) and T_{e0} (see Table 3) into Equation (16). The T_{bp0} can also be obtained by substituting E_{pO} (listed in Tables 1 and 2) and T_{p0} (see Table 3) into the Equation (17). The values of T_{be0} and T_{bp0} for NC, CuO(b)-NC, CuO(f)-NC, Al/CuO(b)-NC, and Al/CuO(f)-NC are also listed in Table 3. Generally, the value of T_b (T_{be0} or T_{bp0}) is one of the most important evaluation parameters for thermal safety, which represent the degree of difficulty of the transition from thermal decomposition to thermal explosion. The higher the value of T_b, the transition can take place easier.

$$T_{be0\ or\ bp0} = \frac{E_{eO\ or\ pO} - \sqrt{E_{eO\ or\ pO}^2 - 4E_{eO\ or\ pO}RT_{e0\ or\ p0}}}{2R} \tag{17}$$

While using $T = T_{p0}$, $E_a = E_K$, and $A = A_K$, the values of activation entropy ($\Delta S^{\neq}$), activation enthalpy ($\Delta H^{\neq}$), and activation of activation free energy ($\Delta G^{\neq}$) of the main exothermic decomposition reaction of the NC, CuO(b)-NC, CuO(f)-NC, Al/CuO(b)-NC, and Al/CuO(f)-NC are obtained by Equation (18) to Equation (20), as listed in Table 3.

$$A \exp\left(-\frac{E_a}{RT}\right) = \frac{k_B T}{h} \exp\left(-\frac{\Delta G^{\neq}}{RT}\right) \tag{18}$$

$$\Delta H^{\neq} = E_a - RT \tag{19}$$

$$\Delta G^{\neq} = \Delta H^{\neq} - T\Delta S^{\neq} \tag{20}$$

where k_B is the Boltzmann constant, 1.38066×10^{-23} J·K^{-1}; h is the Planck constant, 6.626×10^{-34} J·s.

According to the previous studies [27,28,46], the thermal decomposition reaction of NC is a typical competition reaction between the O-NO$_2$ bond rupture and the decomposition of polymer skeleton products decomposition. The NO$_2$ gas is the initial decomposition product in the first step of high-temperature pyrolysis, similar to RDX. However, the NO$_2$ that is produced by RDX pyrolysis is derived from the rupture of the N-NO$_2$ bond [47]. However, the O-NO$_2$ bond rupture is deemed to be the first step, resulting in the release of NO$_2$. The NO$_2$ stagnates in the polymer skeleton and then reacts with the RO• radical or its decomposition products to produce the NO, NO$_2$, CO$_2$, CO, H$_2$O, N$_2$O, HCHO, HCOOH, etc. Finally, the secondary autocatalytic reaction is significantly strengthened.

From the above calculation results, it could be derived that the addition of CuO and Al/CuO nanothermites can reduce the apparent activation energy (E_a), pre-exponential factor (A), onset temperature (T_{e0}), thermal decomposition peak temperature (T_{p0}), critical thermal ignition temperature (T_{be0}), and the critical temperatures of thermal explosion (T_{bp0}), as compared with single-component NC. This indicates that the CuO and Al/CuO nanothermites can accelerate the thermal decomposition of NC. When compared with the CuO-NC composite system, the E_a, A, T_{e0}, T_{p0}, and T_{be0} values of the Al/CuO-NC composite system are all decreased. It indicates that the addition of Al nanopowders can increase the reactive sites and then accelerate the thermal decomposition of NC because of the high specific surface area of Al nanopowders or CuO. The physical adsorption might occur between the Al nanopowders or CuO and decomposition product of nitrocellulose. This may further effect the autocatalytic decomposition reaction of nitrocellulose.

From Tables 1 and 2, one can find that the E_a values of the thermal decomposition processes of Al/CuO(b)-NC and Al/CuO(f)-NC are 181.1 kJ·mol^{-1} and 180.6 kJ·mol^{-1}, and the A values are $10^{17.4}$ s^{-1} and $10^{17.3}$ s^{-1}, respectively. The little difference of E_a and A between the two nanothermites, indicating that the addition of Al/CuO(b) and Al/CuO(f) may reduce the thermal decomposition energy barrier of the composites and make it easy to decompose. However, the self-accelerated decomposition temperature (T_{SADT}) of Al/CuO(f)-NC is 152.9 °C, which is 16.4 °C lower than that of Al/CuO(b)-NC. Additionally, the thermal decomposition peak temperature (T_{p0}) of Al/CuO(f)-NC is 7.4 °C higher than the Al/CuO(b)-NC. This shows that the decomposition reaction of Al/CuO(f)-NC is stronger than that of Al/CuO(b)-NC, but the time-consumption of the decomposition process of Al/CuO(f)-NC is long. From the results of the critical thermal ignition temperature (T_{be0}) and the critical temperatures of thermal explosion (T_{bp0}), one can get that the T_{be0} of Al/CuO(f)-NC is 17.3 °C lower than the Al/CuO(b)-NC, but T_{pe0} of Al/CuO(f)-NC is higher than the Al/CuO(b)-NC. This indicates that the Al/CuO(f)-NC is more easily ignited, but its decomposition rate is relatively slow. Accordingly, the Al/CuO(f)-NC has a much higher thermal stability than that of the Al/CuO(b)-NC.

4. Conclusions

Two kinds of nanosize CuO with different morphologies and sizes were prepared via the hydrothermal method by adjusting the reaction temperature and time. The ultrasonic composite method was used to combine the aluminum nanopowders with the as-prepared bamboo leaf-shaped CuO(b) and flaky-shaped CuO(f) to obtain the corresponding Al/CuO(b) and Al/CuO(f) nanothermites. Based on the non-isothermal decomposition kinetics, the catalytic effect of the different morphologies of CuO and corresponding nanothermites on the thermal decomposition properties of NC was investigated. The additions of CuO(b), Al/CuO(b), CuO(f), and Al/CuO(f) do not change the thermal decomposition mechanism of NC. The catalytic effect of Al/CuO nanothermites to NC is better than the CuO, and the Al/CuO(b)-NC, Al/CuO(f)-NC is easier to decompose than the CuO(b)-NC and CuO(f)-NC. The results of compatibility and thermal safety analysis show that the CuO and Al/CuO catalysts have good compatibility with NC, and the catalysts can be used safely. However, the agglomeration of Al/CuO(f) is relatively serious. The self-accelerated decomposition temperature (T_{SADT}) and the critical thermal ignition temperature (T_{be0}) of Al/CuO(f)-NC are the lowest, while the thermal decomposition peak temperature (T_{p0}) and the critical temperatures of thermal explosion (T_{bp0}) are higher. It shows that the Al/CuO(f)-NC composite material is more easily ignited after being integrated with NC and Al/CuO(f), and the thermal stability during the thermal decomposition process is also better. The CuO(b), Al/CuO(b), CuO(f), and Al/CuO(f) as the catalysts has a wide application prospect in solid propellants.

Author Contributions: Conceptualization, E.Y., N.Z. and H.M.; methodology, E.Y. and N.Z.; validation, Z.Q., H.L., T.A. and F.Z.; formal analysis, S.X. and J.Y.; investigation, E.Y.; writing—original draft preparation, E.Y. and N.Z.; writing—review and editing, N.Z., Z.Q., H.M., H.L. and F.Z.; supervision, F.Z. and N.Z. All authors have read and agreed to the published version of the manuscript.

Funding: This research was supported by the National Natural Science Foundation of China (No. 21473130 and 21905224), Doctoral Scientific Research Foundation of Xi'an University of Technology (No. 109-451117004), Natural Science Basic Research Plan in Shaanxi Province of China (No. 2018JQ2077), and Shaanxi Provincial Department of Education (No. 19JK0595).

Acknowledgments: The authors would like to gratefully thank Luigi T. Deluca for his valuable comments and suggestions to improve the quality of this article.

Conflicts of Interest: The authors declare no conflict of interest.

References

1. Jia, S.; Wang, Y.; Liu, X.; Zhao, S.; Zhao, W.; Huang, Y.; Li, Z.; Lin, Z. Hierarchically porous CuO nano-labyrinths as binder-free anodes for long-life and high-rate lithium ion batteries. *Nano Energy* **2019**, *59*, 229–236. [CrossRef]
2. Tan, G.; Wu, F.; Yuan, Y.; Chen, R.; Zhao, T.; Yao, Y.; Qian, J.; Liu, J.; Ye, Y.; Shahbazian-Yassar, R.; et al. Freestanding three-dimensional core–shell nanoarrays for lithium-ion battery anodes. *Nat. Commun.* **2016**, *7*, 11774. [CrossRef] [PubMed]
3. Nakate, U.T.; Lee, G.H.; Ahmad, R.; Patil, P.; Hahn, Y.-B.; Yu, Y.T.; Suh, E.-k. Nano-bitter gourd like structured CuO for enhanced hydrogen gas sensor application. *Int. J. Hydrog. Energy* **2018**, *43*, 22705–22714. [CrossRef]
4. Du, X.; Zhang, Y.; Si, F.; Yao, C.; Du, M.; Hussain, I.; Kim, H.; Huang, S.; Lin, Z.; Hayat, W. Persulfate non-radical activation by nano-CuO for efficient removal of chlorinated organic compounds: Reduced graphene oxide-assisted and CuO (0 0 1) facet-dependent. *Chem. Eng. J.* **2019**, *356*, 178–189. [CrossRef]
5. Wang, L.; Hou, J.; Liu, S.; Carrier, A.J.; Guo, T.; Liang, Q.; Oakley, D.; Zhang, X. CuO nanoparticles as haloperoxidase-mimics: Chloride-accelerated heterogeneous Cu-Fenton chemistry for H_2O_2 and glucose sensing. *Sensor Actuat. B-CH.* **2019**, *287*, 180–184. [CrossRef]
6. Wang, Z.; Qureshi, N.; Yasin, S.; Mukhin, A.; Ressouche, E.; Zherlitsyn, S.; Skourski, Y.; Geshev, J.; Ivanov, V.; Gospodinov, M.; et al. Magnetoelectric effect and phase transitions in CuO in external magnetic fields. *Nat. Commun.* **2016**, *7*, 10295. [CrossRef]
7. Pandas, H.M.; Fazli, M. Preparation and application of La_2O_3 and CuO nano particles as catalysts for ammonium perchlorate thermal decomposition. *Propell. Explos. Pyrot.* **2018**, *43*, 1096–1104. [CrossRef]
8. Dolgachev, V.; Khaneft, A.; Mitrofanov, A. Ignition of organic explosive materials by a copper oxide film absorbing a laser pulse. *Propell. Explos. Pyrot.* **2018**, *43*, 992–998. [CrossRef]
9. Hu, Y.; Yang, S.; Tao, B.; Liu, X.; Lin, K.; Yang, Y.; Fan, R.; Xia, D.; Hao, D. Catalytic decomposition of ammonium perchlorate on hollow mesoporous CuO microspheres. *Vacuum* **2019**, *159*, 105–111. [CrossRef]
10. Zhu, Y.; Zhou, X.; Xu, J.; Ma, X.; Ye, Y.; Yang, G.; Zhang, K. In situ preparation of explosive embedded CuO/Al/CL20 nanoenergetic composite with enhanced reactivity. *Chem. Eng. J.* **2018**, *354*, 885–895. [CrossRef]

11. Sharma, J.K.; Srivastava, P.; Singh, G.; Akhtar, M.S.; Ameen, S. Catalytic thermal decomposition of ammonium perchlorate and combustion of composite solid propellants over green synthesized CuO nanoparticles. *Thermochim. Acta* **2015**, *614*, 110–115. [CrossRef]

12. Chowdhury, S.; Sullivan, K.; Piekiel, N.; Zhou, L.; Zachariah, M.R. Diffusive vs explosive reaction at the nanoscale. *J. Phys. Chem C* **2010**, *114*, 9191–9195. [CrossRef]

13. Glavier, L.; Nicollet, A.; Jouot, F.; Martin, B.; Barberon, J.; Renaud, L.; Rossi, C. Nanothermite/RDX-based miniature device for impact ignition of high explosives. *Propell. Explos. Pyrot.* **2018**, *42*, 308–317. [CrossRef]

14. Yin, Y.; Li, X. Al/CuO composite coatings with nanorods structure assembled by electrophoretic deposition for enhancing energy released. *Vacuum* **2019**, *163*, 216–223. [CrossRef]

15. Kim, K.J.; Cho, M.H.; Kim, J.H.; Kim, S.H. Effect of paraffin wax on combustion properties and surface protection of Al/CuO-based nanoenergetic composite pellets. *Combust. Flame* **2018**, *198*, 169–175. [CrossRef]

16. Courty, L.; Lagrange, J.-F.; Gillard, P.; Boulnois, C. Laser ignition of a low vulnerability propellant based on nitrocellulose: Effects of Ar and N_2 surrounding atmospheres. *Propell. Explos. Pyrot.* **2018**, *43*, 986–991. [CrossRef]

17. Wei, R.; He, Y.; Zhang, Z.; He, J.; Yuen, R.; Wang, J. Effect of different humectants on the thermal stability and fire hazard of nitrocellulose. *J. Therm. Anal. Calorim.* **2018**, *133*, 1291–1307. [CrossRef]

18. Liu, J.; Chen, M. A simplified method to predict the heat release rate of industrial nitrocellulose materials. *Appl. Sci.* **2018**, *8*, 910. [CrossRef]

19. Tarchoun, A.F.; Trache, D.; Klapötke, T.M.; Chelouche, S.; Derradji, M.; Bessa, W.; Mezroua, A. A promising energetic polymer from *Posidonia oceanica* brown algae: Synthesis, characterization, and kinetic modeling. *Macromol. Chem. Phys.* **2019**, *220*, 1900358. [CrossRef]

20. Trache, D.; Tarchoun, A.F. Analytical methods for stability assessment of nitrate esters-based propellants. *Crit. Rev. Anal. Chem.* **2019**, *49*, 415–438. [CrossRef]

21. Betzler, F.M.; Klapötke, T.M.; Sproll, S. Energetic nitrogen-rich polymers based on cellulose. *Cent. Eur. J. Energy Mater.* **2011**, *8*, 157–171.

22. Luo, L.; Jin, B.; Xiao, Y.; Zhang, Q.; Chai, Z.; Huang, Q.; Chu, S.; Peng, R. Study on the isothermal decomposition kinetics and mechanism of nitrocellulose. *Polym. Test.* **2019**, *75*, 337–343. [CrossRef]

23. Trache, D.; Maggi, F.; Palmucci, I.; DeLuca, L.T. Thermal behavior and decomposition kinetics of composite solid propellants in the presence of amide burning rate suppressants. *J. Therm. Anal. Calorim.* **2018**, *132*, 1601–1615. [CrossRef]

24. Trache, D.; Tarchoun, A.F. Stabilizers for nitrate ester-based energetic materials and their mechanism of action: A state-of-the-art review. *J. Mater. Sci.* **2018**, *53*, 100–123. [CrossRef]

25. Chelouche, S.; Trache, D.; Tarchoun, A.F.; Khimeche, K. Effect of organic eutectic on nitrocellulose stability during artificial aging. *J. Energy Mater.* **2019**, *37*, 387–406. [CrossRef]

26. Chelouche, S.; Trache, D.; Tarchoun, A.F.; Abdelaziz, A.; Khimeche, K.; Mezroua, A. Organic eutectic mixture as efficient stabilizer for nitrocellulose: Kinetic modeling and stability assessment. *Thermochim. Acta* **2019**, *673*, 78–91. [CrossRef]

27. Zhao, N.; Li, J.; Gong, H.; An, T.; Zhao, F.; Yang, A.; Hu, R.; Ma, H. Effects of α-Fe_2O_3 nanoparticles on the thermal behavior and non-isothermal decomposition kinetics of nitrocellulose. *J. Anal. Appl. Pyrol.* **2016**, *120*, 165–173. [CrossRef]

28. Guo, Y.; Zhao, N.; Zhang, T.; Gong, H.; Ma, H.; An, T.; Zhao, F.; Hu, R. Compatibility and thermal decomposition mechanism of nitrocellulose/Cr_2O_3 nanoparticles studied using DSC and TG-FTIR. *RSC Adv.* **2019**, *9*, 3927–3937. [CrossRef]

29. Chelouche, S.; Trache, D.; Tarchoun, A.F.; Abdelaziz, A.; Khimeche, K. Compatibility assessment and decomposition kinetics of nitrocellulose with eutectic mixture of organic stabilizers. *J. Energy Mater.* **2020**, *38*, 48–67. [CrossRef]

30. Trache, D.; Tarchoun, A.F.; Chelouche, S.; Khimeche, K. New insights on the compatibility of nitrocellulose with aniline-based compounds. *Propell. Explos. Pyrot.* **2019**, *44*, 970–979. [CrossRef]

31. Yousef, M.A.; Hudson, M.K.; Berry, B.C. Study on the compatibility of azo-tetrazolate high-energy materials using DSC. *J. Therm. Anal. Calorim.* **2018**, *133*, 1481–1490. [CrossRef]

32. Huang, H.; Shi, Y.; Yang, J.; Li, B. Compatibility study of dihydroxylammonium 5,5'-bistetrazole-1,1'-diolate (TKX-50) with some energetic materials and inert materials. *J. Energy Mater.* **2015**, *33*, 66–72. [CrossRef]

33. Liu, Z. *Thermal Analyses for Energetic Materials*; National Defense Industry Press: Beijing, China, 2008; pp. 21–22.
34. An, T.; Zhao, F.; Pei, Q.; Xiao, L.; Xu, S.; Gao, H.; Xing, X. Preparation, characterization and combustion catalytic activity of nanopartical super thermites. *Chin. J. Inorg. Chem.* **2011**, *27*, 231–238.
35. Zhao, F.; Yi, J.; An, T.; Wang, Y.; Hong, W. *Combustion Catalysts for Solid Propellant*; National Defense Industry Press: Beijing, China, 2016; pp. 90–120.
36. Prentice, D.; Pantoya, M.L.; Clapsaddle, B.J. Effect of nanocomposite synthesis on the combustion performance of a ternary thermite. *J. Phys. Chem. B* **2005**, *109*, 20180–20185. [CrossRef] [PubMed]
37. Shende, R.; Subramanian, S.; Hasan, S.; Apperson, S.; Thiruvengadathan, R.; Gangopadhyay, K.; Gangopadhyay, S.; Redner, P.; Kapoor, D.; Nicolich, S.; et al. Nanoenergetic composites of CuO nanorods, nanowires, and Al-nanoparticles. *Propell. Explos. Pyrot.* **2008**, *33*, 122–130. [CrossRef]
38. Criado, J.M.; Sánchez-Jiménez, P.E.; Pérez-Maqueda, L.A. Critical study of the isoconversional methods of kinetic analysis. *J. Therm. Anal. Calorim.* **2008**, *92*, 199–203. [CrossRef]
39. Ma, H.; Song, J.; Hu, R.; Pan, Q.; Wang, Y. Non-isothermal decomposition kinetics, thermal behavior and computational detonation properties on 4-amino-1,2,4-triazol-5-one (ATO). *J. Anal. Appl. Pyrol.* **2008**, *83*, 145–150. [CrossRef]
40. Vyazovkin, S.; Burnham, A.K.; Criado, J.M.; Pérez-Maqueda, L.A.; Popescu, C.; Sbirrazzuoli, N. ICTAC Kinetics Committee recommendations for performing kinetic computations on thermal analysis data. *Thermochim. Acta* **2011**, *520*, 1–19. [CrossRef]
41. Trache, D. Comments on "Thermal degradation behavior of hypochlorite-oxidized starch nanocrystals under different oxidized levels". *Carbohydr. Polym.* **2016**, *151*, 535–537. [CrossRef]
42. Sánchez-Jiménez, P.E.; Pérez-Maqueda, L.A.; Perejón, A.; Criado, J.M. Nanoclay nucleation effect in the thermal stabilization of a polymer nanocomposite: A kinetic mechanism change. *J. Phys. Chem. C* **2012**, *116*, 11797–11807. [CrossRef]
43. Zhang, T.; Hu, R.; Xie, Y.; Li, F. The estimation of critical temperatures of thermal explosion for energetic materials using non-isothermal DSC. *Thermochim. Acta* **1994**, *244*, 171–176.
44. Yi, J.; Zhao, F.; Wang, B.; Liu, Q.; Zhou, C.; Hu, R.; Ren, Y.; Xu, S.; Xu, K.; Ren, X. Thermal behaviors, nonisothermal decomposition reaction kinetics, thermal safety and burning rates of BTATz-CMDB propellant. *J. Hazard. Mater.* **2010**, *181*, 432–439. [CrossRef] [PubMed]
45. Muravyev, N.V.; Kiselev, V.G. Cheaper, faster, or better: Are simple estimations of safety parameters of hazardous materials reliable? *J. Hazard. Mater.* **2017**, *334*, 267–270. [CrossRef] [PubMed]
46. Hu, R.; Gao, S.; Zhao, F.; Shi, Q.; Zhang, T.; Zhang, J. *Thermal Analysis Kinetics*, 2nd ed.; Science Press: Beijing, China, 2008; pp. 148–167.
47. Peng, L.; Yao, Q.; Wang, J.; Li, Z.; Zhu, Q.; Li, X. Pyrolysis of RDX and its derivatives via reactive molecular dynamics simulations. *Acta Phys. Chim. Sin.* **2017**, *33*, 745–754.

 *nanomaterials*

Article

Hydrothermal Synthesis of Hematite Nanoparticles Decorated on Carbon Mesospheres and Their Synergetic Action on the Thermal Decomposition of Nitrocellulose

Abdenacer Benhammada [1,2], Djalal Trache [1,*], Mohamed Kesraoui [1] and Salim Chelouche [1]

[1] UER Procédés Energétiques, Ecole Militaire Polytechnique, BP 17, Bordj El-Bahri, Algiers 16046, Algeria; nbenhammada@yahoo.fr (A.B.); kesraoui.mohamed@gmail.com (M.K.); salim.chelouche@gmail.com (S.C.)

[2] Ecole Nationale Préparatoire Aux Etudes d'Ingénieur Badji Mokhtar, ENPEI, BP 5, Rouiba, Algiers 16013, Algeria

* Correspondence: djalaltrache@gmail.com

Received: 27 March 2020; Accepted: 24 April 2020; Published: 18 May 2020

Abstract: In this study, carbon mesospheres (CMS) and iron oxide nanoparticles decorated on carbon mesospheres (Fe_2O_3-CMS) were effectively synthesized by a direct and simple hydrothermal approach. α-Fe_2O_3 nanoparticles have been successfully dispersed in situ on a CMS surface. The nanoparticles obtained have been characterized by employing different analytical techniques encompassing Fourier transform infrared (FTIR) spectroscopy, Raman spectroscopy, X-ray diffraction (XRD) and scanning electron microscopy (SEM). The produced carbon mesospheres, mostly spherical in shape, exhibited an average size of 334.5 nm, whereas that of Fe_2O_3 supported on CMS is at around 80 nm. The catalytic effect of the nanocatalyst on the thermal behavior of cellulose nitrate (NC) was investigated by utilizing differential scanning calorimetry (DSC). The determination of kinetic parameters has been carried out using four isoconversional kinetic methods based on DSC data obtained at various heating rates. It is demonstrated that Fe_2O_3-CMS have a minor influence on the decomposition temperature of NC, while a noticeable diminution of the activation energy is acquired. In contrast, pure CMS have a slight stabilizing effect with an increase of apparent activation energy. Furthermore, the decomposition reaction mechanism of NC is affected by the introduction of the nano-catalyst. Lastly, we can infer that Fe_2O_3-CMS may be securely employed as an effective catalyst for the thermal decomposition of NC.

Keywords: carbon mesosphere; Fe_2O_3; supported nanoparticles; nitrocellulose; thermal decomposition; kinetics

---

1. Introduction

Cellulose nitrate, known as nitrocellulose (NC), is one of a main components of gun powder and solid propellants [1–3]. It has been widely investigated owing its thermal decomposition features, such as the decomposition temperature, activation energy, and reaction decomposition mechanism, notably influence the combustion behavior and/or performance characteristics of NC-based formulations [4,5]. It is recognized that the burning efficacy of solid propellants is closely dependent in the decomposition behavior of NC as well. Hence, tailoring the thermal decomposition of NC allows tuning the combustion properties of propellants containing NC. On the other hand, NC-based formulations can exhibit a low thermal stability because of the rupture of O–NO_2 of NC even at ordinary conditions, which can cause the deterioration of their prominent characteristics and subsequently restrain their performance and safe and reliable service-life [6,7]. This situation can be prevented by incorporating either stabilizers or other additives. Thus, the investigation of the safety characteristics, namely the thermal stability, of energetic materials such as NC are indispensable for practical applications.

A few years ago, it was revealed that the incorporation of nanomaterials to NC is an efficient approach to enhance its thermal decomposition by tailoring the decomposition process and/or the activation energy without altering the safety and thermal compatibility [8,9]. Such nanomaterials may comprise metal oxides, metal nanoparticles (NPs), organometallic compounds, metallic composites, energetic nano-catalysts, and carbon nanomaterials [10–12]. Various metal oxide NPs have been tested as NC additives. The effect of CuO [13], Fe_2O_3 [14,15] nanoboron [16], bismuth oxide [17] on the thermal decomposition of nitrocellulose has been assessed and prominent results have been reported.

Hematite (α-Fe_2O_3) nanoparticles, one of the most stable phases of iron oxide which is an n-type semiconductor [18], have received a particular attention owing to their interesting properties and their wide range of application fields in various industrial reactions such as catalysis and biotechnology [19], lithium-ion batteries [20], gas sensing, magnetic memory, biological uses and degradation of organic contaminants [18]. α-Fe_2O_3 nanoparticles can be synthesized with various shapes like nano-platelets, nano-belts, nano-rods, nano-cubes, and nanotubes utilizing miscellaneous physicochemical methodologies [20–24]. For energetic materials applications, α-Fe_2O_3 nanoparticles have been used to increase the ammonium perchlorate thermal decomposition [21,25]. Zhao et al. have confirmed that Fe_2O_3 nanoparticles may be used safely with NC without affecting the kinetic thermal decomposition model [14,15]. Nevertheless, Fe_2O_3 reduces the activation energy and critical temperature of thermal explosion of NC and has a good catalytic effect by promoting the O-NO_2 bond cleavage.

However, the catalytic performance of metal oxide nanoparticles such as Fe_2O_3 depends closely not only on the particles' size and shape, but also on their distribution and dispersion. Particles with nanoscale size range are prone to aggregate because of the importance of surface energy, which will generate lowed available surface areas and reduce the catalytic efficiencies [26]. Therefore, a good dispersion of these nanomaterials using catalytic supports has drawn more attention from the scientific community. Such an efficient approach may reduce the self-aggregation drawbacks of nanoparticles and allows exploring and fully benefiting from the unique physicochemical properties of nanoparticles compared to bulk materials. To produce well-dispersed nanoparticles, numerous substances have been employed to resolve the intractable issues. Recently, carbon-based catalytic supports with outstanding features such as large surface area, chemical stability and tailorable electrical and thermal conductivity, have been revealed as useful supporting material for metal and metal oxides nanoparticles [27]. It was revealed that various metal oxide nanoparticles attached on carbon-based supports not only avoided aggregation but also improved catalytic, thermal, magnetic, and optoelectronic characteristics [28].

Several carbon catalytic supports have been reported such as graphene, graphene oxide, carbon nanotubes, fullerene and carbon mesospheres (CMS), and have been comprehensively employed in various fields [29–33]. CMS are a kind of carbon material, which have some specific characteristics owing to their spherical shape, such as excellent mechanical strength, high packing density, and large specific surface area [34]. Commonly, they can be readily produced by a hydrothermal carbonization procedure of organic materials like glucose [35]. CMS have attracted much interest as a catalytic support owing to the uniformity and homogeneity of their nanoparticles [36]. Their porous nanostructure and high specific area allow a large loading of metal oxide nanoparticles [37]. Nanoparticles supported on carbon mesospheres have been widely used in various field. For instance, CuO NPs supported on CMS have been used for sensing application and super capacitors [36,38], whereas ZnO NPs find utilization as photocatalysts [22], catalysis [39], and other applications [40].

To the best of the authors' knowledge, a research gap still exists in investigation the effect of Fe_2O_3 nanoparticles supported on carbon mesospheres as a catalyst for the thermal decomposition of nitrocellulose. Thus, this work deals with the synthesis of carbon mesosphere CMS as catalytic support and α-Fe_2O_3 nanoparticles decorated on carbon mesospheres, respectively. Then, the catalytic effect of the synthesized materials has been evaluated using differential scanning calorimetry (DSC). The kinetic parameters, i.e., the activation energy (Ea), the pre-exponential factor (A) and reaction model were computed through the isoconversional analysis using four kinetic methods, explicitly, *it*-FWO (iterative Flynn–Wall–Ozawa), *it*-KAS (iterative Kissinger–Akahira–Sunose), TAS (Trache–Abdelaziz–Siwani), and Vyazovkin's equation.

2. Experiment and Methods

2.1. Materials

Nitrocellulose with nitrogen content of 12.56% was produced using the methodology mentioned in our recent works [2,41,42]. Different analytical chemicals comprising glucose ($C_6H_{12}O_6$) for CMS preparation, iron chloride ($FeCl_3 \cdot 6H_2O$), as iron precursor, and ammonia NH_4OH as reducing agent for iron oxide synthesis, were provided by VWR chemicals (100 Matsonford Road, Radnor, PA, USA) and used without further purification. Absolute ethanol and distilled water have been used to purify the obtained catalyst.

2.2. Carbon Mesospheres (CMS) and Fe_2O_3-CMS Preparation

CMS were prepared using a simple hydrothermal treatment of glucose as precursor as reported elsewhere [43], with slight modifications. Typically, a 0.5 M solution of glucose was kept in an autoclave introduced in an oven at 180 °C for 6 h and then cooled at room temperature. A black residue was recovered by centrifugation and washed three times with absolute ethanol and distilled water. Finally, the product obtained was dried at 60 °C in an oven for 8 h.

Fe_2O_3-CMS composite was prepared by a hydrothermal method as follows. First, 1 g of the prepared CMS was dispersed with vigorous stirring in 80 mL of distilled water. An iron solution was simultaneously prepared using iron chloride as precursor in distilled water (5 g, 100 mL). The two solutions were then mixed under stirring for 45 min. After that, 34% ammonia solution was dropwise added to a stirred mixture until a pH of 8. The obtained solution was then incorporated in Teflon-sealed autoclave and heated at 180 °C for 24 h. After being cooled at room temperature, the dispersion was centrifuged, washed several times using distilled water and absolute ethanol, and dried in oven at 60 °C for 8 h, and stored for further characterizations.

2.3. Preparation of CMS-Nitrocellulose (NC) and α-Fe_2O_3-CMS-NC Composites

In order to obtain a good dispersion of the prepared nano-catalysts within the NC matrix, NC-catalyst films were prepared via a dissolution method. In a typical experiment, after being dried in an oven at 60 °C for 24 h, 0.5 g of nitrocellulose was dissolved in 30 mL of acetone under stirring. Then, 25 mg of catalyst, with mass ratio NC:catalyst of (95%:5%), was gradually added under stirring. After a sonication for 20 min, the obtained colloidal mixtures were consistently spread in glass Petri dishes at room temperature until the total elimination of acetone, forming thin films of NC-CMS and NC-Fe_2O_3-CMS, respectively. A pure NC film was also produced with similar method without the addition of catalyst. The experimental procedure was schematized in Figure 1.

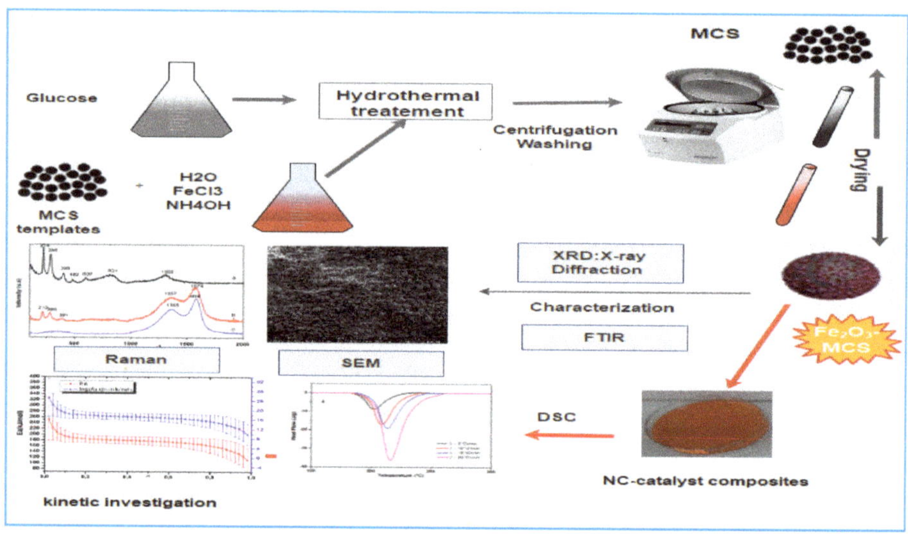

Figure 1. Schematic illustration of preparation and characterization of carbon mesosphere (CMS) and Fe$_2$O$_3$-CMS composites.

2.4. Samples Characterization

2.4.1. Raman Spectroscopy and Fourier Transform Infrared Spectroscopy Analyses

As an imperative nondestructive analytical tool to investigate the chemical composition and the structure of a variety of materials, Raman spectroscopy has been proved as an appropriate technique for the characterization of nanomaterials since it allows detecting characteristic vibrations with low intensities [44,45]. These vibrational features of the produced nanoparticles were determined by employing Raman spectroscopy (Thermo Scientific DXR, Waltham, MA, USA). The chemical functions of the prepared catalysts were investigated using Fast Fourier infrared transform (Bruker-Vertex 70, Rudolf-Plank-Str., Ettlingen, Germany) in attenuated total reflectance (ATR) mode in the wavenumber range 400–4000 cm^{-1}.

2.4.2. Structural and Morphological Investigations

The phase purity of CMS and Fe$_2$O$_3$-CMS was assessed by using PANalytical X'Pert PRO X-ray diffractometer (XRD, Westborough, MA, USA) at 40 mA and 45 kV with Cu anode Kα radiation (λ = 1.54 Å) from 20 to 70° (2θ) at a Step Size of 0.0170. A FEG JSM 7100F TTLS scanning electron microscope (SEM) (JEOL, Leuvensesteenweg, Zaventem, Belgium) was employed to determine the morphology and the particles size of the obtained nanoparticles. The micrographs were acquired with an accelerating voltage of 2 kV. To guarantee reproducible data, more than 50 nanoparticles were used. The particle size of the catalysts was estimated by Image J software (National Institutes of Health by an employee of the Federal Government, MD, USA).

2.4.3. Thermal Analysis

The influence of the incorporated nanocatalysts on the thermal decomposition of NC was assessed by a Perkin Elmer differential scanning calorimeter (DSC, Waltham, MA, USA). For each measurement, 0.8–1 mg of fine cut film is place in a closed aluminum pan. The DSC experiments were realized within the temperature range of 50–300 °C at various heating rates (5, 10, 15 and 20 °C/min) under nitrogen atmosphere (20 cm^{-3}/min). Analysis uncertainties are lower than 0.2 °C for the temperature.

2.5. Kinetic Parameters Determination

A few years ago, the International Confederation for Thermal Analysis and Calorimetry (ICTAC) kinetics committee has claimed that the isoconversional methodology is the utmost appropriate methodology to study the kinetic of thermally stimulated reactions [46]. Based on the use of multiple heating rates rather than isothermal methods, this allows consistent kinetic triplets to be obtained, encompassing, the activation energy, the pre-exponential factor and the most suitable reaction model [47,48].

The reaction rate of solid-state thermal decomposition can be written in terms of T and α as [46,49]:

$$\frac{d\alpha}{dt} = k(T)f(\alpha) \tag{1}$$

where α, t, T, $k(T)$ and $f(\alpha)$ refer, respectively, to the extent of conversion ($0 < \alpha < 1$), the time, temperature, the rate constant, and the reaction mathematical function model that denotes the reaction mechanism. The values of α are experimentally determined from the DSC data as the ratio of the current physical feature change to the total change of this property in the process. Using DSC analysis, α is given as:

$$\alpha = \frac{\int_{t0}^{t} \frac{dH}{dt}\, dt}{\int_{t0}^{t\infty} \frac{dH}{dt}\, dt} = \frac{\Delta H}{\Delta H \text{tot}} \tag{2}$$

where $\frac{dH}{dt}$ is the heat flow, ΔH is the current heat change and ΔH_{tot} is the total heat change determined by DSC.

The substitution of $k(t)$ by its expression from Arrhenius equation leads to:

$$\frac{d\alpha}{dt} = A \exp\left(-\frac{Ea}{RT}\right)f(\alpha) \tag{3}$$

where A is the pre-exponential factor, E_a is the apparent activation energy of the decomposition reaction and R is the universal gas constant.

In the case of multiple heating rate programs, the introduction of heating rate β ($\beta = \frac{dT}{dt}$) transforms Equation (3) to:

$$\frac{d\alpha}{dT} = \frac{A}{\beta} \exp\left(-\frac{Ea}{RT}\right)f(\alpha) \tag{4}$$

By integration, one obtains the integral form of the reaction model $g(\alpha)$, and the 41 forms are reported by Trache et al. [50]:

$$g(\alpha) = \int_{0}^{\alpha} \frac{d\alpha}{f(\alpha)} = \frac{A}{\beta} \int_{0}^{T} \exp\left(-\frac{Ea}{RT}\right)dT \tag{5}$$

As the integral of the temperature dependence part of Equation (5) does not have an analytical solution for an arbitrary temperature program, different approximate equations have been suggested in the literature in order to carry out the kinetic analysis leading to approximate integral methods such as Doyle [51], Coats–Redfern [52] and Senum and Yang [53].

In the present work, to evaluate the kinetics parameters, we have employed four isoconversional methods, i.e., *it*-KAS [54], *it*-FWO [54], TAS [50] and Vyazovkin's equation (VYA/CE) [55]. The details of these methods were given in our previous works [42,50].

3. Results and Discussion

3.1. Characterization of CMS and Fe$_2$O$_3$-CMS

Figure 2A shows the FTIR spectra of the prepared carbon mesospheres and iron oxide nanoparticles decorated on carbon mesospheres, as well as that of the commercial hematite (α-Fe$_2$O$_3$). For the CMS spectrum, the peak absorption at 1703 and 1613 cm^{-1} could be attributed to C=O and C=C vibrations, respectively. The broad absorption peak around 3400 cm^{-1} and the band in the range of 1000–1300 cm^{-1} are attributed to the stretching vibrational modes of C–OH bond and O–H bending [43], indicating the existence of large amount of hydroxyl groups on the surface of CMS.

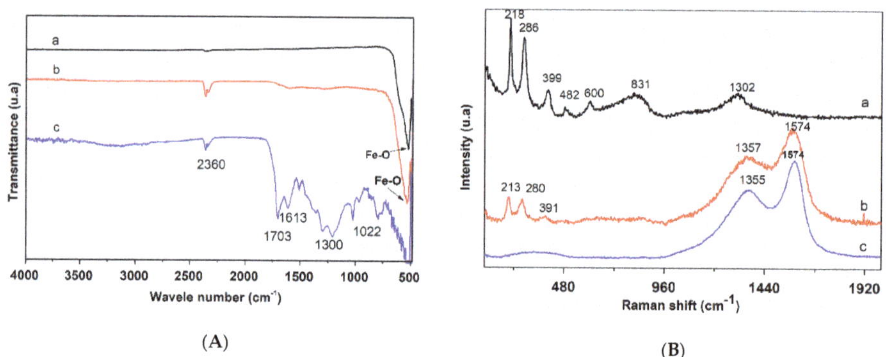

Figure 2. (**A**) Fourier transform infrared (FTIR) spectra of: (a) commercial hematite, (b) Fe$_2$O$_3$-CMS, (c) CMS, (**B**) Raman spectra of: (a) commercial hematite, (b) α-Fe$_2$O$_3$-CMS, (c) CMS.

After being decorated with iron oxide nanoparticles, the band at 1703 cm^{-1} has disappeared demonstrating the cleavage of C=C bonds. Furthermore, the absorption peak intensities of the band at 3400 cm^{-1} and the band in the range of 1000–1300 cm^{-1} are significantly decreased, revealing the formation of metal-oxygen (Fe–O) bonds [56–58]. This finding can be confirmed by the appearance of an absorption peak at around 523 cm^{-1}, which is attributed to stretching vibrational modes of metal-oxygen (Fe–O) bonds. Similar absorption bands have been also found in the pure commercial hematite at 523 cm^{-1}. An absorption peak at 2358 cm^{-1} for the three curves corresponded to the asymmetric stretching of the adsorbed CO$_2$ during sample synthesis [23]. These results indicate the existence of α-Fe$_2$O$_3$ nanoparticles on the surface of carbon mesosphere as well. On the other hand, Raman spectra of the prepared catalyst as well as the commercial hematite specimen are displayed in Figure 2B. The spectra of CMS and Fe$_2$O$_3$-CMS displayed two characteristic bands. The first at 1357 cm^{-1}, assigned to the D band, is related to structural defects, whereas the second at 1574 cm^{-1} corresponded to the G band, representing the graphitic structure in carbon materials [38,59]. It has been reported that the D/G ratio of band intensities represents the graphitic structure with respect to the structure disorder [60,61]. These structural defects are mainly due to surface groups containing oxygen [62]. In addition, the bands at 213 and 480 cm^{-1} (218 and 482 cm^{-1} for the commercial hematite) belonged to two A$_{1g}$ symmetry species. However, the peaks at 280 and 391 cm^{-1} (286 and 399 cm^{-1} for the commercial hematite) were assigned to Eg symmetry as a characteristic Raman phonon bands for Fe$_2$O$_3$ These bands were observed with a slight attenuation for Fe$_2$O$_3$-CMS [24,63]. A red shift for Fe$_2$O$_3$-CMS Raman peaks, in comparison with the commercial sample, is detected which is due to the nanoparticles size reduction. Furthermore, these results indicate the production of -Fe$_2$O$_3$ nanoparticles on the surface of CMS.

Figure 3 shows the diffraction peaks of the synthesized catalyst. The XRD pattern of carbon mesospheres indicates the amorphous character of the synthesized mesospheres with one large diffraction peak at 2θ = 24° [36]. The peaks at 24.13, 33.15, 35.62, 40.87, 49.40, 53.98, 57.53, 62.41, 64.04

and 71.92°, corresponding to (012), (104), (110), (113), (024), (116), (122), (214) and (300) planes of Fe_2O_3, respectively, match well the rhombohedral Fe_2O_3 hematite with a space group R3c and unit cell parameters a = 5.038 Å and c = 13.772 Å (JCPDS Card No. 33-664) [24]. These peaks are found for both Fe_2O_3 nanoparticles and Fe_2O_3-MSC. Besides, the XRD spectra reveal that the peaks of Fe_2O_3 nanoparticles are more intense and sharper compared to those of Fe_2O_3-MSC, indicating the high crystallinity of iron oxide [64], and show a strong preferential orientation of (104) and (110) planes [23]. The average crystallite size diameter D for the prepared Fe_2O_3 nanoparticles, is determined from the diffraction peak widths, employing Debye–Scherrer's equation:

$$D = \frac{k.\lambda}{\beta \cos\theta} \tag{6}$$

with D: crystallite size diameter, k shape factor (k = 0.94), λ: Cu-Kα anode radiation wavelength (λ = 1.54 Å), β_{hkl}: full width at half maxima value (FWHM) in radians, and θ the scattering angle. The computed average crystallite size diameter D is indicated to be in the range of 33 nm.

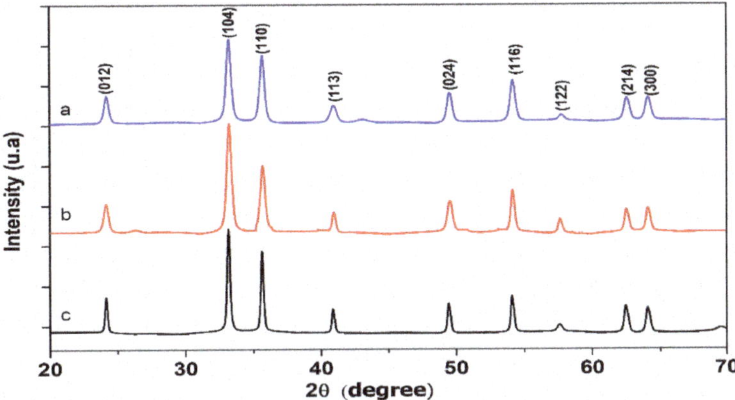

Figure 3. X-ray diffraction (XRD) patterns of (**a**) Fe_2O_3, (**b**), Fe_2O_3-CMS and (**c**) CMS.

The XRD result of Fe_2O_3-MSC indicates also the presence of both iron oxide nanoparticles and carbon mesospheres.

Figure 4 and Figure S1 show the morphology of the prepared CMS and Fe_2O_3-CMS. The SEM images of CMS (a and b) indicate a uniform and homogenous spherical shape with a particle size of about 334.5 nm (Figure S2 and Table S1). From micrographs c, d, and e, the Fe_2O_3 nanoparticles could easily be observed on the surface of carbon mesospheres with a relatively uniform distribution. The particle size of the supported Fe_2O_3 nanoparticles is around 80 nm (Figure S3 and Table S2). These results further indicate that the extern surface of carbon mesospheres act as a template for growing the iron oxide nanoparticles.

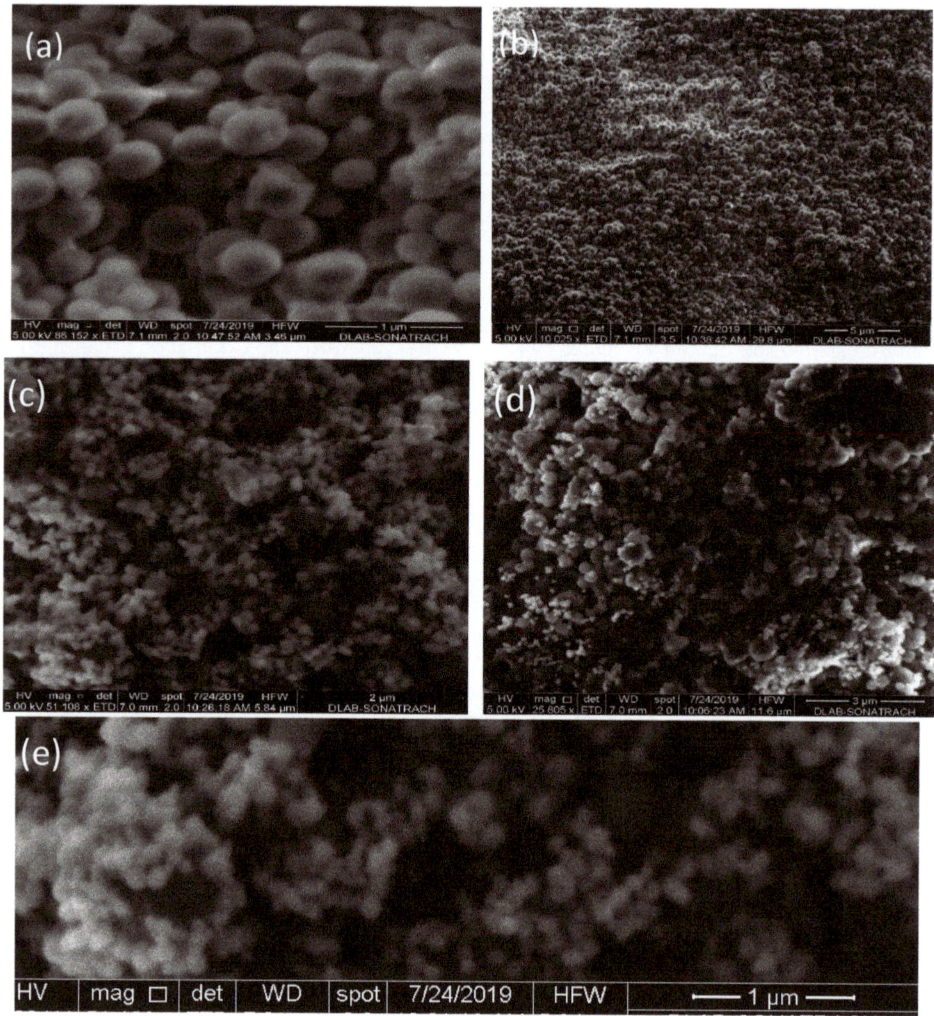

Figure 4. Scanning electron microscope (SEM) images of (**a**,**b**) CMS and (**c**–**e**) Fe$_2$O$_3$-CMS.

3.2. Thermal Analysis

To evaluate different material combinations and ensure the safety during production and storage of energetic materials, compatibly is an important parameter that should be taken into account [65]. Among the various techniques used to evaluate compatibility, DSC thermal analysis is widely employed owing its outstanding features [6,7,42,66]. In order to investigate the compatibility and the effect of the prepared catalysts on the thermal decomposition of nitrocellulose, DSC analyses have been performed at different heating rate and the obtained thermograms are given in Figure 5. The three systems (Pure NC, NC-CMS, NC-Fe$_2$O$_3$-CMS) present the same trend, indeed, one exothermic peak is observed and corresponded to the decomposition of NC [67,68]. With the increase of the heating rate, the peak temperature shifts to higher values. Such results are in good agreement with other works [69]. According to the values of the onset and peak temperatures, given in Table 1, one can observe that the introduction of catalyst has slightly increased the peak temperature. Indeed, for $\beta = 10$ °C/min (Figure 5), pure NC decomposes 1.1 °C earlier than NC-CMS and 1.2 °C earlier than NC-Fe$_2$O$_3$-CMS.

Considering the NC system with and without CMS/CMS-Fe$_2$O$_3$, the peak temperatures of DSC curves increase with the addition of CMS and CMS-Fe$_2$O$_3$, respectively. As the heating rate increases, the exothermic peak becomes sharper indicating a faster chemical reaction [70].

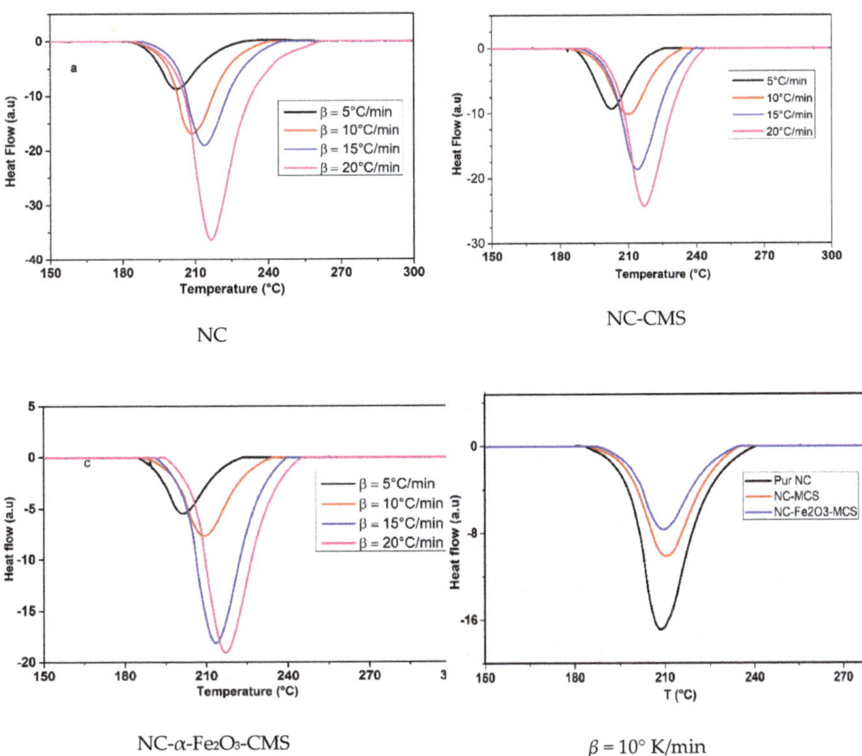

Figure 5. Differential scanning calorimetry (DSC) thermograms of pure nitrocellulose (NC), NC-CMS and NC-Fe$_2$O$_3$-CMS composites.

Table 1. The decomposition temperatures and heat release at various heating rates for the different samples.

Samples	β (°C/min)	T_{onset} (°C) [b]	T_{peak} (°C) [a]	ΔH (J/g)
NC	5	193.3	202.3	−1214
	10	198.8	208.6	−1324
	15	202.9	213.7	−1427
	20	205.5	216.3	−1769
NC-CMS	5	192.4	202.5	−1284
	10	196.8	209.8	−1465
	15	201.4	213.8	−1479
	20	204.5	216.9	−1784
NC-Fe$_2$O$_3$-CMS	5	194.2	202.7	−1595
	10	196.8	209.8	−1900
	15	200.7	213.6	−2707
	20	203.6	216.9	−3092

[a] Uncertainty u associated with the onset decomposition temperature is $u(T_{\text{onset}}) = \pm0.4$ K. [b] Uncertainty u associated with the top decomposition temperature is $u(T_{\text{peak}}) = \pm0.2$ K.

Moreover, From Table 1, it can be inferred that the shifts in the decomposition temperature ($\beta = 5\,°\text{C/min}$) for NC + MCS and NC+ Fe_2O_3-MCS are 0.3 K and 0.5 K, respectively. These shift values are small enough to conclude that the two additives are compatible with NC, even though the STANAG 4147 [71] standard recommends DSC experiments performed on mixtures prepared in 1:1 (*w/w*) at $\beta = 2\,°\text{C/min}$. These results indicate the high compatibility of CMS and Fe_2O_3-CMS with NC [72], and accordingly Fe_2O_3-CMS may be employed as nanocatalyst in the production of NC-based propellant formulations [14].

3.3. Kinetic Parameters

Exploring the DSC values obtained at various heating rates, *it*-KAS, TAS, *it*-FWO and VYA/CE have been used to investigate the thermal decomposition kinetics and thus, evaluate the kinetic parameters, i.e., the activation energy E_a, the pre-exponential factor A and the most probable decomposition model $g(\alpha)$. Numeric calculations were carried out using a MATLAB interface.

The kinetic parameters as well as their corresponding confidence intervals evolution determined by the different isoconversional methods for NC and NC-Fe_2O_3-CMS are depicted in Figures 6 and 7, respectively. Likewise, the mean values of activation energy, pre-exponential factor and the most probable reaction mechanism $g(\alpha)$ for the three studied systems are given in Table 2.

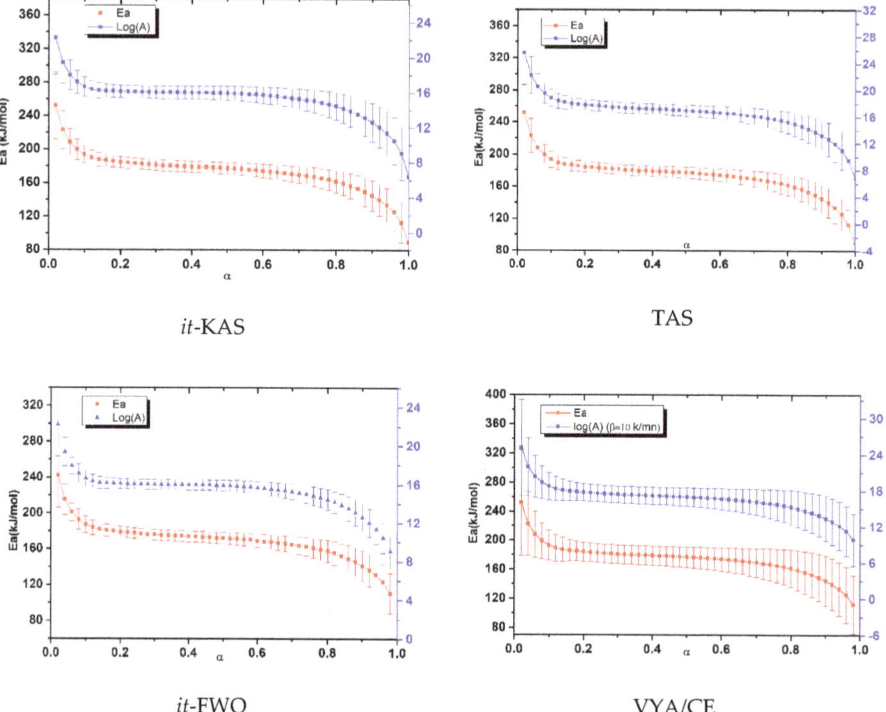

Figure 6. Arrhenius parameters and their associated confidence intervals evolution with respect to α determined by the different kinetic methods for pure NC.

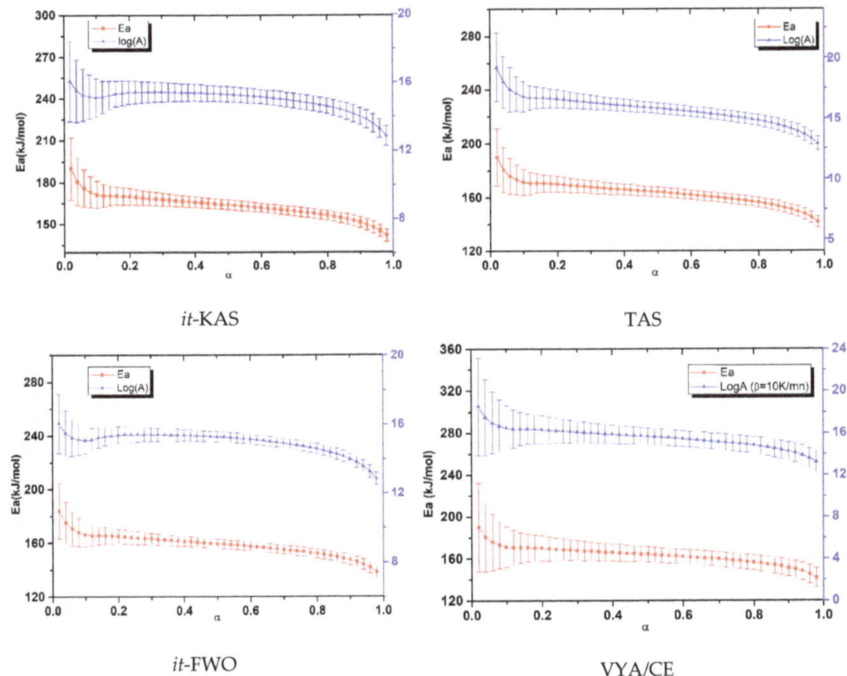

Figure 7. Arrhenius parameters and their associated confidence intervals evolution with respect to α determined by the different kinetic methods for NC-Fe$_2$O$_3$-CMS.

The accuracy of the obtained activation energy and the pre-exponential factor values for *it*-KAS, TAS and *it*-FWO is confirmed with the linear correlation coefficient (R^2), which is found to be in the range (0.95626 to 0.99993). Furthermore, the obtained values of the activation energy and pre-exponential factor allowed us to check that the used isoconversional methods provided close values of E_a and A with a relative deviation of 15.85% and 23.35% for NC, 8.09% and 11.54% for NC-CMS and 7.88 and 11.60% for NC-Fe$_2$O$_3$-CMS, respectively.

The error bars obtained for both E_a and log(A) are very close, indicating the accuracy of implemented computations [69]. The differences that appeared are undoubtedly caused by the different approximations used by the employed kinetic methods.

Moreover, Figures 8 and 9, show respectively, the evolution of a E_a and log(A) against α using the four isoconversional methods for NC, NC-CMS and NC-Fe$_2$O$_3$-CMS. The results obtained show the same trends for E_a and log(A) evolution with a slight difference between models. Indeed, for the four employed isoconversional models, the values obtained seem to be close to each other with slight inferior values for *it*-FWO kinetic method. On the other hand, for the extent of conversion between 0 and 0.01, E_a and log(A) turn out to be more important what is attributed to the cleavage of O-NO$_2$ linkages in nitrocellulose and the liberation of NO$_2$ chemical groups. Then, the two parameters decrease until the end of the reaction. This behavior could be attributed to the autocatalytic parallel reactions, which can generate reactive species that may accelerate the thermolysis and hydrolysis processes [67,68]. As a very strong oxidizing agent, NO$_2$ stagnates in the polymer skeleton and then reacts with the RO$^\bullet$ radicals or their degradation products, causing the opening of the NC anhydroglucopyranose rings to generate other released gases [15,73]. Moreover, the values of E_a and log(A) obtained are within the range of 158.278–176.137 kJ/mol for E_a and 15.2281–17.2619 (s^{-1}) for log(A) which correspond to the common rage values of energetic materials [14].

Table 2. The kinetic parameters of the investigated samples.

System		Isoconversional Method Kinetic Parameters		
		E_a (kJ/mol)	Log(A(S-1))	Reaction Model: g(α)
NC	it-KAS	170 ± 2	16 ± 3	$D_4 = 1 - \left(\frac{2}{3}\right)\alpha - (1-\alpha)^{\frac{2}{3}}$
	it-FWO	170 ± 2	16 ± 3	$D_4 = 1 - \left(\frac{2}{3}\right)\alpha - (1-\alpha)^{\frac{2}{3}}$
	TAS	170 ± 2	16 ± 2	$G_7 = \left[1 - (1-\alpha)2^{\frac{1}{2}}\right]^{1/2}$
	VYA/CE β = 5 °C/min	170 ± 4	16 ± 5	
	β = 10 °C/min		16 ± 5	
	β = 15 °C/min		17 ± 5	
	β = 20 °C/min		17 ± 5	
NC-CMS	it-KAS	180 ± 1	17 ± 2	$A_2 = -\ln(1-\alpha)^{\frac{1}{2}}$
	it-FWO	170 ± 1	17 ± 1	$A_2 = -\ln(1-\alpha)^{\frac{1}{2}}$
	TAS	180 ± 1	16 ± 1	$F_{3/4} = 1 - (1-\alpha)^{\frac{1}{4}},$ $R_3 = F_{2/3} = 1 - (1-\alpha)^{\frac{1}{3}}$
	VYA/CE β = 5 °C/min	180 ± 2	17 ± 3	
	β = 10 °C/min		17 ± 3	
	β = 15 °C/min		17 ± 3	
	β = 20 °C/min		17 ± 3	
NC-Fe$_2$O$_3$-CMS	it-KAS	160 ± 1	15 ± 1	$P_2 = \alpha^2$
	it-FWO	160 ± 1	15 ± 1	$P_2 = \alpha^2$
	TAS	160 ± 1	16 ± 1	$P_{1/3} = \alpha^{\frac{1}{3}},\ P_{1/4} = \alpha^{\frac{1}{4}}$
	VYA/CE β = 5 °C/min	160 ± 2	15 ± 2	
	β = 10 °C/min		16 ± 2	
	β = 15 °C/min		16 ± 2	
	β = 20 °C/min		16 ± 2	

D_4, Three-dimensional diffusion (Ginstling–Brounshtein); $F_{3/4}$, Chemical reaction; A_2, Random nucleation (Avrami–Erofeev); $P_{1/3}$ and $P_{1/4}$, nucleation (Power low); G_7, Other kinetic equations with unjustified mechanism (TAS); P_2, nucleation (parabolic low).

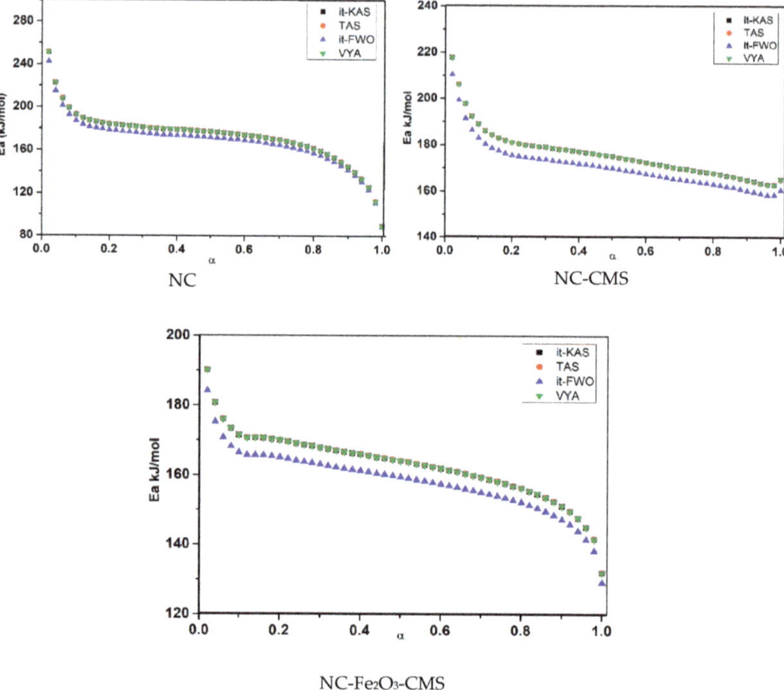

Figure 8. Activation energy (*E*a) evolution for the investigated systems by using different isoconversional methods.

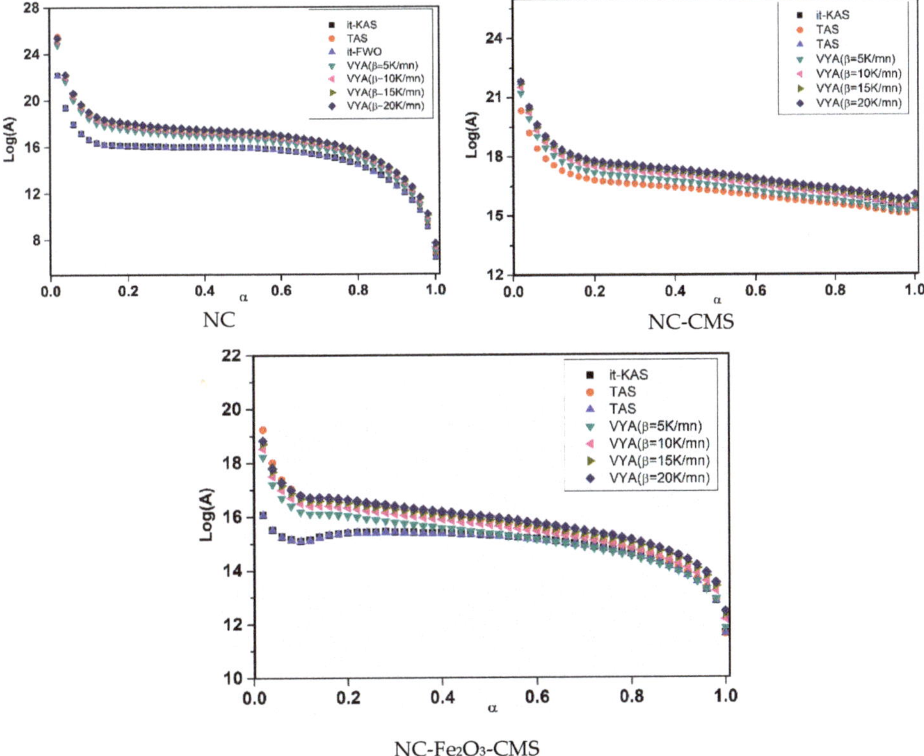

Figure 9. Pre-exponential factor evolution) for the investigated systems by using the different kinetic methods.

Regarding the catalytic activity of the incorporated nanocatalyst, one can observe that the addition of carbon mesospheres, acting as catalytic support, has slightly increased the activation energy based on the four used kinetic models, and thus may have a stabilizing effect on the thermal decomposition of NC. Other carbon nanomaterials such as carbon nanotubes, nanodiamond, and graphene oxide (GO) have been revealed to play stabilizing effect on some organic energetic materials owing to their great thermal conductivity [28]. In our case, CMS facilitates the heat transfer from the reaction zone to the unburned portion of NC, which sustains the propagation of the exothermic reaction. The improvement in heat conductivity would result in less heat accumulation and low hotspots formation, which are important factors determining the sensitivity and stability of energetic materials [74]. Moreover, as shown in Table 1, CMS increases slightly the energy release of NC (ΔH (J/g)) through the improvement of the contact between fuel/oxidizer species of NC due to the better dispersion of CMS within NC matrix. Thiruvengadathan et al. have demonstrated a similar effect for nanothermites supplemented with GO [75].

After the decoration of CMS with Fe_2O_3 nanoparticles, this trend has changed and the activation energy decreases by 12.9 kJ/mol, indicating the good catalytic activity of this additive on the decomposition behavior of nitrocellulose. It was reported that the presence of GO-based catalyst slightly improved the thermal stability (higher decomposition temperature) of 1,3,5,7-tetranitro-1,3,5,7-tetrazocane (HMX) and decreased the activation energy [76]. In another work, Chen et al. have evaluated the effect of GO-Ni on the thermal decomposition of triaminoguanidine nitrate and found a similar trend, where an increase of thermostability but lower energy barrier for the thermal decomposition have been mentioned [70]. On the other hand, the agglomeration problem of nano-sized catalysts is excluded in the presence of CMS as the supported dispersion media, resulting in better contact of Fe_2O_3 nanoparticles with NC fibers. The Fe_2O_3-CMS would show strong catalytic effect only in gas-phase reaction after initial decomposition of the NC at higher temperature (increased T_{peak}), resulting in lowering activation energy of decomposition. The catalytic effect of Fe_2O_3-CMS on the NC themolysis reaction occurs mostly in the gas phase, where the reaction could be accelerated upon initial decomposition of the NC fibers. The presence of CMS would, however, prevent the initial decomposition of NC due to its high thermal conductivity. Similar trend has been recently reported by He et al., who investigated the thermal decomposition of the 1,3,5-trinitro-1,3,5-trizocane (RDX) supplemented with GO-based catalyst [77]. Furthermore, as displayed in Table 1, Fe_2O_3-CMS increases sensibly the energy release of NC through the improvement of the contact between fuel/oxidizer species of NC due to the better dispersion of Fe_2O_3-CMS within NC matrix.

On the other hand, the same trend of both activation energy and pre-exponential factor with respect to α is well represented by the compensation effect. Therefore, in our work, the compensation effect has been investigated using TAS and VYA/CE models for different heating rates of by plotting $Log(A)$ as a function of E_a allowing us to compute the compensation parameters as displayed in Table 3. The obtained values of linear correlation coefficient (R^2) confirm the good compensation effect between $Log(A)$ and E_a.

Table 3. Compensation parameters obtained by the Trache–Abdelaziz–Siwani (TAS) method and Vyazovkin's equation (VYA/CE's).

System			$Log\,A = a\,E_a + b$		
			a (mol/J)	b	R^2
NC	TAS		$0.2653 \pm 5 \times 10^{-4}$	-7.959 ± 0.088	0.99991
	VYA/CE	$\beta = 5\,°C/min$	$0.2518 \pm 2 \times 10^{-5}$	-6.225 ± 0.008	0.99936
		$\beta = 10\,°C/min$	$0.2518 \pm 2 \times 10^{-5}$	-5.532 ± 0.008	0.99936
		$\beta = 15\,°C/min$	$0.2518 \pm 2 \times 10^{-5}$	-5.126 ± 0.007	0.99936
		$\beta = 20\,°C/min$	$0.2518 \pm 3 \times 10^{-5}$	-4.839 ± 0.011	0.99936
NC-CMS	TAS		$0.2202 \pm 1 \times 10^{-5}$	-1.164 ± 0.020	0.99816
	VYA/CE	$\beta = 5\,°C/min$	$0.2522 \pm 3 \times 10^{-5}$	-6.062 ± 0.013	0.999515
		$\beta = 10\,°C/min$	$0.2522 \pm 2 \times 10^{-5}$	-5.371 ± 0.006	0.999515
		$\beta = 15\,°C/min$	$0.2522 \pm 3 \times 10^{-5}$	-4.965 ± 0.013	0.999515
		$\beta = 20\,°C/min$	$0.2522 \pm 3 \times 10^{-5}$	-4.677 ± 0.011	0.999515
NC-Fe_2O_3-CMS	TAS		$0.3024 \pm 2 \times 10^{-5}$	13.128 ± 0.026	0.99956
	VYA/CE	$\beta = 5\,°C/min$	$0.2528 \pm 2 \times 10^{-5}$	-6.073 ± 0.076	0.999541
		$\beta = 10\,°C/min$	$0.2528 \pm 2 \times 10^{-5}$	-5.380 ± 0.071	0.999541
		$\beta = 15\,°C/min$	$0.2528 \pm 2 \times 10^{-5}$	-4.975 ± 0.065	0.999541
		$\beta = 20\,°C/min$	$0.2528 \pm 2 \times 10^{-5}$	-4.687 ± 0.061	0.999541

Regarding the reaction model, among the 41 available, the evolution of the integral reaction mechanism values *vis* α displayed in Figure 10 and the most probable models $g(\alpha)$ are reported in Table 3. Various models can be attributed to NC decomposition either pure or with catalyst, which is dependent on the chosen kinetic method. NC decomposes according to $G_7 = \left[1 - (1 - \alpha)^{\frac{1}{2}}\right]^{1/2}$ with the TAS model and the three-dimensional diffusion using *it*-KAS and *it*-FWO. The incorporation of carbon mesospheres could stabilize NC through the decrease of the heat accumulation and preventing the hotspots formation during the decomposition. Therefore, CMS could change the decomposition mechanism from G_7 model to others. With *it*-KAS and *it*-FWO, NC-CMS decomposes according to a random nucleation mechanism (Avrami–Erofeev) $A_2 = -\ln(1 - \alpha)^{\frac{1}{2}}$ and a chemical reaction for the TAS model. Similar behavior has been reported by Sánchez-Jiménez, showing that the addition of clay nano-flakes produced a change of the thermal degradation mechanism towards a nucleation and growth model [78]. The authors indicated that such mechanism change is likely to be responsible for the increased stability.

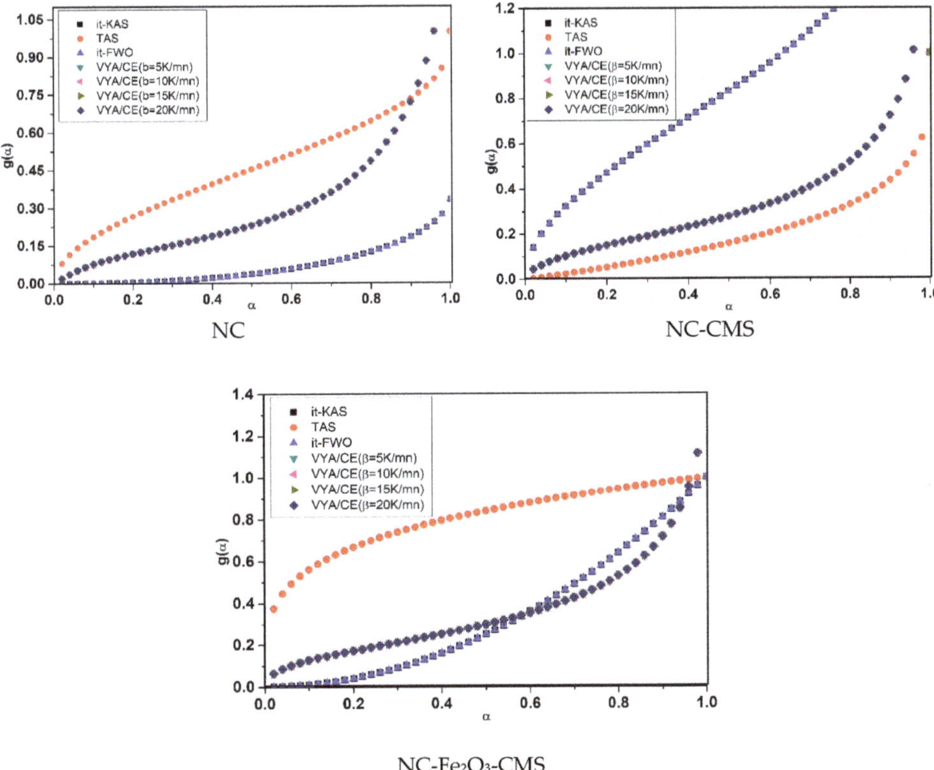

Figure 10. Integral reaction model evolution for the different systems studied.

For NC-Fe$_2$O$_3$-CMS, it decomposes according to a nucleation power and parabolic low (P$_2$ = α^2P$_{1/3}$ = $\alpha^{\frac{1}{3}}$) for the three isoconversional models, revealing that Fe$_2$O$_3$ behaves as the catalytic center for which the decomposition depends considerably on the nucleation sites of Fe$_2$O$_3$. Similar result has been reported by Shen et al., who determined the decomposition mechanism of triaminoguanidine nitrate supplemented with GO-Ni [70].

Recently, our research group revealed that NC containing Fe$_2$O$_3$ nanoparticles using iron chloride as precursor decomposes following the random nucleation mechanism (Avrami–Erofeev) [4]. In other works, chemical reaction mechanism F$_{3/4}$ = $1 - (1 - \alpha)^{\frac{1}{4}}$ has been assumed for pure NC and NC-Fe$_2$O$_3$ composite utilizing iron chloride as precursor [14], and for NC-Al/Fe$_2$O$_3$ mixture [15]. More recently, Chellouche et al. [42] attributed a nucleation (parabolic law) for nitrocellulose. Such differences in the kinetic could be assigned to the difference in the composition that influence the NC stability and its decomposition behavior [79] as well as the difference in the employed analytical tool and the kinetic methodologies.

4. Conclusions

From the foregoing experiments and modeling, the following conclusions regarding preparation of the nano-catalyst as well as its catalytic effect on NC decomposition may be drawn:

(1)   Carbon mesospheres CMS and iron oxide nanoparticles decorated on carbon mesosphere (Fe$_2$O$_3$-CMS) have been effectively produced by a hydrothermal method, and easily mixed with nitrocellulose to obtain NC-catalyst composites. The DSC analyses proved a high compatibility of

the as prepared Fe_2O_3-CMS with a slight increase of the temperature of decomposition, suggesting the safety use of such NC-Fe_2O_3-CMS composite.

(2)   The kinetic methods employed to determine the kinetics parameters revealed that the activation energy decreased by 12.9 kJ/mol with the presence of supported Fe_2O_3 nanoparticles, whereas the addition of CMS alone has no catalytic activity on the decomposition behavior of NC. Furthermore, the non-isothermal decomposition of nitrocellulose has been modeled. While TAS allowed the same kinetic reaction model ($G_7 = \left[1 - (1 - \alpha)^{\frac{1}{2}}\right]^{1/2}$ for pure NC, the *it*-KAS and *it*-FWO models provided the same decomposition model D4 of three-dimensional diffusion (Ginstling–Brounshtein). It is demonstrated that NC-Fe_2O_3-CMS decomposes according to the nucleation power and parabolic low ($P_2 = \alpha^2$, $P_{1/3} = \alpha^{\frac{1}{3}}$) for the three isoconversional models. The addition of carbon mesospheres, which does not significantly change the activation energy, affects the reaction model, where for *it*-KAS and *it*-FWO isoconversional models, NC-CMS decomposes according to the random nucleation mechanism (Avrami–Erofeev) $A_2 = -\ln(1 - \alpha)^{\frac{1}{2}}$, whereas a chemical reaction is obtained by the TAS model.

(3)   The Fe_2O_3-CMS catalyst presents interesting features due to the stabilization effect before the decomposition point of energetic ingredient (NC), which is of great importance for the safety of energetic materials for long-term storage. Once the decomposition occurs, Fe_2O_3-CMS would accelerate the reactions and result in faster decomposition and higher energy releases.

Supplementary Materials: The following are available online at http://www.mdpi.com/2079-4991/10/5/968/s1, Figure S1: SEM images treated with ImageJ (a, b) CMS, (c, d) CMS-Fe_2O_3; Figure S2: Particle size distribution using ImageJ software for MCS; Figure S3: Particle size distribution using ImageJ software for Fe_2O_3-MCS; Table S1: Statistics on columns of particle size distribution for MCS; Table S2: Statistics on columns of particle size distribution for Fe_2O_3-MCS.

Author Contributions: Conceptualization, A.B. and D.T.; Methodology, A.B., D.T. and M.K.; Investigation, A.B. and M.K.; Resources, A.B., D.T. and M.K.; Writing—original draft preparation, A.B.; Writing-review and editing, D.T. and S.C.; Supervision, D.T. All authors have read and agreed to the published version of the manuscript.

Funding: This work was financially supported by the Ecole Militaire Polytechnique.

Acknowledgments: The authors would like to gratefully thank Luigi T. Deluca for his valuable comments and suggestions to improve the quality of this article.

Conflicts of Interest: The authors declare no conflict of interest.

References

1.   Wei, W.; Cui, B.; Jiang, X.; Lu, L. The catalytic effect of nio on thermal decomposition of nitrocellulose. *J. Therm. Anal. Calorim.* **2010**, *102*, 863–866. [CrossRef]

2.   Trache, D.; Khimeche, K.; Mezroua, A.; Benziane, M. Physicochemical properties of microcrystalline nitrocellulose from alfa grass fibres and its thermal stability. *J. Therm. Anal. Calorim.* **2016**, *124*, 1485–1496. [CrossRef]

3.   Tarchoun, A.F.; Trache, D.; Klapötke, T.M.; Chelouche, S.; Derradji, M.; Bessa, W.; Mezroua, A. A promising energetic polymer from posidonia oceanica brown algae: Synthesis, characterization, and kinetic modeling. *Macromol. Chem. Phys.* **2019**, *220*, 1900358. [CrossRef]

4.   Benhammada, A.; Trache, D.; Kesraoui, M.; Tarchoun, A.F.; Chelouche, S.; Mezroua, A. Synthesis and characterization of α-fe2o3 nanoparticles from different precursors and their catalytic effect on the thermal decomposition of nitrocellulose. *Thermochim. Acta* **2020**, *686*, 178570. [CrossRef]

5.   Zhang, T.; Zhao, N.; Li, J.; Gong, H.; An, T.; Zhao, F.; Ma, H. Thermal behavior of nitrocellulose-based superthermites: Effects of nano-fe 2 o 3 with three morphologies. *RSC Adv.* **2017**, *7*, 23583–23590. [CrossRef]

6.   Chelouche, S.; Trache, D.; Tarchoun, A.F.; Abdelaziz, A.; Khimeche, K. Compatibility assessment and decomposition kinetics of nitrocellulose with eutectic mixture of organic stabilizers. *J. Energetic Mater.* **2020**, *38*, 48–67. [CrossRef]

7.   Trache, D.; Tarchoun, A.F. Analytical methods for stability assessment of nitrate esters-based propellants. *Crit. Rev. Anal. Chem.* **2019**, *49*, 415–438. [CrossRef]

8. Yang, X.; Wang, Y.; Li, Y.; Li, Z.; Song, T.; Liu, X.; Hao, J. Thermal stability and mechanical properties of hybrid materials based on nitrocellulose grafted by aminopropylisobutyl polyhedral oligomeric silsesquioxane. *Polimery* **2017**, *62*, 576–587. [CrossRef]

9. Ayoman, E.; Hosseini, S.G. Synthesis of cuo nanopowders by high-energy ball-milling method and investigation of their catalytic activity on thermal decomposition of ammonium perchlorate particles. *J. Therm. Anal. Calorim.* **2016**, *123*, 1213–1224. [CrossRef]

10. Xu, J.; Li, D.; Chen, Y.; Tan, L.; Kou, B.; Wan, F.; Jiang, W.; Li, F. Constructing sheet-on-sheet structured graphitic carbon nitride/reduced graphene oxide/layered mno2 ternary nanocomposite with outstanding catalytic properties on thermal decomposition of ammonium perchlorate. *Nanomaterials* **2017**, *7*, 450. [CrossRef]

11. Zhao, S.; Ma, D. Preparation of $CoFe_2O_4$ nanocrystallites by solvothermal process and its catalytic activity on the thermal decomposition of ammonium perchlorate. *J. Nanomater.* **2010**, *2010*. [CrossRef]

12. Yan, Q.-L.; Zhao, F.-Q.; Kuo, K.K.; Zhang, X.-H.; Zeman, S.; DeLuca, L.T. Catalytic effects of nano additives on decomposition and combustion of rdx-, hmx-, and ap-based energetic compositions. *Prog. Energy Combust. Sci.* **2016**, *57*, 75–136. [CrossRef]

13. Mahajan, R.; Makashir, P.; Agrawal, J. Combustion behaviour of nitrocellulose and its complexes with copper oxide. Hot stage microscopic studies. *J. Therm. Anal. Calorim.* **2001**, *65*, 935–942. [CrossRef]

14. Zhao, N.; Li, J.; Gong, H.; An, T.; Zhao, F.; Yang, A.; Hu, R.; Ma, H. Effects of α-fe2o3 nanoparticles on the thermal behavior and non-isothermal decomposition kinetics of nitrocellulose. *J. Anal. Appl. Pyrolysis* **2016**, *120*, 165–173. [CrossRef]

15. Zhao, N.; Li, J.; Zhao, F.; An, T.; Hu, R.; Ma, H. Combustion catalyst: Nano-fe2o3 and nano-thermite al/fe2o3 with different shapes. *Dev. Combust. Technol.* **2016**, *325*. [CrossRef]

16. Li, Y.; Yang, H.; Hong, Y.; Yang, Y.; Cheng, Y.; Chen, H. Electrospun nanofiber-based nanoboron/nitrocellulose composite and their reactive properties. *J. Therm. Anal. Calorim.* **2017**, *130*, 1063–1068. [CrossRef]

17. Maraden, A.; Zeman, S.; Zigmund, J.; Mokhtar, M. Physical and thermal impact of lead free ballistic modifiers. *Thermochim. Acta* **2018**, *662*, 16–22. [CrossRef]

18. Satheesh, M.; Paloly, A.R.; Krishna Sagar, C.; Suresh, K.; Bushiri, M.J. Improved coercivity of solvothermally grown hematite (α-fe2o3) and hematite/graphene oxide nanocomposites (α-fe2o3/go) at low temperature. *Phys. Status Solidi A* **2018**, *215*, 1700705. [CrossRef]

19. Zanganeh, S.; Hutter, G.; Spitler, R.; Lenkov, O.; Mahmoudi, M.; Shaw, A.; Pajarinen, J.S.; Nejadnik, H.; Goodman, S.; Moseley, M. Iron oxide nanoparticles inhibit tumour growth by inducing pro-inflammatory macrophage polarization in tumour tissues. *Nat. Nanotechnol.* **2016**, *11*, 986. [CrossRef]

20. Yu, W.J.; Zhang, L.; Hou, P.X.; Li, F.; Liu, C.; Cheng, H.M. High reversible lithium storage capacity and structural changes of fe2o3 nanoparticles confined inside carbon nanotubes. *Adv. Energy Mater.* **2016**, *6*, 1501755. [CrossRef]

21. Xu, H.; Wang, X.; Zhang, L. Selective preparation of nanorods and micro-octahedrons of fe2o3 and their catalytic performances for thermal decomposition of ammonium perchlorate. *Powder Technol.* **2008**, *185*, 176–180. [CrossRef]

22. Wang, X.; Hu, P.; Fangli, Y.; Yu, L. Preparation and characterization of zno hollow spheres and zno– carbon composite materials using colloidal carbon spheres as templates. *J. Phys. Chem. C* **2007**, *111*, 6706–6712. [CrossRef]

23. Ayachi, A.A.; Mechakra, H.; Silvan, M.M.; Boudjaadar, S.; Achour, S. Monodisperse α-fe2o3 nanoplatelets: Synthesis and characterization. *Ceram. Int.* **2015**, *41*, 2228–2233. [CrossRef]

24. Su, X.; Yu, C.; Qiang, C. Synthesis of α-fe2o3 nanobelts and nanoflakes by thermal oxidation and study to their magnetic properties. *Appl. Surf. Sci.* **2011**, *257*, 9014–9018. [CrossRef]

25. Zhang, Y.; Liu, X.; Nie, J.; Yu, L.; Zhong, Y.; Huang, C. Improve the catalytic activity of α-fe2o3 particles in decomposition of ammonium perchlorate by coating amorphous carbon on their surface. *J. Solid State Chem.* **2011**, *184*, 387–390. [CrossRef]

26. Liu, B.; Wang, W.; Wang, J.; Zhang, Y.; Xu, K.; Zhao, F. Preparation and catalytic activities of cufe 2 o 4 nanoparticles assembled with graphene oxide for rdx thermal decomposition. *J. Nanoparticle Res.* **2019**, *21*, 48. [CrossRef]

27. Wang, W.; Guo, S.; Zhang, D.; Yang, Z. One-pot hydrothermal synthesis of reduced graphene oxide/zinc ferrite nanohybrids and its catalytic activity on the thermal decomposition of ammonium perchlorate. *J. Saudi Chem. Soc.* **2019**, *23*, 133–140. [CrossRef]

28. Yan, Q.-L.; Gozin, M.; Zhao, F.-Q.; Cohen, A.; Pang, S.-P. Highly energetic compositions based on functionalized carbon nanomaterials. *Nanoscale* **2016**, *8*, 4799–4851. [CrossRef]

29. Song, J.; Xu, L.; Zhou, C.; Xing, R.; Dai, Q.; Liu, D.; Song, H. Synthesis of graphene oxide based cuo nanoparticles composite electrode for highly enhanced nonenzymatic glucose detection. *ACS Appl. Mater. Interfaces* **2013**, *5*, 12928–12934. [CrossRef]

30. Xu, Z.-X.; Xu, G.-S.; Fu, X.-Q.; Wang, Q. The mechanism of nano-cuo and cufe2o4 catalyzed thermal decomposition of ammonium nitrate. *Nanomater. Nanotechnol.* **2016**, *6*, 1847980416681699. [CrossRef]

31. Xu, S.; Du, A.J.; Liu, J.; Ng, J.; Sun, D.D. Highly efficient cuo incorporated tio2 nanotube photocatalyst for hydrogen production from water. *Int. J. Hydrog. Energy* **2011**, *36*, 6560–6568. [CrossRef]

32. Chen, M.; Wang, L.; Yang, H.; Zhao, S.; Xu, H.; Wu, G. Nanocarbon/oxide composite catalysts for bifunctional oxygen reduction and evolution in reversible alkaline fuel cells: A mini review. *J. Power Sources* **2018**, *375*, 277–290. [CrossRef]

33. Li, S.; Pasc, A.; Fierro, V.; Celzard, A. Hollow carbon spheres, synthesis and applications–a review. *J. Mater. Chem. A* **2016**, *4*, 12686–12713. [CrossRef]

34. Wang, Y.; Cao, X.; Sun, S.; Zhang, R.; Shi, Q.; Zheng, L.; Sun, R. Carbon microspheres prepared from the hemicelluloses-rich pre-hydrolysis liquor for contaminant removal. *Carbohydr. Polym.* **2019**, *213*, 296–303. [CrossRef] [PubMed]

35. Titirici, M.-M.; Antonietti, M. Chemistry and materials options of sustainable carbon materials made by hydrothermal carbonization. *Chem. Soc. Rev.* **2010**, *39*, 103–116. [CrossRef] [PubMed]

36. Zhang, J.; Ma, J.; Zhang, S.; Wang, W.; Chen, Z. A highly sensitive nonenzymatic glucose sensor based on cuo nanoparticles decorated carbon spheres. *Sens. Actuators B* **2015**, *211*, 385–391. [CrossRef]

37. Tang, S.; Vongehr, S.; Meng, X. Controllable incorporation of ag and ag–au nanoparticles in carbon spheres for tunable optical and catalytic properties. *J. Mater. Chem.* **2010**, *20*, 5436–5445. [CrossRef]

38. Fan, Y.; Liu, P.-F.; Yang, Z.-J. Cuo nanoparticles supported on carbon microspheres as electrode material for supercapacitors. *Ionics* **2015**, *21*, 185–190. [CrossRef]

39. Demir-Cakan, R.; Makowski, P.; Antonietti, M.; Goettmann, F.; Titirici, M.-M. Hydrothermal synthesis of imidazole functionalized carbon spheres and their application in catalysis. *Catal. Today* **2010**, *150*, 115–118. [CrossRef]

40. Liu, J.; Wickramaratne, N.P.; Qiao, S.Z.; Jaroniec, M. Molecular-based design and emerging applications of nanoporous carbon spheres. *Nat. Mater.* **2015**, *14*, 763–774. [CrossRef] [PubMed]

41. Trache, D.; Donnot, A.; Khimeche, K.; Benelmir, R.; Brosse, N. Physico-chemical properties and thermal stability of microcrystalline cellulose isolated from alfa fibres. *Carbohydr. Polym.* **2014**, *104*, 223–230. [CrossRef] [PubMed]

42. Chelouche, S.; Trache, D.; Tarchoun, A.F.; Abdelaziz, A.; Khimeche, K.; Mezroua, A. Organic eutectic mixture as efficient stabilizer for nitrocellulose: Kinetic modeling and stability assessment. *Thermochim. Acta* **2019**, *673*, 78–91. [CrossRef]

43. Miao, Z.-H.; Wang, H.; Yang, H.; Li, Z.; Zhen, L.; Xu, C.-Y. Glucose-derived carbonaceous nanospheres for photoacoustic imaging and photothermal therapy. *ACS Appl. Mater. Interfaces* **2016**, *8*, 15904–15910. [CrossRef] [PubMed]

44. Gouadec, G.; Colomban, P. Raman spectroscopy of nanomaterials: How spectra relate to disorder, particle size and mechanical properties. *Prog. Cryst. Growth Charact. Mater.* **2007**, *53*, 1–56. [CrossRef]

45. Rashad, M.; Rüsing, M.; Berth, G.; Lischka, K.; Pawlis, A. Cuo and co 3 o 4 nanoparticles: Synthesis, characterizations, and raman spectroscopy. *J. Nanomater.* **2013**, *2013*, 82. [CrossRef]

46. Vyazovkin, S.; Burnham, A.K.; Criado, J.M.; Pérez-Maqueda, L.A.; Popescu, C.; Sbirrazzuoli, N. Ictac kinetics committee recommendations for performing kinetic computations on thermal analysis data. *Thermochim. Acta* **2011**, *520*, 1–19. [CrossRef]

47. Vyazovkin, S. Some basics en route to isoconversional methodology. In *Isoconversional Kinetics of Thermally Stimulated Processes*; Springer: Berlin/Heidelberg, Germany, 2015; pp. 1–25.

48. Trache, D. Comments on "thermal degradation behavior of hypochlorite-oxidized starch nanocrystals under different oxidized levels". *Carbohydr. Polym.* **2016**, *151*, 535–537. [CrossRef]

49. Vyazovkin, S. *Isoconversional Kinetics of Thermally Stimulated Processes*; Springer: Berlin/Heidelberg, Germany, 2015.

50. Trache, D.; Abdelaziz, A.; Siouani, B. A simple and linear isoconversional method to determine the pre-exponential factors and the mathematical reaction mechanism functions. *J. Therm. Anal. Calorim.* **2017**, *128*, 335–348. [CrossRef]

51. Doyle, C.D. Kinetic analysis of thermogravimetric data. *J. Appl. Polym. Sci.* **1961**, *5*, 285–292. [CrossRef]

52. Coats, A.W.; Redfern, J. Kinetic parameters from thermogravimetric data. *Nature* **1964**, *201*, 68–69. [CrossRef]

53. Senum, G.; Yang, R. Rational approximations of the integral of the arrhenius function. *J. Therm. Anal. Calorim.* **1977**, *11*, 445–447. [CrossRef]

54. Liqing, L.; Donghua, C. Application of iso-temperature method of multiple rate to kinetic analysis. *J. Therm. Anal. Calorim.* **2004**, *78*, 283–293. [CrossRef]

55. Vyazovkin, S.; Dollimore, D. Linear and nonlinear procedures in isoconversional computations of the activation energy of nonisothermal reactions in solids. *J. Chem. Inf. Comput. Sci.* **1996**, *36*, 42–45. [CrossRef]

56. Dong, Q.; Kumada, N.; Yonesaki, Y.; Takei, T.; Kinomura, N.; Wang, D. Template-free hydrothermal synthesis of hollow hematite microspheres. *J. Mater. Sci.* **2010**, *45*, 5685–5691. [CrossRef]

57. Woo, K.; Lee, H.J.; Ahn, J.P.; Park, Y.S. Sol–gel mediated synthesis of fe2o3 nanorods. *Adv. Mater.* **2003**, *15*, 1761–1764. [CrossRef]

58. Xu, Y.; Yang, S.; Zhang, G.; Sun, Y.; Gao, D.; Sun, Y. Uniform hematite α-fe2o3 nanoparticles: Morphology, size-controlled hydrothermal synthesis and formation mechanism. *Mater. Lett.* **2011**, *65*, 1911–1914. [CrossRef]

59. Hanafi, S.; Trache, D.; He, W.; Xie, W.; Mezroua, A.; Yan, Q.-L. Thermostable energetic coordination polymers based on functionalized go and their catalytic effects on the decomposition of ap and rdx. *J. Phys. Chem. C* **2020**, *124*, 5182–5195. [CrossRef]

60. Zhu, Y.-G.; Cao, G.-S.; Sun, C.-Y.; Xie, J.; Liu, S.-Y.; Zhu, T.-J.; Zhao, X.; Yang, H.Y. Design and synthesis of nio nanoflakes/graphene nanocomposite as high performance electrodes of pseudocapacitor. *RSC Adv.* **2013**, *3*, 19409–19415. [CrossRef]

61. Liu, T.; Shao, G.; Ji, M.; Ma, Z. Composites of olive-like manganese oxalate on graphene sheets for supercapacitor electrodes. *Ionics* **2014**, *20*, 145–149. [CrossRef]

62. Chen, Z.; Hou, Z.; Xu, W.; Chen, Y.; Li, Z.; Chen, L.; Wang, W. Ultrafine cuo nanoparticles decorated activated tube-like carbon as advanced anode for lithium-ion batteries. *Electrochim. Acta* **2019**, *296*, 206–213. [CrossRef]

63. Nasibulin, A.G.; Rackauskas, S.; Jiang, H.; Tian, Y.; Mudimela, P.R.; Shandakov, S.D.; Nasibulina, L.I.; Jani, S.; Kauppinen, E.I. Simple and rapid synthesis of α-fe 2 o 3 nanowires under ambient conditions. *Nano Res.* **2009**, *2*, 373–379. [CrossRef]

64. Zak, A.K.; Majid, W.A.; Abrishami, M.E.; Yousefi, R. X-ray analysis of zno nanoparticles by williamson–hall and size–strain plot methods. *Solid State Sci.* **2011**, *13*, 251–256.

65. Yan, Q.-L.; Xiao-Jiang, L.; La-Ying, Z.; Ji-Zhen, L.; Hong-Li, L.; Zi-Ru, L. Compatibility study of trans-1, 4, 5, 8-tetranitro-1, 4, 5, 8-tetraazadecalin (tnad) with some energetic components and inert materials. *J. Hazard. Mater.* **2008**, *160*, 529–534. [CrossRef] [PubMed]

66. Trache, D.; Tarchoun, A.F.; Chelouche, S.; Khimeche, K. New insights on the compatibility of nitrocellulose with aniline-based compounds. *Propellants Explos. Pyrotech.* **2019**, *44*, 970–979. [CrossRef]

67. Katoh, K.; Ito, S.; Kawaguchi, S.; Higashi, E.; Nakano, K.; Ogata, Y.; Wada, Y. Effect of heating rate on the thermal behavior of nitrocellulose. *J. Therm. Anal. Calorim.* **2009**, *100*, 303–308. [CrossRef]

68. Trache, D.; Tarchoun, A.F. Stabilizers for nitrate ester-based energetic materials and their mechanism of action: A state-of-the-art review. *J. Mater. Sci.* **2018**, *53*, 100–123. [CrossRef]

69. Chelouche, S.; Trache, D.; Tarchoun, A.F.; Khimeche, K. Effect of organic eutectic on nitrocellulose stability during artificial aging. *J. Energetic Mater.* **2019**, 1–20. [CrossRef]

70. Chen, S.; He, W.; Luo, C.-J.; An, T.; Chen, J.; Yang, Y.; Liu, P.-J.; Yan, Q.-L. Thermal behavior of graphene oxide and its stabilization effects on transition metal complexes of triaminoguanidine. *J. Hazard. Mater.* **2019**, *368*, 404–411. [CrossRef]

71. Agreement, S. 4147: *Chemical Compatibility of Ammunition Components with Explosives (Non-Nuclear Applications)*; NATO Military Agency for Standardization: Brussels, Belgium, 2001.

72. An, T.; Zhao, F.; Gao, H.; Ma, H.; Hao, H.; Yi, J.; Yang, Y. Preparation of super thermites and their compatibilities with db propellants components. *J. Mater. Eng.* **2011**, *1*, 23–28.

73. Bohn, M.A.; Volk, F. Aging behavior of propellants investigated by heat generation, stabilizer consumption, and molar mass degradation. *Propellants Explos. Pyrotech.* **1992**, *17*, 171–178. [CrossRef]

74. Jain, S.; Park, W.; Chen, Y.; Qiao, L. Flame speed enhancement of a nitrocellulose monopropellant using graphene microstructures. *J. Appl. Phys.* **2016**, *120*, 174902. [CrossRef]

75. Thiruvengadathan, R.; Chung, S.W.; Basuray, S.; Balasubramanian, B.; Staley, C.S.; Gangopadhyay, K.; Gangopadhyay, S. A versatile self-assembly approach toward high performance nanoenergetic composite using functionalized graphene. *Langmuir* **2014**, *30*, 6556–6564. [CrossRef] [PubMed]

76. Zeng, G.Y.; Zhou, J.H.; Lin, C.M. The Thermal Decomposition Behavior of Graphene Oxide/Hmx Composites. In *Key Engineering Materials*; Trans Tech Publ: Zurich, Switzerland, 2015; pp. 110–114.

77. He, W.; Guo, J.-H.; Cao, C.-K.; Liu, X.-K.; Lv, J.-Y.; Chen, S.-W.; Liu, P.-J.; Yan, Q.-L. Catalytic reactivity of graphene oxide stabilized transition metal complexes of triaminoguanidine on thermolysis of rdx. *J. Phys. Chem. C* **2018**, *122*, 14714–14724. [CrossRef]

78. Sánchez-Jiménez, P.E.; Pérez-Maqueda, L.A.; Perejón, A.; Criado, J.M. Nanoclay nucleation effect in the thermal stabilization of a polymer nanocomposite: A kinetic mechanism change. *J. Phys. Chem. C* **2012**, *116*, 11797–11807. [CrossRef]

79. Pourmortazavi, S.; Hosseini, S.; Rahimi-Nasrabadi, M.; Hajimirsadeghi, S.; Momenian, H. Effect of nitrate content on thermal decomposition of nitrocellulose. *J. Hazard. Mater.* **2009**, *162*, 1141–1144. [CrossRef]

Review

Al-Based Nano-Sized Composite Energetic Materials (Nano-CEMs): Preparation, Characterization, and Performance

Weiqiang Pang [1,2,*], Xuezhong Fan [1], Ke Wang [1], Yimin Chao [3], Huixiang Xu [1,2], Zhao Qin [2] and Fengqi Zhao [2]

1   The Third Department, Xi'an Modern Chemistry Research Institute, Xi'an 710065, China; xuezhongfan@126.com (X.F.); zhuzhangmangxiewk@163.com (K.W.); xhx204@163.com (H.X.)

2   Science and Technology on Combustion and Explosion Laboratory, Xi'an Modern Chemistry Research Institute, Xi'an 710065, China; qzhao087@hotmail.com (Z.Q.); zhaofqi@163.com (F.Z.)

3   School of Chemistry, University of East Anglia, Norwich Research Park, Norwich NR4 7TJ, UK; y.chao@uea.ac.uk

*   Correspondence: nwpu_pwq@163.com; Tel.: +86-029-88291063

Received: 21 April 2020; Accepted: 25 May 2020; Published: 29 May 2020

Abstract: As one of the new types of functional materials, nano-sized composite energetic materials (nano-CEMs) possess many advantages and broad application prospects in the research field of explosives and propellants. The recent progress in the preparation and performance characterization of Al-based nano-CEMs has been reviewed. The preparation methods and properties of Al-based nano-CEMs are emphatically analyzed. Special emphasis is focused on the improved performances of Al-based nano-CEMs, which are different from those of conventional micro-sized composite energetic materials (micro-CEMs), such as thermal decomposition and hazardous properties. The existing problems and challenges for the future work on Al-based nano-CEMs are discussed.

Keywords: composite energetic materials; nano-sized particles; Al-based; morphology performance; thermal decomposition; hazardous properties

---

1. Introduction

Nano-sized composite energetic materials (nano-CEMs), owing to their large specific surface area, high surface energy, and high surface activity, can realize the nanometer of energetic materials in the process of preparation, storage, and application [1,2]. Moreover, nano-CEMs can integrate the characteristics of each component, thus the new types of energetic materials with various characteristics can be obtained, which can reduce the size of reactants in composites to nano-scale, and then increase the contact interfacial area effectively between materials, so as to solve the problem of slow transmission speed of micro-sized composite energetic materials (micro-CEMs) owing to large particles (chemical reaction rate, mass transfer, and heat transfer), and obtain energetic materials with energy density and a high energy release rate [3,4]. In addition, the energy density of nano-sized energetic composites prepared by the sol–gel method is two times the monomolecular energetic material [5]. The mass transfer speed between the oxidant and the aluminum powder in the pore is fast and the reaction is sensitive. The burning speed is more than 1000 times compared with that of the similar ordinary oxides. Therefore, nano-CEMs as a new type of energetic material have attracted wide interests in the research community. Al is widely used to increase the energy and raise the flame temperature in rocket propellants because of its high combustion enthalpy, easy availability, low toxicity, and good stability in energetic applications, such as propellants, pyrotechnics, and explosives. Especially, nano-sized Al (nAl) is broadly exploited to improve performance incrementing the burning rate and combustion efficiency of energetic systems, leading to shorter ignition delays and shorter agglomerate burning times with respect to propellants containing μAl [6–9]. In this paper, the developments and achievements in the preparation and performance of Al-based nano-CEMs are reviewed. The preparation methods and properties of Al-based nano-CEMs are emphatically analyzed, and the existing problems and challenges in the future work are reviewed.

2. Preparation Methods of Nano-CEMs

Nano-CEMs, being composed of one or more principal energetic components and some additives or binders, have been widely used in military. The uniform dispersing of energetic materials with nano-sized components can display the surface activities, improving the performance of traditional energetic materials and enhancing their practical applications. There are many methods to prepare the nano-CEMs, such as the solid phase method, liquid phase method, and gas phase method (according to the state of raw materials), as well as physical and chemical methods (according to the prepared techniques) [1,5,10]; while there is a big difference for the preparation methods of nano-CEMs mentioned above, here, the mail methods on their preparation were reviewed.

In the last two decades, nano-sized metal matrix composites have witnessed tremendous growth, and particulate-reinforced nanocomposites have been extensively employed in the propulsion field for their tremendous advantages. As we know, non-homogeneous particle dispersion and poor interface bonding are the main drawbacks of conventional manufacturing techniques. A critical review of nanocomposite manufacturing processes is presented; the distinction between ex situ and in situ processes is discussed in some detail. Moreover, in situ gas/liquid processes are elaborated and their advantages are discussed [11]. It provides the reader with an overview of nanocomposite manufacturing methods along with a clear understanding of advantages and disadvantages.

2.1. Sol–Gel Method

The sol–gel method is one of the most important preparation methods for nano-CEMs, so the preparation of Al-based nano-CEMs will focus on this method. On the one hand, this method can control the components accurately, and provide the feasibility for obtaining the nano-sized materials with controllable and uniform density; one the other hand, it allows the components to mix directly at the nano-scale.

The method is mainly used to prepare the metal oxide-based nano-CEMs. In order to make the oxidant or fuel into sol, other components are added to form gel. After the gel was formed, the porous aerogels with low density could be obtained by supercritical fluid extraction, and the xerogels with high density could be obtained under slow evaporation and at a certain pressure. Then, the sol–gel forms the main framework, and fuel and the other components are filled in it. The metal oxide sol can be obtained using hydrolyze metal salts and alkoxy metals. Figure 1 shows the typical process for sol–gel method preparation.

$$M_aX_b + bH_2O \rightarrow M_a(H_2O)_b + bX \tag{1}$$

$$M_a(H_2O)_b + 2b[\text{proton scavnger}] \rightarrow M_aO_b + 2b[\text{proton scavenger-H}] \tag{2}$$

$$M_a(OR)_b + 5H_2O \rightarrow Ma(OH)_b + bROH \tag{3}$$

$$M_a(OH)_b \rightarrow M_aO_{1/2b} + 1/2bH_2O \tag{4}$$

Nano-sized composite energetic materials of Al/Fe$_2$O$_3$, RDX (hexogen, cyclotrimethylenetrinitramine)/RF (resorcinol-formaldehyde), and RDX/SiO$_2$ were prepared by this method [13–15]. Compared with the nano-sized composite particles by means of the mechanical mill method, the thermal decomposition temperature can be forwarded by addition of nano-sized RDX/RF composite energetic materials by the sol–gel method; the mechanical sensitivity reduced as well, indicating that it has great potential application in micro initiating explosive devices.

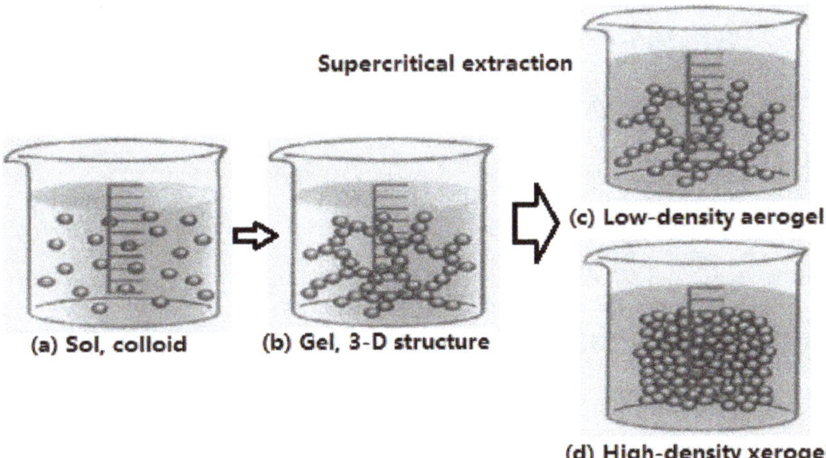

Figure 1. Typical process for sol–gel method preparation. (**a**) Sol, colloid; (**b**) Gel, 3D structure; (**c**) Low-density aerogel; (**d**) High-density xerogel. Reproduced from [12], with permission from Elsevier, 2001.

2.2. Solvent/Nonsolvent Method

The solvent/nonsolvent method for nano-CEMs' preparation is based on crystallography. Nano-sized cuprous dichromate particles coated with perchlorate crystal were prepared, and the particle size can be effectively controlled by adjusting the mass ratio of solvent to nonsolvent. The catalytic effect of nano-CEMs prepared by this method on ammonium perchlorate can be greatly improved, and the thermal decomposition reaction temperature with ammonium perchlorate was obviously decreased, thus the intensity of the thermal decomposition reaction was enhanced [16]. Nano-sized Al/AP (ammonium perchlorate) composite energetic materials were prepared with the solvent/nonsolvent method, the second thermal decomposition temperature decreased after addition of prepared nano particles, while the total heat release increased [17]. Nano-sized AP/CC (copper chromite) and AP/Fe$_2$O$_3$ composite energetic materials were also prepared by the same method, which has good efficacy [18].

2.3. High-Energy Ball Milling Method

The high-energy ball milling method, which makes the hard ball impact, grind, and stir the raw material intensively according to the rotation and vibration actions, and makes it into nano-sized particles. Nano-sized Al/B/Fe$_2$O$_3$, AP/CuO, AP/CC (copper chromite), and Al/RDX composite energetic materials were prepared by this method [19,20]. The formation of nano-sized AP/CC composite particles can increase the contact area between ammonium perchlorate and copper dichromate, improving the dispersion and uniformity of the prepared particles, thus enhancing the catalysis effect of copper dichromate and reducing the thermal decomposition temperature of AP (For nano-sized Al/B/Fe$_2$O$_3$ composition, see Section 3.2.1).

2.4. Spray Drying Method

The spray drying method is the process of atomizing small particles and drying them with solution, suspension, emulsion, sol, and so on. It is the main process for preparing submicron and nano-sized particles in industry, and its equipment is simple, low cost, and highly efficient. HMX-(octogen, cyclotetramethylenetetranitramine), RDX-, and CL-20 (hexanitrohexaazaisowurtzitane)-based nano-sized composite energetic materials were prepared by means of the spray drying method, and the process was optimized [21]. Compared with the original particles, the average particle size of prepared HMX-based composite is 30–100 nm, the activation energy and thermal sensitivity are much lower than that of pure HMX, and the impact sensitivity of sample with NC (nitrocellulose) as the binder has little change. For the RDX-based nano-sized composite, the addition of different binders can greatly reduce the impact sensitivity of RDX. For example, with F_{2602} as the binder, the activation energy of nano-sized CL-20/F_{2602} composite increased, the impact sensitivity decreased significantly, while the thermal sensitivity had little change.

2.5. Other Methods

Hydrothermal reactions have been widely recognized in natural and artificial systems. Particularly, the formation, production, alteration, and decomposition of all substances and materials in natural systems are always related to the action of water (aqueous solutions) at higher temperatures under different pressures. Hydrothermal/solvothermal is another significant and proposed method to prepare nano-sized energetic materials. The feature and future of hydrothermal/solvothermal reactions for synthesis/preparation of nano-materials with desired shapes, sizes, and structures were elaborated [22].

The nano-sized Al/Al_2O_3(p) composites were fabricated from Al-K_2ZrF_6-$Na_2B_4O_7$ system by sonochemistry in situ reaction. The fabrication mechanisms, including high intensity ultrasonic influence on microstructures and reinforcement particles–aluminum matrix interface, were investigated by X-ray diffraction (XRD), scanning electron microscopy (SEM), and transmission electron microscopy (TEM), and the reaction mechanisms were investigated [23]. The results show that the component of the as-prepared composites is Al_2O_3 reinforcement. Al_2O_3 particles are uniformly distributed in the aluminum matrix, the morphologies of Al_2O_3 particles present in a nearly sphere-shape, the sizes are in the range of 20–100 nm, and the interfaces are net and no interfacial outgrowth is observed. The sonochemistry method was reviewed for use in the fabrication of nanomaterials [24].

Freeze-drying is another one of the effective methods to prepare the nano-sized composites. Al/AP nano-sized composites containing nano-sized aluminum powder with low density were prepared by means of the freeze-drying process [25,26]; this method has been used in solid rocket propellants. The reduction method, EEW (electro-exploding wire), CVD (chemical vapor deposition), and PCD (photo catalytic degradation) are well known methods to prepare nano-sized energetic materials. Different from the traditional mixture of reductant and oxidant, the porous metal matrix composite has nano-sized holes and an extremely high specific surface area, and the contact area between oxidant and reductant is large, which displays different performances of explosion characteristics [27–30].

3. Characteristics and Performance of Nano-CEMs

3.1. Al-Based Binary System

3.1.1. Nano-Sized Al/Fe_2O_3 Energetic Composite

Nano-sized Al(s)/Fe_2O_3 composites are readily produced from a solution of Fe(III) salt by adding an organic epoxide and metal powder, and nano-CEMs formed with Fe_2O_3 as metal oxide and Al as fuel, that is, super thermite, have an extremely fast chemical reaction and the burning rate can reach up to 1000 times that of ordinary thermite, which can thus generate an extremely fast reaction wave and release huge energies [4,10,12]. Nano-sized Al/Fe_2O_3 composites were prepared by Gash et al., and the gel structure was observed by high resolution transmission electron microscopy (HRTEM) [21]. It was found that the nano-sized Al/Fe_2O_3 energetic composites were composed of 3–10 nm Fe_2O_3 clusters that are in intimate contact with ultrafine Al particles with 25 nm in diameter, and the Al particles have an oxide coating with thickness of ~5 nm. This value agrees well with the pristine ultrafine grain Al powder, indicating that the sol–gel synthetic method and processing does not significantly perturb the metal fuel. Nonetheless, both qualitative and quantitative characterizations have shown that these materials are indeed energetics, while the materials described here are relatively insensitive to standard impact, electrostatic spark, and friction tests. Qualitatively, it does appear that these nano-sized energetic composites burn faster, which are more sensitive to thermal ignition than their conventional counterparts, and the aerogel materials are more sensitive to ignition than those of xerogels.

Nano-sized Al/Fe_2O_3 energetic composite was prepared by means of the sol–gel method as well as with the supercritical fluid drying method in China [26,27,31,32]; SEM photographs are listed in Figure 2. The experimental results showed the structure of nano-sized Al/Fe_2O_3 composite energetic material was a significant fiber network. The average particle size of Al in the Al/Fe_2O_3 composite was 40 nm. The specific surface area of the Al/Fe_2O_3 composite was 147.9 m^2/g, which is much smaller than that of blank Fe_2O_3 aerogel. Its average pore size was 8 nm and the distribution of pore size was more uniform. From the results, it was believed that the sol–gel method can provide processing advantages over the conventional methods in the areas of cost, purity, homogeneity, and safety at the very least, and potentially act as energetic materials with interesting and special properties.

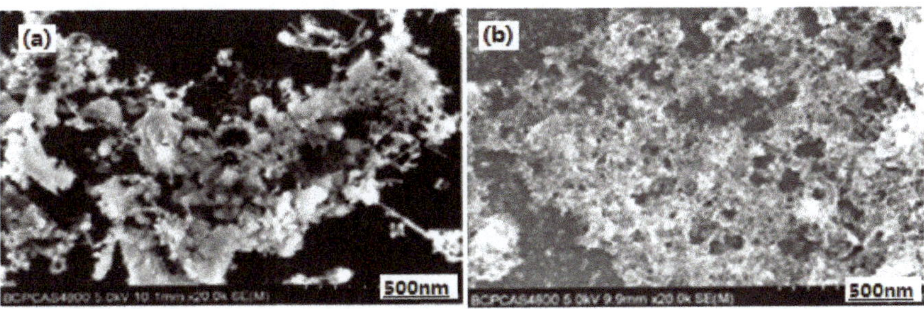

Figure 2. Scanning electron microscopy (SEM) photographs of Fe_2O_3 aerogel (**a**) and Al/Fe_2O_3 composite energetic material (**b**). Reproduced from [31], with permission from Huo Zha Yao Xue Bao, 2010.

3.1.2. Nano-Sized Al/NC Energetic Composite

In order to utilize the full potentials of the existing energetic materials, nano-sized Al/nitrocellulose (Al/NC) composite materials were prepared with the mass ratio of nAl to NC of 0:10, 1:10, 3:10, 5:10, 7:10, and 9:10, respectively, through the sol–gel and supercritical CO_2 drying methods. The structures of nano-CEMs were characterized by SEM and DSC methods. The physical mixed Al/NC and NC aerogel/Al mixture with the same mass ratio as that of nano-sized Al/NC composites was prepared as the compared composition at the same time. SEM images of different composite samples are shown in Figure 3. Al elements distribution in nano-sized Al/NC energetic composite by EDS (energy dispersive spectroscopy) analysis are shown in Figure 4. The experimental results show that the nano-sized Al/NC composite materials belong to mesoporous material and nAl powder is well-distributed in the gel matrix. The specific surface area of the Al/NC particle decreases with the increasing Al powder mass fraction. When the mass ratio of Al/NC is 5:10, the decomposition heat per unit mass of NC increases from 1689.21 $J \cdot g^{-1}$ to 2408.07 $J \cdot g^{-1}$.

3.1.3. Nano-Sized Al/MnO_2 Energetic Composite

Owing to the close contact area and uniform distribution between metal fuel and oxidant, nano-structured energetic materials have better energy release performance and ignition performance. Nano-structured energetic composite composed of MnO_2 nano sheets with root-embedded Al on silicon wafer was prepared [34]. Figure 5 shows the SEM photos of MnO_2 and nano-sized Al/MnO_2 composite particles. As shown, the thickness of MnO_2 grown on the silicon substrate is 900 nm, and the thickness of each sheet is less than 10 nm. The uniform distribution of MnO_2 nano-flakes on the whole silicon substrate indicates that this method can realize the growth of MnO_2 on the silicon wafer with a large surface area (Figure 5a,c). Moreover, Al can deposit on the porous MnO_2 nano sheet successfully, and there was some Al powder at the interface of $Si–MnO_2$, indicating that Al could fill in the porous structure of MnO_2 nano sheet and come into close contact with the MnO_2 nano sheet (Figure 5b). Additionally, the deposited Al is "rooted" in the porous MnO_2 and "grows" into a rod-shaped structure, while the surplus rod-shaped Al aggregates and forms a densed vertical "coral" structure (Figure 5d).

Figure 3. SEM images of different composite samples. (**a**) NC aerogel; (**b**) Al powder; (**c**) Al/NC mixture; (**d**) NC aerogel/Al mixture; and (**e**) Al/NC nano-sized composite. Reproduced from [33], with permission from Han Neng Cai Liao, 2013.

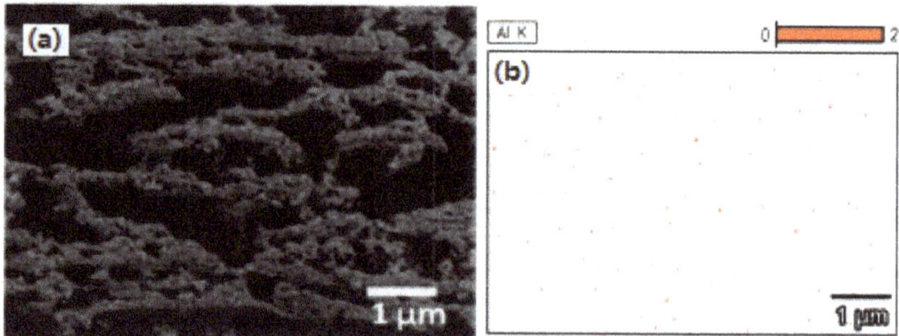

Figure 4. Al elements distribution of Al/NC nano-composite (Al/NC = 1:2) by EDS analysis. (**a**) Selected area of Al/NC nano-sized composite; (**b**) Al elements' distribution. Reproduced from [33], with permission from Han Neng Cai Liao, 2013.

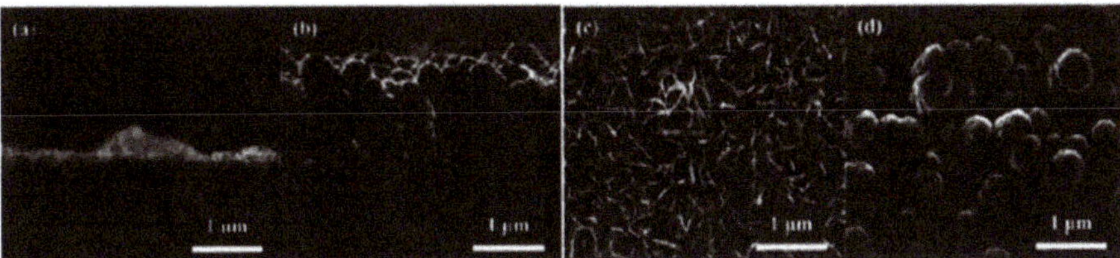

Figure 5. The cross-sectional and top-view SEM photos of MnO_2 and MnO_2/Al particles. (**a**) Cross-sectional of MnO_2; (**b**) cross-sectional of MnO_2/Al; (**c**) top-view of MnO_2; and (**d**) top-view of MnO_2/Al. Reproduced from [34], with permission from Initiators & Pyrotechnics, 2018.

Additionally, Al powder with 500 nm, 1000 nm, and 1500 nm thickness was deposited on MnO_2 with 900 nm thickness, as samples S900-500, S900-1000, and S900-1500 were conducted (Figure 6). Thermal analysis shows that the nano-structured Al/MnO_2 energetic materials have a lower onset reaction temperature (510 °C) and relatively higher heat of reaction (2380 J·g^{-1}) before Al melting. Hydrophobicity is achieved by coating a layer of fluorocarbon, indicating the possible long-term storage ability.

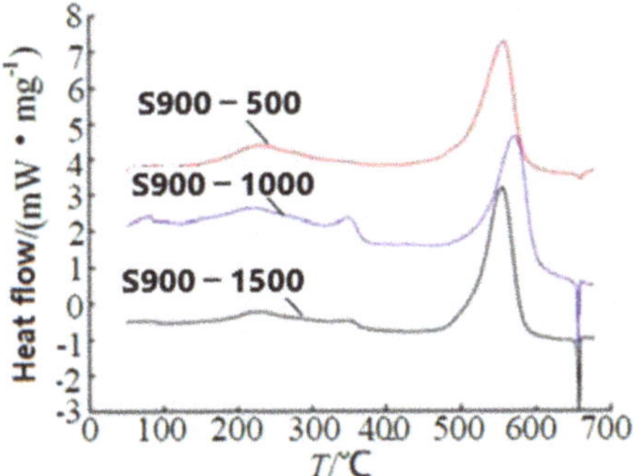

Figure 6. DSC curves of 900 nm thickness MnO_2 samples with nominal 500 nm, 1000 nm, and 1500 nm Al. Reproduced from [34], with permission from Initiators & Pyrotechnics, 2018.

In order to improve the long-term storage stability, fluorocarbon compound was deposited on the surface of nano-sized Al/MnO_2 composite, which have nano texture structures. Figure 7 shows the SEM photos of Al/MnO_2/fluorocarbon composite. It can be seen that the structure of Al/MnO_2 remained unchanged after the fluorocarbon coating, and the surface of the coated composite is better than that of coating before.

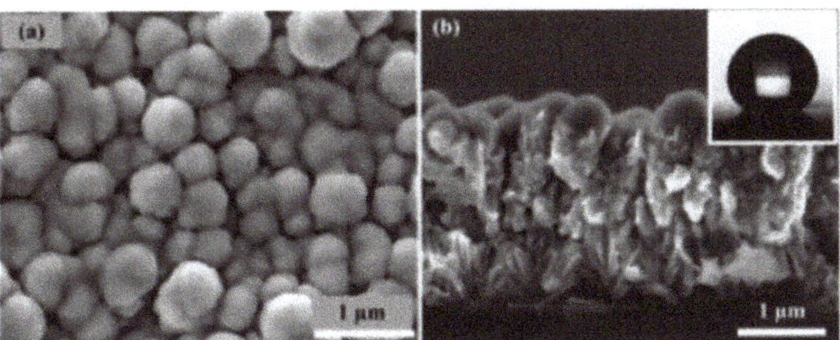

Figure 7. SEM photos of MnO_2/Al/fluorocarbon composite. (**a**) Cross-sectional; (**b**) top-view. Reproduced from [34], with permission from Initiators & Pyrotechnics, 2018.

3.1.4. Nano-Sized Al/CuO Energetic Composite

CuO is a good catalyst on the decomposition and combustion of solid propellant, as we know. In order to significantly enhance the energetic characteristics and long-term storage stability, two types of super-hydrophobic nano-sized Al/CuO core/shell structured particles were prepared onto Cu foils, with CuO nano-rods or nano-tubes as the core and Al as the shell, combined with the solution chemistry method, magnetron sputtering, and surface treatments. Both nano-sized materials exhibit excellent thermal behaviors, especially for the hollow tube structured Al/CuO nano-tubes. Moreover, the functionalized nano-sized Al/CuO particles possess long-term storage stability for as high as 88% of energy left after being exposed in the air for one month, benefitting from the enhanced resistibility to the humid environment. Combustion analysis of functionalized Al/CuO nanotubes is performed and the results reveal that the steady combustion process with the flame propagation speed is 100 m·s^{-1}. This study can provide new ideas for maintaining the activity of nano-CEMs through surface functionalization and significativity for practical applications [35].

A method is disclosed for producing an energetic metastable nano-composite material. Under pre-selected milling conditions, a mixture of powdered components is reactively milled. These components will spontaneously react at a known duration of the pre-selected milling conditions [36]. The milled powder is recovered as a highly reactive nanostructured composite for subsequent use by controllably initiating destabilization thereof. On the basis of this method, several thermite powders with a nominal composition of 2Al·3CuO were prepared by arrested reactive milling using different milling times [37]. The powder particles comprise a fully dense Al matrix with CuO inclusions. The dimensions of the CuO inclusions are around 100 nm and are effectively the same for all prepared powders. The number of inclusions per unit of mass of the composite particles increases with the increased milling times, while unattached CuO particles are present in the samples prepared with shorter milling times. The redox reaction kinetics in the Al/CuO thermites was described, and the multistep reaction model remains valid for composite powders prepared with different milling conditions, and thus characterized by different aluminum/copper oxide ratios. One point should be note that, to accurately describe the rate of redox reactions in specific nanocomposite powders, the model must account for the size of CuO inclusions and the number of inclusions per unit of mass in the nanocomposite particle.

3.1.5. Nano-Sized Al/Co$_3$O$_4$ Composite

The energetic performance of nano-CEMs depends on the interfacial diffusion and mass transfer during the reacted process. However, the development of the desired structure to significantly enhance the reactivity still remains challenging for researchers. Co$_3$O$_4$ micro-spheres with 3D porous hollow were designed and prepared, in which gas-blowing agents (air) and maximized interfacial interactions were introduced to enhance the mass transport and reduce the diffusion distance between the oxidizer and fuel (aluminum) [38]. A low-onset decomposition temperature (423 °C) and high heat output (3118 J·g^{-1}) were observed, resulting from porous and hollow nano-structure of Co$_3$O$_4$ micro-spheres. Furthermore, 3D hierarchical Al/Co$_3$O$_4$ arrays were directly fabricated on the silicon substrate, which was fully compatible with silicon-based microelectromechanical systems to achieve functional nano-sized energetics-on-a-chip. This approach can provide a simple and efficient way to fabricate the 3D ordered nano-sized energetic arrays with superior reactivity, and it has potential applications in micro-energetic devices.

3.1.6. Nano-Sized Al/MoO$_3$ Energetic Composite

A monolayer of two-dimensional (2D) MoO$_3$ with 3–4 µm thickness was prepared by the ultrasonic dispersion method, and then the nano-sized Al/MoO$_3$ composite with a high interface contact area by combining Al nanoparticles with 80 nm in diameter was prepared [39]. The combustion data show that the burning rate reaches (1730 ± 98.1) m·s^{-1} for their high interface contact area, the increasing rate of pressure is (3.49 ± 0.31) MPa·s^{-1}, and the temperature in the reacted area is 3253 K [40]. This is the highest value for the nano-sized Al/MoO$_3$ CEMs reported so far. At the same time, the Al/Fe$_2$O$_3$ and other metal oxides-based energetic materials were coated on Al/Monel (Monel nickel-based alloy), while Ni/Al, Ni/Si, and other laminated structural materials were prepared as well using the sol–gel method and multi-layer injection method; see details in [41].

Moreover, low-temperature redox reactions in nano-thermites prepared by arrested reactive milling (ARM) were described by the Cabrera–Mott (CM) mechanism, in which the growth of very thin oxide layers is accelerated by an electric field induced across such layers. A reaction mechanism combining the initial CM step with the following oxidation steps identified earlier for oxidation of Al and representing growth and phase transitions in various polymorphs of alumina was proposed and shown to be valid for different Al/CuO nanocomposite powders prepared by ARM. Al/MoO$_3$ nanocomposite was prepared by the similar multi-step reaction mechanism [42]. The powder particles comprise a fully dense Al matrix with nano-sized MoO$_3$ inclusions. A weight loss representing dehydration of the composite material was observed at low-temperature thermogravimetry (TG) measurements. Kinetics of the initial exothermic redox reaction observed in both DSC and micro-calorimetry experiments could be successfully described by the CM mechanism. Reactions at elevated temperatures were also well described by the previously identified sequence of steps associated with phase changes and oxidation of various alumina polymorphs. A modification to the CM kinetics was necessary to describe an intermediate range of temperatures, between approximately 400 and 600 K. This modification is suggested to represent a temperature induced change in the structure of a very thin precursor aluminum oxide layer separating Al and MoO$_3$.

3.1.7. Nano-Sized Al/Ni Energetic Composite

As we know, under the stimulation of external conditions (such as heating, current, and laser), Al/Ni composite can produce solid–solid exothermic reaction with a high temperature, and maintain the self-sustaining reacted process, forming the compounds between these two elements. Further, it has been used in the welding and ignition research fields [43]. On the basis of the high temperature and exothermic reaction of Al/Ni composite with a large specific surface area of nano-sized particles, a new nano-structured Al/Ni energetic composite on silicon (Si) substrate using the template method were prepared [44]. Figure 8 shows the SEM photo and DSC curve of the Al/Ni nano-sized composite. As can be seen, Ni nano-sized rods (50 nm in diameter) grow vertically on the substrate, and Al covers the surface of Ni nano-sized rods and is partially embedded into the array of Ni nano-sized rods to form an embedded nano-sized structure (Figure 8a). This kind of embedded nano-sized structure can increase the contact area between the reactants (Al and Ni) and increase their solid–solid reaction rate. The ignition of nano-sized Al/Ni composite was examined by a single pulsed laser irradiation technique, and the results show that the nano-sized Al/Ni composite produces sparks for several milliseconds, which is advantageous to its explosion application. The 50% firing energy of the Al/Ni energetic composite is 36.28 mJ, with an energy density of 46.22 mJ·mm^{-2}.

The core–shell shaped carbon coated nano-sized aluminum (nAl) in the atmosphere of CH$_4$ and Ar gases by means of laser induction combined heating method was prepared [5]. It was shown that its core is crystalline aluminum and its shell is graphite like carbon. Most of the particles are spherical with a diameter of 20–60 nm (thickness is 3–8 nm). During the formation of core–shell shaped nano-sized Al/C particles, owing to the low density of carbon and its poor solubility with aluminum, carbon atoms do not dissolve in the interior of the nanoparticles, but they are deposited on the surface of the nanoparticles.

Additionally, nano-sized Al–Ni energetic material with 90–100 nm particles is fabricated and the experimental results of its properties are presented [45]. High values of specific energy and the rate of its release make it possible to use this material as a heat release element in thermoelectric power generation devices. It has been demonstrated experimentally that it is possible to maintain a voltage value higher than 1 V for 45 s as a result of combustion of a 3 g Al–Ni sample, and that using a simple DC–DC converter will allow charging supercapacitors or accumulators.

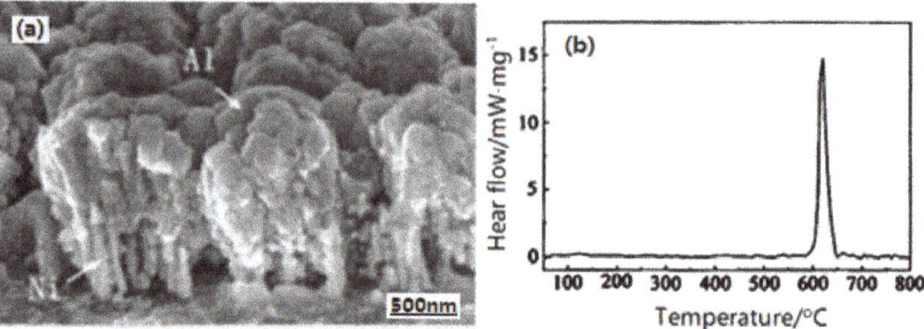

Figure 8. SEM and DSC curve of Al/Ni nano-sized composite. (**a**) SEM of Al/Ni nano-sized composite; (**b**) DSC curve of Al/Ni nano-sized composite. Reproduced from [44], with permission from Huo Zha Yao Xue Bao, 2012.

3.1.8. Nano-Sized Al/Fatty Acid Energetic Composite

Aluminum-oleic acid composite nanoparticles with a mean diameter of 85 nm were successfully prepared by means of a wet chemical process [46]. The Al/oleic acid molar ratio affects the thickness of the oleic acid layer on Al nanoparticles. Al electrodes can be formed by firing an Al nanoparticle paste film at 600 °C, and the firing temperature is about 300 °C lower than that required for micrometer-sized Al particles. The electrode formed from commercial Al nanoaparticles is not electrically conducted because of the oxide layer Al nanoparticles; however, the film from the prepared Al nanoparticles for an Al/oleic acid molar ratio of 1:0.05 has a minimum specific resistance of 5.6 mΩ·cm.

3.1.9. Nano-Sized Al/Polymer Energetic Composite

Polyimide (PI) composites with high dielectric permittivity have received a great deal of attention in embedded capacitors and energy-storage devices owing to its excellent thermal stability and good mechanical properties. Nano-aluminum (Al) particles were introduced into PI to prepare promising PI/nano-Al composites [47]. The results indicate that the dielectric constant of the composite films increased with the increase of nano-Al contents, and the highest dielectric constant was 15.7 for a composite film incorporating 15 wt % nano-Al. The effects of mixture doping concentration on volume resistivity and loss tangent are analyzed. The correlation effects of the Al nanoparticles on the different factors that influence the dielectric performance in the PI matrix such as microstructure, resistivity, and interface of the composites were discussed in detail. This composite film would thus possess potential application in flexible energy-storage devices.

3.2. Al-Based Ternary System

3.2.1. Nano-Sized Al/B/Fe$_2$O$_3$ Energetic Composite

In order to reduce the acidic impurity of B$_2$O$_3$ and H$_3$BO$_3$ on the surface of amorphous boron powder and improve the efficiency of boron powder application, the nano-sized Al/B/Fe$_2$O$_3$ energetic composites were prepared through the high energy ball milling [48]. It was found in Figure 9 that the particle size of raw Al and B is micro-scale; with the increase of milling time, the average particle size of Al/B bimetal composite decreases. This is because of the large plastic deformation of the two mixed powders in the composite at the initial stage of ball milling. In the physical mixed samples, the spherical particles of aluminum can still be seen. Boron powder and iron oxide are simply adsorbed onto the surface of aluminum powder, indicating that aluminum powder, iron oxide, and boron powder are simply blended, and the morphology of the grains has not changed. For the Al/B/Fe$_2$O$_3$ tri-metal composite, there are no spherical aluminum particles, flake boron powder, and flocculent iron oxide, which shows that these three components have been combined and the average particle size is about 90 nm. The heat release of the prepared nano-sized Al/B/Fe$_2$O$_3$ composite increases with an increase in the mass fraction of B in the composition, as can be seen from the DSC curves in Figure 10.

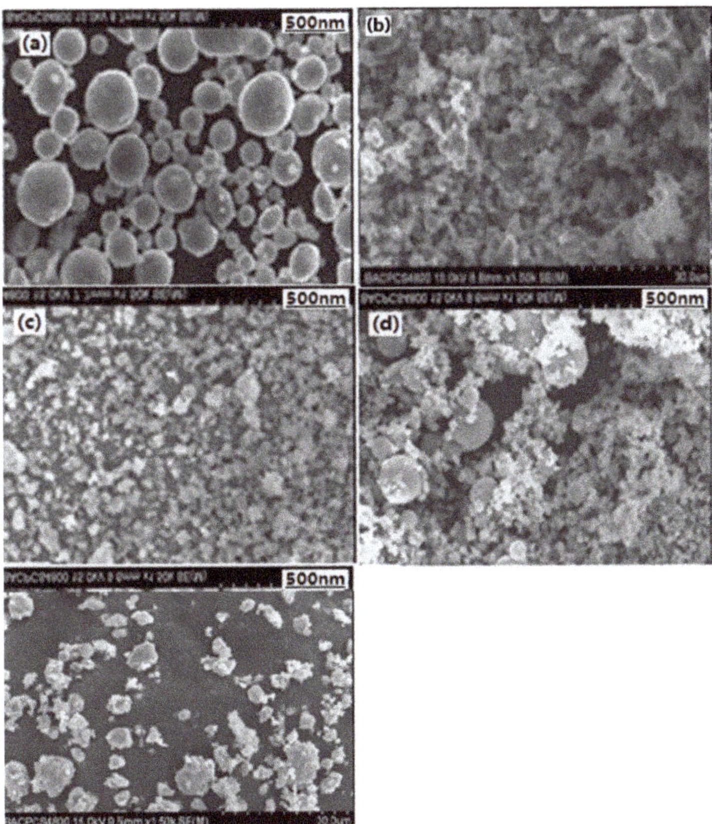

Figure 9. SEM photographs of different samples. (**a**) Raw Al; (**b**) raw B; (**c**) Al/B composite; (**d**) physical mixed; and (**e**) ball milling. Reproduced from [48], with permission from Gu Ti Huo Jian Ji Shu, 2014.

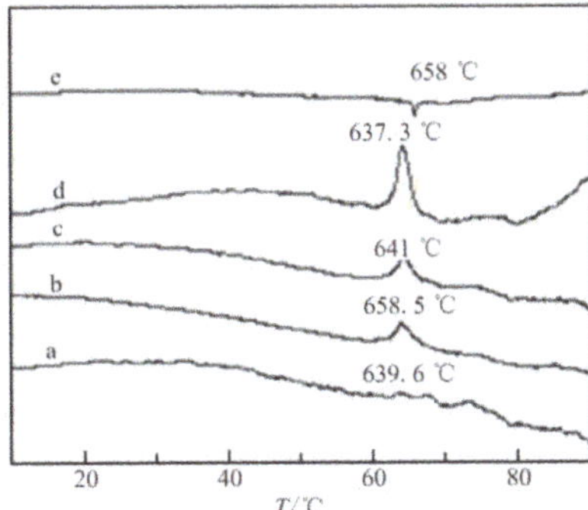

Figure 10. DSC of Al/B/Fe$_2$O$_3$ nano-sized composites. Reproduced from [48], with permission from Gu Ti Huo Jian Ji Shu, 2014.

3.2.2. Nano-Sized Al/CuO/KClO$_4$ Energetic Composite

A solvent/non-solvent synthetic approach was utilized in preparing nano-sized Al/CuO/KClO$_4$ composite by coating Al/CuO particles with a layer of nanoscale oxidizer KClO$_4$. The results reveal that, after ball milling and the chemical synthesis process, the phase compositions were not changed. Scanning electron microscopy (SEM) images show that these energetic nano-sized composites consist of small clusters of Al/CuO that are in intimate contact with a continuous and clear-cut KClO$_4$ layer (100–400 nm). High K/Cl intensity on the perimeter of the nano-sized particles and high Al/Cu mass fraction in the interior powerfully demonstrated the Al/CuO/KClO$_4$ core–shell nanostructure. Electrical ignition experiments and pressure cell test prove that these nano-sized energetic composites are more sensitive to ignition with a much higher burning rate than that of traditional formulations (conventional counterparts). The TG and DSC results show that the burning rate of these energetic nano-sized composites nearly tripled [49].

3.2.3. Nano-Sized Al/B/Ni Energetic Composite

In order to improve the dispersion of nano-sized catalyst particles in the propellant composition and make full use of its catalytic performance, nano-sized Al/B/Ni composites were prepared [50]. It can be seen that the surface of raw Al is smooth, and their average particle size is 20–30 nm (Figure 11a). The coated particles on the surface of nano-sized composite particles are uniform and continuous, and the smooth surface of core aluminum particles is not exposed, thus the coating is relatively complete, and most of their average particle sizes are 20–30 nm, while a few are 50 nm (Figure 11b). Moreover, the proportion of Ni/B alloy particles in the composite particles in the presence of original unit particles increased significantly, and the proportion of agglomerated particles decreased significantly, indicating that the preparation of nano-sized composite particles can improve its dispersion performance in the composition. It can be also found that a thin coating film is attached on the surface of aluminum powder, and the coating layer is uniform and continuous. There are two obscure rings (Figure 11d), which is similar to Figure 11c.

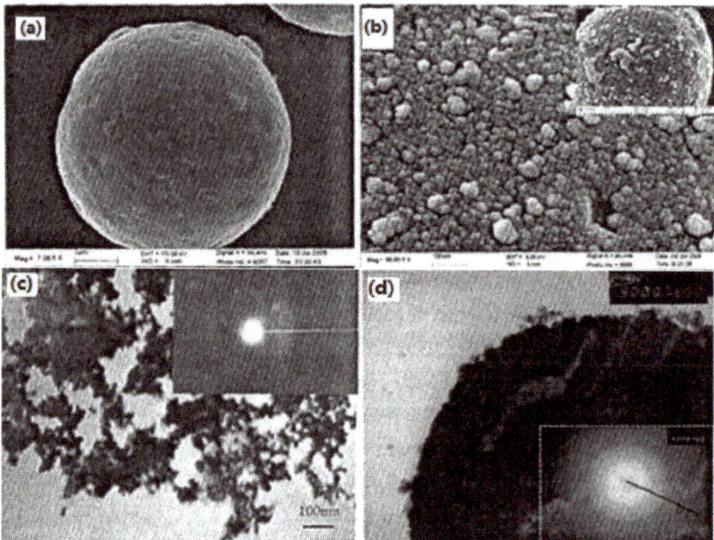

Figure 11. SEM and transmission electron microscopy (TEM) photographs of different samples. (**a**) Raw Al powder; (**b**) Al/B/Ni nano-sized composite particles. Reproduced from [36], with permission from Elsevier, 2016; (**c**) Ni/B nano-sized alloy; and (**d**) Al/B/Ni nano-sized composite particles. Reproduced from [50], with permission from Han Neng Cai Liao, 2009.

A certain proportion of nano-sized Ni/B amorphous alloy and AP are ground and compounded on a planetary ball mill in the presence of a small amount of anhydrous ethanol, and then dried to obtain the nano-sized Ni/B composite particles and AP. Figure 12 shows the DTA (differential thermal analysis) curves of AP with different mass ratios of nano-sized Al/B/Ni composite particles, and the mass fraction data refers to the percentage of Ni/B in Ni/B/Al and AP composites, which is 5%, 10%, and 12%, respectively. It is shown that, with the increase of the mass fraction of nano-sized Ni/B particles, the exothermic peak temperature of AP decreases, indicating that the catalytic effect increased with an increase in the mass fraction of the nano-sized composite. However, when the mass fraction is more than 10%, the exothermic peak temperature cannot continue to decline, and the decomposition speed decreases, indicating that the catalytic effect does not continue to increase with its mass fraction. Moreover, when the mass fraction of nano-sized Ni/B is 12%, the exothermic peak temperature of composites is higher than that of pure AP at low-temperature range, indicating that nano-sized Ni/B particle hinders the low-temperature thermal decomposition of AP. Additionally, the minimum (or maximum) value of the quadratic curve equation is used to get the optimum amount of catalyst in the graph, as well as the fitting equation between the decomposition temperature and mass fraction of each sample in Figure 12, which is $y = 1.363x^2 - 21.625x + 475.77$, where the fitting coefficient is 0.97, it was calculated that the optimum mass fraction of Ni/B particles in Al/NiB composite is 7.93%, and the corresponding minimum temperature is 389.98 °C on the theoretical high-temperature decomposition, indicating that the catalytic effect of nano-sized particles has been improved significantly after treatment.

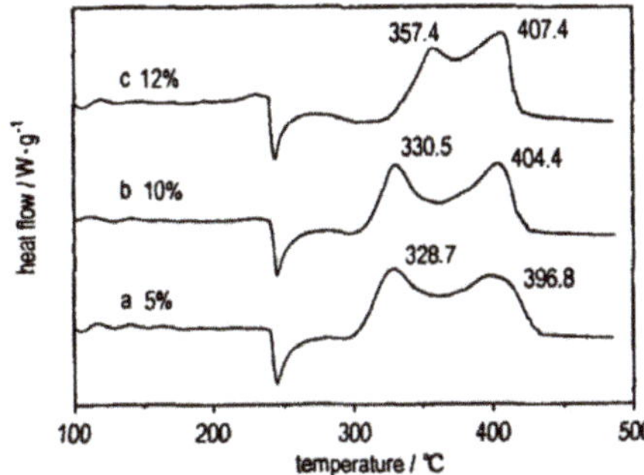

Figure 12. DTA curves of AP with different ratios of nano-sized Al/B/Ni composite particles. Reproduced from [50], with permission from Han Neng Cai Liao, 2009.

3.2.4. Nano-Sized Al/RDX/Fe$_2$O$_3$ Energetic Composite

The sol–gel method is one of the most important techniques for preparation of nano-sized particles, while nano-sized Al/RDX/Fe$_2$O$_3$ composites with a mass fraction of 85% hexogen (RDX), 3.75% aluminum, and 11.25% iron oxide (Fe$_2$O$_3$) were prepared by sol–gel template and supercritical CO$_2$ fluid drying technology, where mechanical sensitivity and detonation velocity were compared with those of pure RDX. Particles were characterized by scanning electron micro-scale (SEM) shown in Figure 13. Fe$_2$O$_3$ aerogel is a complete "honeycomb", which has a relatively uniform pore size, and the average pore size is between 50 nm to 100 nm. When the initial particles are polymerized to form gel, the particles are combined to form a three-dimensional porous aerogel skeleton. After supercritical fluid drying, the liquid in the pores is taken away, forming the honeycomb structure of Fe$_2$O$_3$ (Figure 13a). A small amount of RDX grains are embedded on the surface of the crystal, and they are formed with the evaporation and diffusion of dimethylformamide (DMF) during the drying process. The structure of Fe$_2$O$_3$ colloid is broken and a small amount of RDX grains are exposed (Figure 13b). The nano-sized composites in high-resolution microscopic SEM photos exhibit a spherical or near-spherical shape, and the average particle size is in the range of 100–200 nm (Figure 13c).

Figure 13. SEM photos of different particles. (**a**) Fe_2O_3 gas gel; (**b**) lowresolution microscopic of Al/RDX/Fe_2O_3 composite; and (**c**) high-resolution microscopic of Al/RDX/Fe_2O_3 composite. Reproduced from [51], with permission from Han Neng Cai Liao, 2011.

For this formation mechanism of nano-sized composite, DMF solution of the explosive is filled inside the gel mesh in the process of forming Fe_2O_3 gel, and the nAl powder is evenly dispersed in the gel system by ultrasonic vibration instrument. Using the supercritical fluid drying technology by the super solubility and diffusivity of supercritical CO_2 fluid, RDX rapidly recrystallizes around the nAl powder in the gel hole, which suppressed the further crystallization and growth of RDX effectively. The nano-CEMs were formed from nano-sized RDX and nAl powder encapsulated by Fe_2O_3 colloid.

Under the same experimental conditions, the impact sensitivity and friction sensitivity of nano-sized Al/RDX/Fe_2O_3 composites are much lower than that of pure RDX (Table 1). This can be related to the "hot spot" formation of the samples during the test process. On the one hand, the grains of nano-sized composites are much smaller, its surface structure integrity is good, and there are few defects, which make it difficult to form local "hot spots" inside. On the other hand, nano-sized particles also have a unique lubrication effect, reducing the friction between the particles effectively, thus the smaller size of the "hot spot" decreases the mechanical sensitivity. The detonation velocity of nano-sized composites is 7185 m·s^{-1}, which is higher than that of pure RDX by 615 m·s^{-1}. This may be because of the difference between the nano-sized composites and the micro-scale mixed explosives. The mass transfer rate between the explosives or between explosive and metal for micro-scaled explosives is slow, while nano-scaled energetic transfer between explosive particles would not be affected by the mass transfer rate, which contributes to the high detonation velocity [51].

Table 1. Hazardous properties of RDX and different compositions.

Samples	Impact Sensitivity/cm	Standard Deviation	Friction Sensitivity/%	Detonation Velocity/m·s^{-1}
RDX	22.5	0.04	96	6570
Al/RDX/Fe_2O_3	50.2	0.05	7	7185

Note: impact sensitivity test condition: sample mass is (35 ± 1) mg, drop weight is 2.5 kg, and relative huimidity is ≤80% at room temperature; friction sensitivity test condition: swing angle is (90 ± 1)°, gage pressure is 3.92 MPa, and sample mass is (20 ± 1) mg; detonation velocity test condition: sample shape is Φ10 mm × 10 mm and density is 1.55 g cm^{-3}.

In order to compare the properties with Al/RDX/Fe_2O_3, nano-sized B/RDX/Fe_2O_3 (mass ratio is 2:90:8) composites were prepared by adding RDX and B powder into the gel template of Fe_2O_3, which were prepared by the sol–gel process and dried by supercritical CO_2 fluid drying technology [52]. The results indicate that the average particle size of nanocomposite energetic material is 30–50 nm. Compared with the pure RDX, the onset thermal decomposition temperature of B/RDX/Fe_2O_3 increases by 7 °C, the reaction heat increases by 885 J·g^{-1}, and the impact sensitivity (characteristic drop height, H_{50}) is 40.8 cm.

3.2.5. Nano-Sized Al/RDX/SiO_2 Energetic Composite

With the continuous improvement of the performance requirements for energetic materials, RDX is one of the most widely used single explosives. In order to reduce the mechanical sensitivity of RDX and improve its thermal decomposition performance, three types of Al/RDX/SiO$_2$ with 30%, 50%, and 70% of Al/RDX in the composites were prepared by means of the sol–gel method (Figure 14). In Figure 14a, SiO$_2$ shows a cellular network structure. Granular Al/RDX is filled in only a few holes in the SiO$_2$ skeleton, owing to the small proportion of Al/RDX (Figure 14b). With the increase of the Al/RDX mass fraction, the filling amount of Al/RDX in the SiO$_2$ skeleton increased significantly, and only few holes are not filled in (Figure 14c). All holes in the SiO$_2$ skeleton are filled by Al/RDX, which leads to the obvious accumulation of Al/RDX particles in the skeleton (Figure 14d).

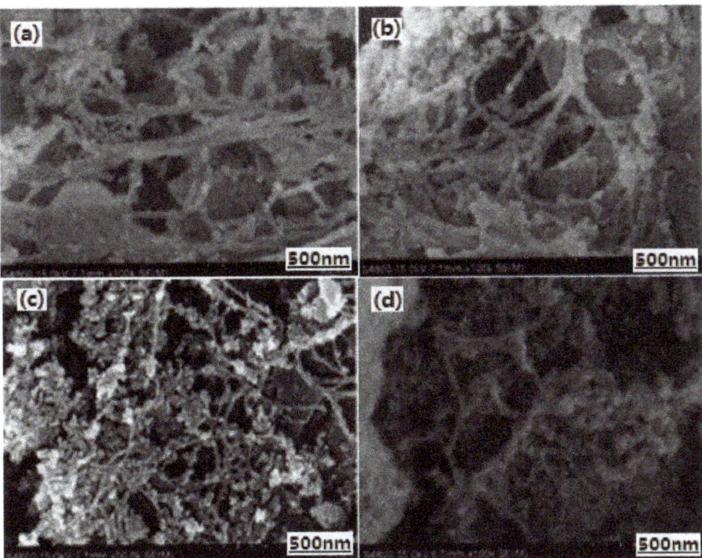

Figure 14. SEM photos of SiO$_2$ xerogel and nano-sized Al/RDX/SiO$_2$ composite particles. (a) SiO$_2$ xerogel; (b) 30%; (c) 50%; (d) 70%. Reproduced from [53], with permission from Han Neng Cai Liao, 2017.

On the basis of the preparation of nano-sized Al/RDX/SiO$_2$ composite particles, its effects on the thermal decomposition of RDX were investigated by DSC and TG techniques at the heating rate of 10 °C·min^{-1} (Figure 15). The melting and thermal decomposition peak temperatures of RDX in the prepared nano-sized Al/RDX/SiO$_2$ composite materials are 1.56–4.49 °C and 18.9–22.4 °C, respectively, which are earlier than those of pure RDX (Figure 15a). With the increase of Al/RDX mass fraction in the SiO$_2$ skeleton, the decomposition peak temperature with heat release of RDX increases gradually in the composite. This may be because, with the increase of Al/RDX mass fraction in the composition, the SiO$_2$ skeleton is relatively reduced, and the agglomeration and accumulation of fillers occur in the skeleton, which results in the decrease of contact area between the reactants, thus the exothermic decomposition peak temperature of RDX in composites slightly increases on the macro-scale. Moreover, with the increase of Al/RDX mass fraction in nano-sized Al/RDX/SiO$_2$ composites, the mass loss of RDX in the composites is gradually delayed (Figure 15b). The reason may be that, with the increasing mass fraction of Al/RDX filler in the SiO$_2$ framework, some holes in the SiO$_2$ framework are filled unevenly, collapse, and grain agglomeration, which reduces the contact area between the reactants. The mass transfer and heat transfer in the chemical reaction process are adversely affected, which is reflected in the TG curves (Figure 15b), that is, the mass loss of RDX in nano-sized Al/RDX/SiO$_2$ composites increased with the increase of the Al/RDX mass fraction.

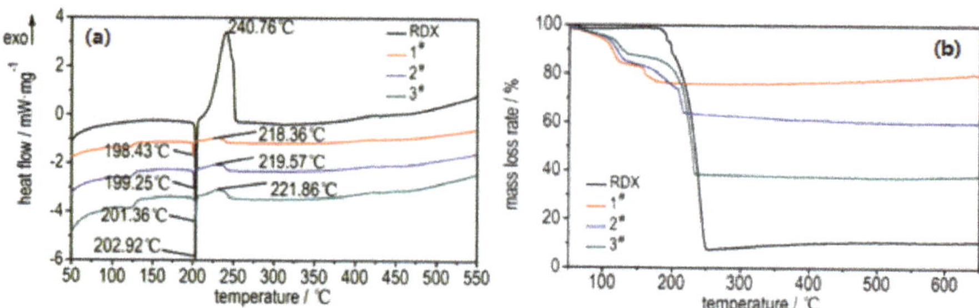

Figure 15. DSC (**a**) and TG (**b**) curves of pure RDX and nano-sized Al/RDX/SiO$_2$ composite particles. 1# is 30%, 2# is 50%, and 3# is 70%. Reproduced from [53], with permission from Han Neng Cai Liao, 2017.

At the same time, the mechanical sensitivities of nano-sized Al/RDX/SiO$_2$ composites were tested, which were compared with the particles by means of mechanical mixed procedure. The friction sensitivity and impact sensitivity of nano-sized Al/RDX/SiO$_2$ composites (26–68%, 27.1–95.6 cm) is much lower than those of pure RDX (10%, 134.9 cm) and mechanical mixed ones (74–85%, 18.6–25.1 cm).

In order to improve the combustion of RDX/Al/SiO$_2$ composites, nano-sized Al/RDX/AP/SiO$_2$ composites were prepared and characterized [54]. As shown in Figure 16a, the morphology of the spherical composite is uniform, the average particle size (higher than 200 nm) and gradation are reasonable, and their accumulation is close. During the dry process, as the solvent evaporates, the gel pores collapse gradually. AP and RDX with solvent diffuse from one gel hole to another gel hole or gel surface (Figure 16b). For the gel complex by the supercritical drying process (Figure 16c), supercritical CO$_2$ rapidly penetrates into the gel hole and extracts the organic solvents instantaneously, so that a large number of AP and RDX can crystallize rapidly in each gel hole. Supercritical CO$_2$ has super solubility and diffusivity, which can effectively inhibit the diffusion and transfer of RDX between adjacent gel pores, thereby maintaining the integrity of the gel void structure.

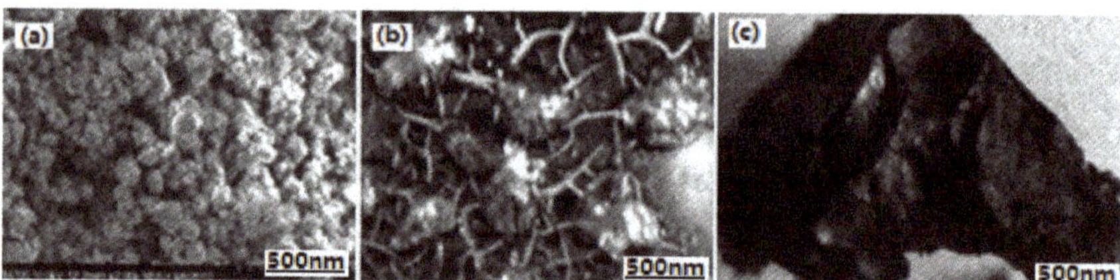

Figure 16. SEM photos of different particles. (**a**) Nano-sized Al/RDX/AP/SiO$_2$ composite; (**b**) gel complex by vacuum freeze drying; and (**c**) gel complex by supercritical drying. Reproduced from [55], with permission from Nanjing University of Science and Technology, 2018.

Moreover, the mechanical sensitivities of nano-sized Al/RDX/AP/SiO$_2$ composites were tested, which were compared with the particles by means of mechanical mixed procedure. The impact sensitivity of nano-sized Al/RDX/SiO$_2$ composites (48.3 cm) is much lower than that of the mechanical mixed one (30.7 cm) [55].

In addition, nano-sized Al/AP composites were prepared by means of the freeze-drying process by Martin et al., nAl powder was added to AP solution, and then the mixture solution was quickly poured into the container filled with liquid nitrogen to obtain the low-density nano-sized Al/AP composites, which improve the combustion performance of propellants [56,57].

3.3. Properties and Applications of Typical Nano-CEMs

At present, most studies use organic solvent as dispersant to disperse nAl powder and oxide. Considering the high activity of nAl and high flammability in the organic solvent, the preparation process is very dangerous. One alternative method is to use water as the dispersant. Because of the high activity of nAl, the nAl powder can still react with water, resulting in the decrease of activity, although the surface is passivated by a layer of oxide [58,59]. It was found that, based on silane or oleic acid, the organic surface coating can effectively inhibit the reaction between nAl powder and water, ensuring its activity in a long period of time.

The impact ignition energy of nano-sized Al/MoO_3 composites is 0.3–1.7 J, which increases with an increase in the particle size. The maximum reaction rate is related to the initial particle size, and the reaction rate increases with the decrease of the particle size. The combustion mechanism and reaction properties of four types of nano-sized Al/MoO_3, Al/WO_3, Al/CuO, and Al/Bi_2O_3 composites were conducted; it was found that the propagation speed depends on the generation of gas and the thermodynamic state of reaction products, and the burning rate decreases with the increase of mixture density, which is contrary to the traditional explosion theory, so the combustion propagation mechanism is different from that of the traditional explosion theory. Nano-sized Al/Cr_2O_3 with 10–15 nm in diameter was prepared, and the relationship between particle size and combustion performance by means of CO_2 laser ignition technique was conducted; the performance is significantly improved compared with that of micro-scaled particles. One interesting point is that nano-sized $Al/WO_3 \cdot H_2O$ composites were prepared by means of the wet chemical method, and its performance was compared with that of nano-sized Al/WO_3 composites; it was found that the energy release rate of reaction of $Al/WO_3 \cdot H_2O$ is significantly improved by the reaction between nAl and H_2O in $WO_3 \cdot H_2O$ to produce hydrogen, which increased the total heat release and energy release rate of the reaction.

Three nanocomposite materials with the same nominal stoichiometric thermite composition of Al/CuO were prepared by three different methods: ultrasonic mixing (USM) of constituent nanopowders, electrospraying (ES), and arrested reactive milling (ARM). The combustion performances of the three prepared types of Al/CuO nanocomposites were conducted [60]. Prepared powders were placed in a 6.7 mm diameter, 0.5 mm deep cavity in a brass substrate and ignited by electro-static discharge. The experiments were performed in air, argon, and helium. The mass of powder removed from the sample holder after ignition was measured in each test. Using a multi-anode photo-multiplier tube coupled with a spectrometer, time-resolved light emission traces produced by the ignited samples were recorded in the range of wavelengths of 373–641 nm. Time-resolved temperatures were then determined by fitting the recorded spectra assuming Planck's black body emission. Temporal pressure generated by ignition events in the enclosed chamber showed that the powder's combustion properties were tied to both their preparation technique as well as the environment in which they were ignited. The agglomeration of nanoparticles hindered the combustion of USM powders, while it was not observed for the ES powders. Lower temperatures and pressures were observed in oxygen-free gas environments for USM and ES powders prepared using starting nano-particles. For the ES powders, the effect of gas environment was less significant, which was interpreted considering that ES materials included gasifying nitrocellulose binder, enhancing heat, and mass transfer between individual Al and CuO particles. Higher pressures and temperatures were observed in inert environments for the ARM-prepared powder.

Addition of reactive nanocomposite powders can increase the burn rate of aluminum, and thus the overall reaction rate of the energetic formulation. Replacing only a small fraction of the fuel by a nanocomposite material can enhance the reaction rate with little change to the thermodynamic performance of the formulation. The nanocomposite materials $Al/3CuO$ and Al/MoO_3 prepared by ARM, a scalable "top-down" technique for manufacturing reactive nano-materials, were added to micron-sized aluminum powder and the mixture was aerosolized and burned in a constant volume chamber with varied oxygen, nitrogen, and methane atmosphere [61]. The resulting pressure traces were recorded and processed to compare different types and amounts of modifiers. Additives of nanocomposite powders of Al/MoO_3 to micron-sized aluminum were found to be effective in increasing both the rate of pressure rise and maximum pressure in the respective constant volume explosion experiments. It was observed that 20 wt % of additive resulted in the best combination of the achieved burn rate and pressure.

The potential application of nano-aluminum/nitrocellulose mesoparticles as an ingredient for solid composite rocket propellants was investigated [62]. The basic strategy is to incorporate nanoaluminum in the form of a micrometer scale particle containing a gas-generator, to enable easier processing and potential benefits resulting from reduced sintering prior to combustion. The mesoparticles were made by electrospray and comprised aluminum nanoparticles (50 nm) and nitrocellulose to form micrometer scale particles. Eighty percent solids loaded composite propellants (AP/HTPB (hydroxilterminated polybutadiene) based) were made with the addition of micrometer sized (2–3 μm) aluminum (10 wt %), and compared directly to propellants made by directly substituting aluminum mesoparticles for traditional micrometer sized particles. The propellant burning rate was relatively insensitive for mesoparticles containing between 5 wt % and 15 wt % nitrocellulose. However, direct comparison between a mesoparticle-based propellant and a propellant containing micrometer scale aluminum particles showed burning rates approximately 35% higher, while having a nearly identical burning rate exponent. High-speed imaging indicates that propellants using mesoparticles have less agglomeration of particles on the propellant surface.

The results show that the burning rate of nano-CEMs is the fastest when the particle size of Al powder and oxide is at nano-metric level, while the burning rate of CEMs composed of micron Al powder and nano-metric oxide is much faster than that of nanometer Al powder and micro-metric level.

4. Conclusions and Future Challenges

As one of new type of functional material, nano-sized composite energetic materials (nano-CEMs) have broad application prospects. However, most of the studies are carried out at the theoretical and laboratory level at present, which is still a long way from engineering and practical application. In the research field of high-energy explosives, the nano-sized composite process can cause the surface of each component of explosive to fully contact each other. On the one hand, it can decrease the agglomeration of ultrafine particles in explosives or improve the physical and chemical properties of explosives. On the other hand, it can also adjust the sensitivity of explosives and improve the safety performance of explosives. In the research field of solid propellants, using the nano-sized composite synthesis technologies, the powder and ultrafine catalyst are prepared into ultrafine composite particles, which can increase the contact area, improve the actual catalytic ability of the catalyst, and then greatly improve the energy release efficiency of the powder [63].

Nano-sized composite energetic materials (nano-CEMs) possess the characteristics of high energy density, high combustion rate of a few kilometers/second, and micron scale critical reaction propagation size. They have shown good application potential in many aspects, such as micro electro mechanical systems (MEMS) devices, anti-infrared decoy materials, and high energy additives. In addition, the combination between oxidants and reducers exhibits unique reaction dynamics characteristics at nano-scale, such as particle size dependence, mass transfer diffusion, energy release, and other mechanisms, which are quite different from the traditional micro-sized scale solid-state reaction. In the past years of study, significant progresses have been made in developing preparation methods, investigating properties, fundamental theory, and applications of nano-CEMs. However, the following difficult challenges to be addressed in the future have been identified.

(1) Understanding of reaction propagation mechanism of nano-CEMs is still in its infancy.

On the basis of the "melting diffusion mechanism" and "metal oxygen turnover mechanism", only the mass diffusion and reaction rate at the interface were considered, which can simply help us to understand the escape of oxygen atom (O) in oxide at the initial stage (ignition stage) of reaction and the process of combining with active metal. In fact, the reaction propagation mechanism is more complicated after ignition during combustion. In the face of this challenge with complex and multiple factors, an effective strategy is to combine the experiment and theoretical model by peeling off the single influencing factor. However, these models are too simple, and there are few experiments to describe the multiple factors [1,2,64–67]. Thus, the understanding of the ignition and reaction propagation mechanism of nano-CEMs still requires the joint efforts of colleagues all over the world.

(2)   The performance breakthrough is also limited by loading density, stability, and other factors.

Although nano-CEMs have high bulk energy density, it is difficult to obtain the ideal output energy, such as high pressure and shock wave owing to the lack of gas release within the reaction products. In order to obtain a high burning rate, it is necessary to heat the air between the powder voids to accelerate the convection and mass transfer, which significantly increases the dependence of reaction speed and loading density. Although the reaction speed can be slightly increased by the design of hollow spheres, branch, and other microstructures, it is difficult to balance the contradiction between the high bulk density and reaction rate of the composition. The reaction speed of the composition could be slightly improved by introducing gas products through nano-CEMs' composition design. However, it will bring new problems, such as stability and storage performance [1,2,37–39].

(3)   Engineering application needs new breakthroughs and deep investigations.

Because most nano-CEMs use highly active nano-sized metals (such as Al, Mg) as reducing agents, how to avoid or reduce the self-oxidation of the highly active metals is the prime problem in engineering applications. For nano-sized aluminum powder, it is easy to agglomerate and influence its dispersion compositions by coating nano Al powder with polymer or other inert materials. In addition, in order to ensure the coating quality, the mass fraction of the inert coating materials is often high, which would reduce the energy density of the composition and increase the complexity of the reaction. In the process of formulation design, it is very difficult to bond the nano-CEMs with high specific surface area using the few mass fractions of binder [1,2,41–45], which is one of the common problems of nano-CEMs in the practical application.

Author Contributions: Conceptualization, W.P. and X.F.; resources, W.P., H.X., and K.W.; data curation, W.P.; writing—Original draft preparation, W.P.; writing—Review and editing, Y.C.; project administration, F.Z. and W.P.; funding acquisition, F.Z. and Z.Q. All authors have read and agreed to the published version of the manuscript.

Funding: This research was funded by "National Natural Science Foundation of China, grant number 21703167" and "Science and Technology on Combustion and Explosion Laboratory Foundation of China, grant number 2019SYSZCJJ".

Acknowledgments: The authors are thankful to Juan Zhao from Analysis and Measurement Center on Energetic Materials, Xi'an Modern Chemistry Research Institute for the particle micrograph experiments and Xiao-ning Ren from Science and Technology on Combustion and Explosion Laboratory, Xi'an Modern Chemistry Research Institute for the thermal decomposition tests. We are specially thankful to Luigi T. DeLuca from Politecnico di Milano, Italy and Djalal Trache from Ecole Militaire Polytechnique, Algeria for providing many helpful suggestions and English statements.

References

1.   He, W.; Liu, P.-J.; He, G.-Q.; Go, M.; Yan, Q.-L. Highly reactive metastable intermixed composites (MICs): Preparations and characterizations. *Adv. Mater.* **2018**, *30*, 1706293. [CrossRef] [PubMed]

2.   Pang, W.-Q.; Fan, X.-Z.; Zhang, Z.-P. *Nano-Sized Metal Powder: Preparation, Characterization and Energetic Application*; National Defense Industry Press: Beijing, China, 2016.

3.   An, T.; Zhao, F.-Q.; Hao, H.-X.; Ma, H.-X.; Yao, E.-G.; Yang, Y.; Tan, Y. Effect of thermites on laser ignition characteristics of double base propellants. *Chin. J. Explos. Propellants* **2011**, *34*, 67–72.

4.   An, T.; Zhao, F.-Q.; Gao, H.-X.; Hao, H.-X.; Ma, H.-X. Sensitivity characteristics of double base propellants containing super thermites. *J. Propuls. Technol.* **2013**, *34*, 129–134.

5.   Dilip, S.; Vigor, Y.; Richard, A.Y. Metal-based nanoenergetic materials: Synthesis, properties, and applications. *Prog. Energy Combust. Sci.* **2017**, *61*, 293–365.

6. Zhang, X.-T.; Song, W.-L.; Guo, L.-G.; Hu, M.-L.; Xie, C.-S. Preparation of carbon-coated Al nanopowders by laser-induction complex heating method. *J. Propuls. Technol.* **2007**, *28*, 333–338.

7. Zhou, J.; Ding, L.; An, J.; Zhu, Y.; Liang, Y. Study on the thermal behaviors of nano-Al based fuel air explosive. *J. Therm. Anal. Calorim.* **2017**, *130*, 1111–1116. [CrossRef]

8. DeLuca, L.T.; Galfetti, L.; Severini, F.; Meda, L.; Marra, G.; Vorozhtsov, A.B.; Sedoi, V.S.; Babuk, V.A. Burning of nano-aluminized composite rocket propellants. *Combust. Explos. Shock Waves* **2005**, *41*, 680–692.

9. Paravan, C.; Verga, A.; Maggi, F.; Galfetti, L. Accelerated ageing of micron- and nano-sized aluminum powders: Metal content, composition and non-isothermal oxidation reactivity. *Acta Astronaut.* **2019**, *158*, 397–406. [CrossRef]

10. Bockmon, B.S.; Son, S.F.; Asay, B.W.; Busse, J.R.; Mang, J.T.; Peterson, P.D.; Pantoya, M.L. Combustion performance of metastable intermolecular composites (MIC). *Cpia Publ.* **2002**, *712*, 613–624.

11. Borgonovo, C.; Apelian, D. Manufacture of Aluminum nanocomposites: A critical review. *Mater. Sci. Forum* **2011**, *678*, 1–22. [CrossRef]

12. Simpson, R.L.; Tillotson, T.M.; Hrubesh, L.W.; Gash, A.E. Nanostructured energetic materials derived from sol-gel chemistry. *J. Non Cryst. Solid* **2001**, *285*, 338–345.

13. Guo, Q.-X.; Nie, F.-D.; Yang, G.-C.; Li, J.-S.; Chu, S.-J. Preparation of RDX/resorcinol-formaldehyde (RF) nano-composite energetic materials by sol-gel method. *Chin. J. Energetic Mater.* **2006**, *14*, 268–271.

14. Zhou, C.; Li, G.-P.; Luo, Y.-J. Preparation of Fe$_2$O$_3$/Al nano composite by sol-gel method. *Chin. J. Explos. Propellants* **2010**, *33*, 1–4.

15. Jiang, X.-B.; Liang, Y.-Q.; Zhang, J.-L.; Chen, J.-S. Preparation of RDX/SiO$_2$ booster membrane by sol-gel method. *Chin. J. Energetic Mater.* **2009**, *17*, 689–694.

16. Gash, A.E.; Tillotson, T.M.; Sateher, J.H., Jr.; Poco, J.F.; Hrubesh, L.W.; Simpson, R.L. Use of epoxides in the sol-gel synthesis of porous Fe$_2$O$_3$ monliths from Fe (III) salts. *Chem. Mater.* **2001**, *13*, 999–1007. [CrossRef]

17. Ma, Z.-Y.; Li, F.-S.; Chen, A.-S.; Song, H.-C. Preparation and characterization of Al/ammonium perchlorate composite particles. *J. Propuls. Technol.* **2004**, *25*, 373–376.

18. Ma, Z.-Y.; Li, F.-S.; Chen, A.-S.; Bai, H.-P. Preparation and thermal decomposition behavior of Fe$_2$O$_3$/ammonium perchlorate composite nanoparticles. *Acta Chim. Sin.* **2004**, *62*, 1252–1255.

19. Xu, G.-C.; Zhang, L.-D. *Nano Composite Materials*; Chemical Industry Press: Beijing, China, 2002.

20. Xie, Q.-D. *Preparation of Carbide and Nanocomposites by High Energy Ball Milling*; Zhejiang University: Zhejiang, China, 2003.

21. Shi, X.-F. *Preparation and Properties of Micro-Spherical Energetic Nanocompoistes Prepared by Spray Drying*; North University of China: Taiyuan, China, 2015.

22. Yusoff, M.S.M.; Paulus, W.; Muslim, M. Synthesis of nano-sized alpha alumina using solvothermal and hydrothermal methods: A comparative study. In Proceedings of the 2012 International Conference on Enabling Science and Nanotechnology (ESciNano), IEEE, Johor Bahru, Malaysia, 5–7 January 2012.

23. Chen, D.; Zhao, Y.; Zhu, H.; Zheng, M.; Chen, G. Microstructure and mechanism of in-situ Al$_2$O$_3$(p)/Al nano-composites synthesized by sonochemistry melt reaction. *Trans. Nonferrous Met. Soc. China* **2012**, *22*, 36–41. [CrossRef]

24. Gedanken, A. Using sonochemistry for the fabrication of nanomaterials. *Ultrason. Sonochem.* **2004**, *11*, 47–55. [CrossRef]

25. Martin Joe, A.; Welch Larry, H. Composition and Method for Preparing Oxidizer Matrix Containing Dispersed Metal Particles. WO Patent WO 2001/38264 Al, 27 August 2001.

26. Dossi, S.; Reina, A.; Maggi, F.; DeLuca, T.L. Innovative metal fuels for solid rocket propulsion. *Int. J. Energ. Mater. Chem. Propul.* **2012**, *11*, 299–322.

27. Sullivan, K.T.; Chiou, W.; Fiore, R.; Zachariah, M.R. In situ microscopy of rapidly heated nano-Al and nano-Al/WO$_3$ thermites. *Appl. Phys. Lett.* **2010**, *97*, 133104.

28. Jian, G.; Piekiel, N.W.; Zachariah, M.R. Time-resolved mass spectrometry of nano-Al and nano-Al/CuO thermite under rapid heating: A mechanistic study. *J. Phys. Chem. C* **2012**, *116*, 26881–26887. [CrossRef]

29. Wang, H.; Zachariah, M.R.; Xie, L.; Rao, G. Ignition and combustion characterization of nano-Al-AP and nano-Al-CuO-AP micro-sized composites produced by electrospray technique. *Energy Procedia* **2015**, *66*, 109–112. [CrossRef]

30. Liu, P.; Liu, J.; Wang, M. Ignition and combustion of nano-sized aluminum particles: A reactive molecular dynamics study. *Combust. Flame* **2019**, *201*, 276–289.

31. Chao, Z.; Guo-Ping, L.; Yun-Jun, L. Preparation and characterization of porous Fe_2O_3/Al composite energetic material. *Chin. J. Explos. Propellants* **2010**, *33*, 1–4.

32. Tillotson, T.M.; Gash, A.E.; Simpson, R.I.; Hrubesh, L.W.; Satcher, J.J.H.; Poco, J.F. Nanostructured energetic materials using sol-gel methodologies. *J. Non Cryst. Solids* **2001**, *85*, 338–345.

33. Jin, M.-M.; Luo, Y.-J. Preparation and characterization of NC/Al nano-composite energetic materials. *Chin. J. Energetic Mater.* **2013**, *21*, 230–234.

34. Zhu, Y.; Ma, X.-X.; Zhang, K.-L. Study on Nano-structured MnO_2/Al Energetic Composite. *Initiators Pyrotech.* **2018**, *1*, 23–27.

35. Ke, X.; Zhou, X.; Gao, H.; Hao, G.Z.; Xiao, L.; Chen, T.; Liu, J.; Jiang, W. Surface functionalized core/shell structured CuO/Al nanothermite with long-term storage stability and steady combustion performance. *Mater. Des.* **2018**, *140*, 179–187. [CrossRef]

36. Dreizin, E.L.; Schoenitz, M. Nano-composite energetic powders prepared by arrested reactive milling. *Thermochim. Acta* **2016**, *636*, 48–56.

37. Williams, R.A.; Schoenitz, M.; Dreizin, E.L. Validation of the thermal oxidation model for Al/CuO nanocomposite powder. *Combust. Sci. Technol.* **2014**, *186*, 46–67. [CrossRef]

38. Wang, J.; Zheng, B.; Qiao, Z.-J.; Chen, J.; Zhang, L.-Y.; Zhang, L.; Li, Z.-Q.; Zhang, X.-Q.; Yang, G.-C. Construct 3D porous hollow Co_3O_4 micro-sphere: A potential oxidizer of nano-energetic materials with superior reactivity. *Appl. Surf. Sci.* **2018**, *442*, 767–772. [CrossRef]

39. Zakiyyan, N.; Wang, A.-Q.; Thiruvengadathan, R.; Staley, C.; Mathai, J.; Gangopadhyay, K.; Maschmann, M.R.; Gangopadhyay, S. Combustion of aluminum nanoparticles and exfoliated 2D molybdenum trioxide composites. *Combust. Flame* **2018**, *187*, 1–10. [CrossRef]

40. Dixon, G.P.; Martin, J.A.; Thompson, D. Lead-Free Precussion Peimer Mixes Based on Metastable Interstitial Composite (MIC) Technology. Patent US 5717159, 10 February 1998.

41. Troy, J.; Barbee, W.; Simpson, R.L.; Gash, A.E.; Joe, S.H., Jr. Nano-Laminate-Based Ignitors. Patent US 2004 0060625, 4 January 2004.

42. Williams, R.A.; Schoenitz, M.; Ermoline, A.; Dreizin, E.L. Low-temperature exothermic reactions in fully-dense Al/MoO_3 nanocomposite powders. *Thermochim. Acta* **2014**, *594*, 1–10. [CrossRef]

43. Wang, J.; Zhang, W.-C.; Shen, R.-Q.; Ye, J.-H.; Wang, X.-W. Research progress of nano thermite. *Chin. J. Explos. Propellants* **2014**, *37*, 1–9.

44. Jin, X.-Y.; Hu, Y.; Shen, R.-Q.; Ye, Y.-H. Preparation and laser ignition studies of Al/Ni energetic nanocomposite. *Explos. Mater.* **2012**, *41*, 12–15.

45. Nemtseva, S.Y.; Lebedev, E.A.; Shaman, Y.P.; Lazarenko, P.I.; Ryazanov, R.M.; Gavrilov, S.A.; Gromov, D.G. Nano-sized Al-Ni energetic powder material for heat release element of thermoelectric device. *J. Phys. Conf. Ser.* **2018**, *1124*. [CrossRef]

46. Lee, H.M.; Yun, J. Preparation of aluminum-oleic acid nano-composite for application to electrode for Si solar cells. *Mater. Trans.* **2011**, *52*, 1222–1227. [CrossRef]

47. Liu, X.; Li, Y.; Liu, Y.; Sun, D.; Guo, W.; Sun, X.; Feng, Y.; Chi, H.; Li, X.; Tian, F.; et al. Performance and microstructure characteristics in polyimide/nano-aluminum composites. *Surf. Coat. Technol.* **2017**, *320*, 103–108. [CrossRef]

48. Shen, L.-H.; Li, G.-P.; Luo, Y.-J.; Gao, K.; Chai, C.-P.; Ge, Z. Preparation of Al/B/Fe_2O_3 nano-composite energetic materials by high energy ball milling. *J. Solid Rocket Technol.* **2014**, *37*, 233–237.

49. Yang, F.; Kang, X.-L.; Luo, J.-S.; Yi, Z.; Tang, Y.-J. Preparation of core-shell structure KClO4@Al/CuO nano energetic material and enhancement of thermal behavior. *Sci. Rep.* **2017**, *7*, 3730. [CrossRef] [PubMed]

50. Yang, Y.; Pan, Z.-H.; Li, L.-X.; Li, Y.-B.; Cao, X.-F. Preparation of nanometer NiB/Al composite and its thermal catalysis effect on AP decomposition. *Chin. J. Energetic Mater.* **2009**, *17*, 446–450.

51. Wang, R.-H.; Zhang, J.-L.; Wang, J.-Y.; Pan, J.; Zhang, J. Preparation and characterization of nano-composite energetic materials Fe_2O_3/Al/RDX. *Chin. J. Energetic Mater.* **2011**, *19*, 739–742.

52. Wang, R.-H.; Jin, R.-Y.; Wang, J.-Y.; Zhang, J.-L.; Wang, D.-J. Preparation of RDX/B/Fe_2O_3 nano-composite energetic material with gel-template method. *Chin. J. Energetic Mater.* **2015**, *23*, 410–414.

53. Zhang, D.-D.; Huang, Y.-S.; Li, R.; Li, M.; Wang, J.-J.; Ge, M.-Z.; Zhang, H.-J.; He, Y.-L. Preparation and properties of RDX/Al/SiO_2 nano-composite energetic materials. *Chin. J. Energetic Mater.* **2017**, *25*, 656–660.

54. Zhang, D.-D. *Preparation and Properties of RDX Based Nanocomposite Energetic Materials*; Nanjing University of Science and Technology: Nanjing, China, 2018.

55. Pan, J.-J.; Zhang, J.-L.; Shang, F.-F.; Song, X.-X. Preparation and characterization of submicron composite energetic materials RDX/AP/Al/SiO$_2$. *Shanxi Chem. Ind.* **2011**, *31*, 15–17.

56. Pang, W.-Q.; DeLuca, T.L.; Fan, X.-Z.; Wang, K.; Li, J.-Q.; Zhao, F.-Q. Progress on modification of high active aluminum powder and combustion agglomeration in chemical propellants. *J. Solid Rocket Technol.* **2019**, *42*, 42–53.

57. Carole, R. *Al-Based Energetic nano Materials: Design, Manufacturing, Properties and Applications*; Wiley: Hoboken, NJ, USA, 2015.

58. Puszynski, J.A. Processing and characterization of aluminum-based nanothermites. *J. Therm. Anal. Calorim.* **2009**, *96*, 677–685. [CrossRef]

59. Puszynski, J.A. *Reactivity of Nano-Sized Aluminum with Metal Oxides and Water Vapor. Materials Research Society Proceedings*; Materials Research Society: Boston, MA, USA, 2003.

60. Monk, I.; Schoenitz, M.; Jacob, R.J.; Dreizin, E.L.; Zachariah, M.R. Combustion characteristics of stoichiometric Al-CuO nanocomposite thermites prepared by different methods. *Combust. Sci. Technol.* **2017**, *189*, 555–574. [CrossRef]

61. Demitrios, S.; Xianjin, J.; Ervin, B.; Edward, L.D. Aluminum burn rate modifiers based on reactive nanocomposite powders. *Propellants Explos. Pyrotech.* **2009**, *35*, 260–267.

62. Young, G.; Wang, H.; Zachariah, M.R. Application of nano-aluminum/nitrocellulose mesoparticles in composite solid rocket propellants. *Propellants Explos. Pyrotech.* **2015**, *43*, 413–418. [CrossRef]

63. Yan, Q.-L.; Zhang, X.-H.; Qi, X.-F.; Liu, M.; Li, X.-J. Preparation and application of nano-sized composite energetic materials. *New Chem. Mater.* **2011**, *39*, 36–39.

64. Zhang, M.; Wang, Y.; Song, X.-L.; Luo, T.-T. Preparation and characterization of NC/PETN nanocomposite. *J. Ordnance Equip. Eng.* **2018**, *39*, 182–186.

65. Weismiller, M.R.; Malchi, J.Y.; Lee, J.G.; Yetter, R.A.; Foley, T.J. Effects of fuel and oxidizer particle dimensions on the propagation of aluminum containing thermites. *Proc. Combust. Inst.* **2011**, *33*, 1989–1996. [CrossRef]

66. Rossi, C. Nano-energetic materials for MEMS: A review. *J. Micro Electro Mech. Syst.* **2007**, *16*, 919–931. [CrossRef]

67. Pang, W.-Q.; DeLuca, L.T.; Gromov, A.; Cumming, A.S. *Innovative Energetic Materials: Properties, Combustion Performance and Application*; Springer: Taramani, India, 2020.

 *nanomaterials*

Review

Review of Experimental Methods for Measuring the Ignition and Combustion Characteristics of Metal Nanoparticles

Vladimir Zarko [1,2,*] and Anatoly Glazunov [2]

[1] Voevodsky Institute of Chemical Kinetics and Combustion, Russian Academy of Sciences, 630090 Novosibirsk, Russia
[2] Laboratory for Designing Elements of Rocket and Space Technology, Tomsk State University, 634050 Tomsk, Russia; gla@niipmm.tsu.ru
* Correspondence: zarko@kinetics.nsc.ru; Tel.: +7-383-333-2292

Received: 20 September 2020; Accepted: 2 October 2020; Published: 12 October 2020

Abstract: Investigations in recent decades have shown that the combustion mechanism of metal particles changes dramatically with diminishing size. Consequently, theoretical description of the ignition and combustion of metal nanoparticles requires additional research. At the same time, to substantiate theoretical models, it is necessary to obtain objective experimental information about characteristics of ignition and combustion processes, which is associated with solving serious technical problems. The presented review analyzes specific features of existing experimental methods implied for studying ignition and combustion of metal nanoparticles. This particularly concerns the methods for correct determination of nanoparticles size, correct description of their heat-exchange parameters, and determining the ignition delay and combustion times. It is stressed that the problem exists of adequate comparison of the data obtained with the use of different techniques of particles' injection into a hot gas zone and the use of different methods of reaction time measurement. Additionally, available in the literature, data are obtained for particles of different material purity and different state of oxide layer. Obviously, it is necessary to characterize in detail all relevant parameters of a particle's material and measurement techniques. It is also necessary to continue developing advanced approaches for obtaining narrow fractions of nanoparticles and for detailed recording of dynamic particles' behavior in a hot gas environment.

Keywords: ignition; metal; nanoparticles; combustion mode; heat transfer; free-molecular; burning time

---

1. Introduction

Metal nanoparticles are known from very distant times; for example, the famous Cup of Lycurgus, colored stained-glass windows in churches, etc. It can be mentioned that scientific studying the properties of nanoparticles (gold and silver) started in the 19th century in fundamental work of M. Faraday [1]. However, intensive research in this area began relatively recently, and this happened due to the development of methods for obtaining nanoparticles by evaporation–condensation [2] and electric wire explosion [3].

The history of metal nanoparticles application in the combustion and explosion processes developed according to a typical scenario for new discoveries: first, a rapid euphoria and great expectations, then a decline in interest, a decrease in activity, followed by a growth of interest and the achievement of real positive results. This situation was described by M. Zachariah in an editorial note published in 2011 [4]. At present, metal nanoparticles are widely used as modifiers of the burning rate of solid propellants, components of pyrotechnic compositions, etc.

It can be noted that nanoparticles are intensively studied as individual objects, the products of production by various technical methods, and also as the combustion products of original macro- and micro-particles of metals. In the first case, there are questions of atomization (dispersion) of individual particles, since aggregation processes at nanoparticles level proceed very fast, and it is necessary to have evidence that the procedure is carried out with a given particle size. Secondly, one has to consider general picture of the metal particles combustion when changing their characteristic size. Research in recent decades has shown that the combustion mechanism of metal particles changes dramatically along with changing their size. This is clearly demonstrated by the diagrams of the burning time dependency on the particle size, Figures 1–3.

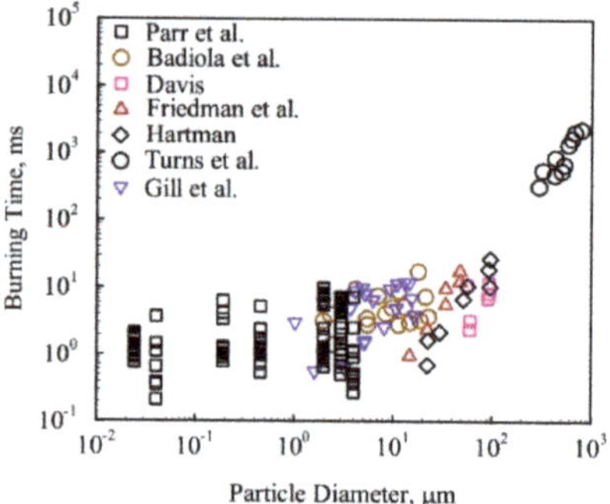

Figure 1. Burning time vs. aluminum particle size exhibiting weak size effect at nano-scales. Reproduced from [5], with permission from Prog. Energy Combust. Sci., 2017.

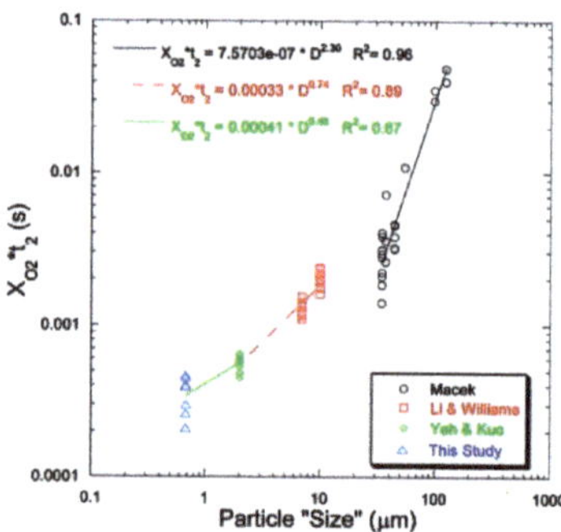

Figure 2. Comparison of the full-fledged combustion times of B particles (Macek [6], Li [7], Yeh [8], this study [9]). Reproduced from [9], with permission from Elsevier, 2009.

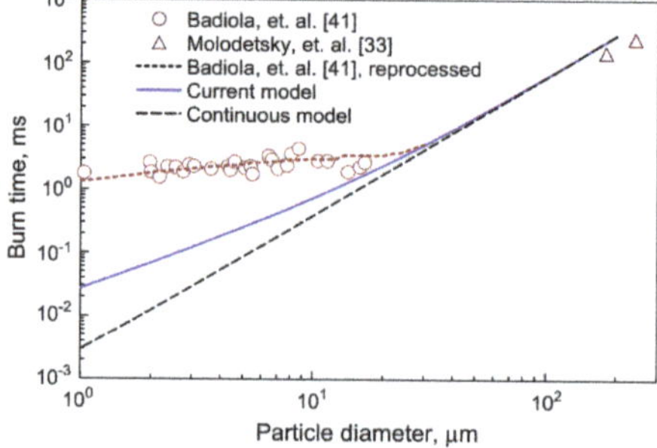

Figure 3. Dependency of the burning time on Zr particles size (Badiola [10], Molodetsky [11]). Reproduced from [12], with permission from Elsevier, 2013.

A sort of generalized information on the burning time t_b dependency of the Al and B particles on size [13] is presented in Figure 4, which is generally true for other metals [11]. Here, one can see that for particles with a size greater than 20–30 microns, a quadratic dependency in the form of $t_b{\sim}D^2$ takes place, then with a decrease in size, the dependency on the particle size weakens and takes the form $t_b{\sim}D^1$, while for submicron particles, it becomes $t_b{\sim}D^n$, where n = 0.3–0.5. The main reason for changing the type of dependency for the burning time is a change in the mechanism of the particle heat exchange with the gas environment [14–16].

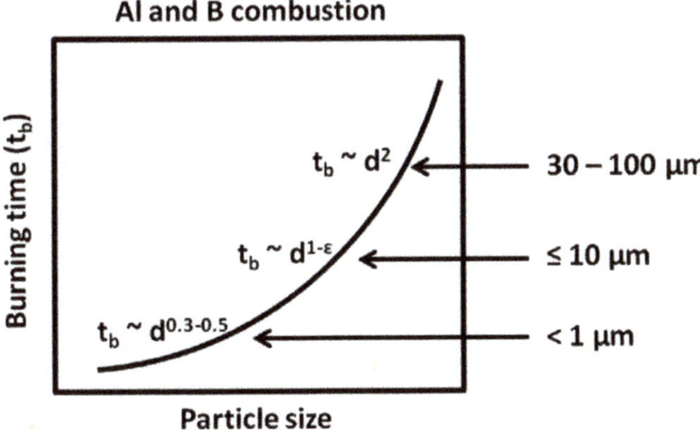

Figure 4. Conceptual figure showing determined experimentally for Al and B the burning time on size dependency. Reproduced from [13], with permission from Elsevier, 2014.

Thus, theoretical description of the processes of ignition and combustion of metal nanoparticles requires additional research. However, to substantiate the theoretical models, it is necessary to obtain objective experimental information about the characteristics of these processes, which is associated with solving serious technical problems. The purpose of the present article is to analyze specific features of experimental methods for studying and recording the characteristics of the ignition and combustion of metal nanoparticles. In particular, this concerns the methods for correct determining the size of nanoparticles, as well as correct accounting for their heat-exchange parameters and recording the ignition delay and combustion times.

2. Combustion Mechanism Variation with Metal Particle Size

The equation shows the qualitative assessment of the prevailing reaction mechanism of a metal particle that one can make via calculation of the Damkohler number D_a representing the ratio of diffusion and chemistry time scales:

$$D_a{\sim}D_p/D_{ox} \tag{1}$$

Here, D is the particle diameter, p the gas pressure, and D_{ox} the oxidizer diffusion coefficient.

At $D_a \gg 1$, a diffusion-controlled process of metal particle oxidation is implemented, whereas at $D_a \ll 1$, the oxidation is governed by chemical kinetics. Equation (1) indicates that the oxidation is characterized by a diffusion-controlled process in the case of large particles and high pressure, while in the case of small particles and low pressures, it is characterized by a kinetically controlled process. Approximate estimates for Equation (1) show [17] that for boron particles (according to [7]), the diffusion-controlled process is realized at sizes over 75 microns at 1 atm and, for aluminum particles (according to [18]), at sizes over 100 microns at 1 atm. Note that these limits decrease in direct proportion to the medium pressure. At the same time, it is important to note that the above estimates are very approximate and more accurate analysis of the heat-exchange modes can be made on the basis of the Knudsen number, which is defined as the ratio of the free molecules path length λ and the particle size $D/2$: $Kn = 2\lambda/D$.

It can be considered as two limiting cases: $Kn \gg 1$ and $Kn \ll 1$ [14]. In the first case, the particle diameter is less than the free gas molecules path length, and this mode of heat exchange is called free-molecular. In the second case, the particle diameter significantly exceeds the free path length, and the heat exchange occurs in the continuous medium mode. It is assumed that the first case corresponds to the values of $Kn > 10$, and the second, when $Kn < 0.01$. At intermediate values $10 > Kn > 0.01$, a transient heat-exchange mode is implemented. The Knudsen number is calculated using Equation (2):

$$Kn = \frac{RT}{\sqrt{2}\pi D_a^2 N_A p D},\tag{2}$$

where R is the universal gas constant, T the temperature, D_a the diameter of the ambient gas molecule, N_A the Avogadro's number, and p the pressure. The particle sizes corresponding to the Knudsen numbers 0.01 and 10 are represented as functions of pressure at temperatures of 300 and 3000 K in Figure 5. It can be seen that at a pressure of 1 atm and a gas temperature of 3000 K, the particle size corresponding to the assumption of a continuous medium is equal to 70 microns. It is reduced by 10 times if the pressure increases to 10 atm or when the temperature falls down to 300 K.

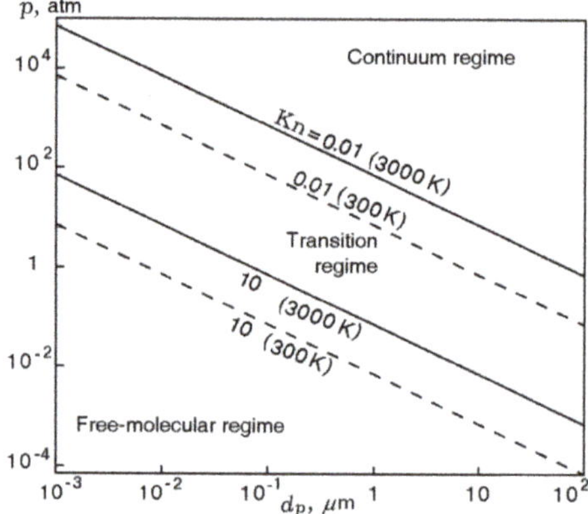

Figure 5. Boundaries between the continuum and free-molecular regimes as a function of particle size at different pressures and temperatures 300 and 3000 K. Reproduced from [14], with permission from Elsevier, 2015.

It is obvious that during the evolution of the size of burning metal particles, the mechanisms of heat exchange with the gas medium change (radiation heat exchange must be considered additionally) and the calculation of the total burning time must be made, accounting the stages. An illustration [19] of the variation of combustion mechanisms versus the aluminum particle size is shown in Figure 6.

Ten micron and larger particles are characterized by a structure with detached diffusion flame and maximum temperature close to the adiabatic flame temperature of Al—the air system. For submicron particles, the reactions practically proceed on the surface of the particles. The temperature of flame is almost equal to the surrounding gas temperature, especially at low pressures. It is not exceeding the aluminum boiling point and decreases when going far from the surface. In the case of intermediate size particles, the flame approaches the surface of a particle and the maximum temperature can be significantly lower of adiabatic one.

In general, the qualitative picture of the burning metal particle depends on the ratio of the metal boiling point and the boiling—dissociation temperature of its oxide. The concept of "Glassman's criterion" [20–22] is known in the literature, which states that if the metal boiling point is lower than its oxide boiling point-dissociation temperature, combustion occurs in the vapor phase. With the reverse ratio of the characteristic temperatures, heterogeneous combustion occurs on the surface of the particles. According to the literature data, the metals Be, Cr, Al, Fe, Hf, Mg, Li, and Ti should burn in pure O_2 at atmospheric pressure in the vapor phase mode forming a diffusion flame. At the same time, Si and Zr should burn heterogeneously [23].

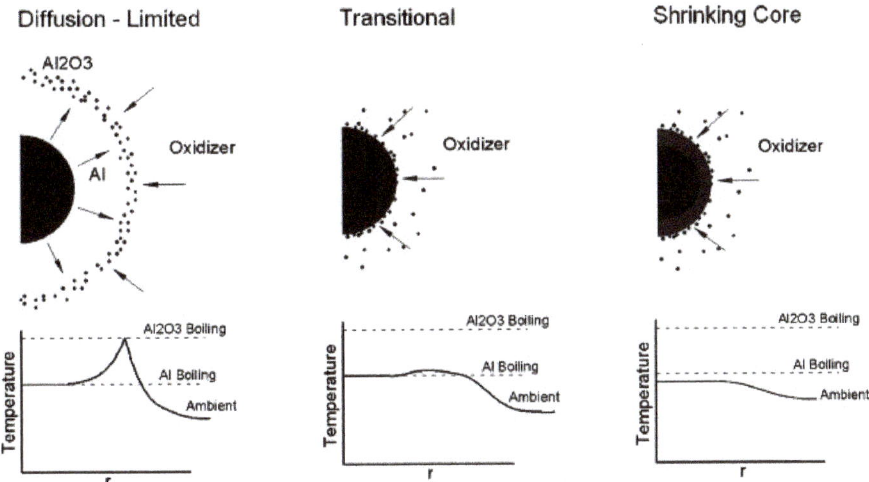

Figure 6. Schematic of the flame structures presenting in aluminum combustion. Reproduced from [19], with permission from Prog. Energy Combust. Sci., 2017.

In the case of metalloid B, the situation is special: due to abnormally low boiling point its oxide vaporizes on the expense of heat feedback from the flame, but the energy is not sufficient to reach the metal boiling point, and potentially possible heterogeneous combustion mechanism is not realized. It is important to emphasize that the necessary condition for the existence of a vapor-phase mechanism of metal combustion is the excess of the flame temperature over the metal boiling point. Therefore, if there exists a noticeable heat loss, the combustion mode of metals with a relatively small difference in the boiling temperatures of metal and oxide (within 400 K; Cr, Fe, Hf, Ti) can be changed from vapor-phase to heterogeneous. Similarly, the combustion mode changes when the composition of the oxidizer changes: aluminum combustion occurs heterogeneously with CO_2 and H_2O vapors (diminished flame temperature). It should be noted that these arguments are qualitative in nature and are valid for aluminum mainly at moderate pressures and temperatures. Particularly, the combustion mode with evaporation can be realized for small particles of aluminum at high pressure and gas temperature.

Thus, in general, the total burning time τ_b^{Σ} of a relatively large size metal particle must be calculated taking into account the changes of heat-exchange modes (from continuum through transition to free-molecular regimes):

$$\tau_b^{\Sigma} = \tau_b^{\text{cont}}\Big|_{D_0}^{D_1} + \tau_b^{\text{trans}}\Big|_{D_1}^{D_2} + \tau_b^{\text{free}}\Big|_{D_2}^{0} \tag{3}$$

Here, D_0 is the initial particle diameter, D_1 the size corresponding to the value of the Knudsen parameter $Kn = 0.01$, and D_2 the size corresponding to the value $Kn = 10.0$. It should be noted that, in available literature, this approach to the calculation of the total burning time of metal particles to date has not been implemented, and in published works, it was assumed by default that the selected burning mechanism is valid from the beginning to the end of a particle's combustion. To the authors' knowledge, there is only one work on the combustion of boron particles, where the attempt to develop such an approach has been made [24].

3. Correct Determination of Nanoparticles Size

Determination of nanoparticles' size is rather difficult technical problem, which also involves the issue of disconnecting the aggregated and sintered particles. In many cases, the available results of such measurements do not look as well-substantiated. In research conducted with Al nanoparticles in the shock wave tube [25], the particles were injected radially into the tube using a pneumatically driven piston. High resolution scanning electron microscope (SEM) was employed to measure for each sample the particle size distribution, and more than 100 diameter measurements from each sample were usually made for obtaining particle size distribution.

Then, the particles were characterized by number average and mass average diameters. In particular, a fraction called "50 nm SkySpring NanoMaterials" had the number average and mass average diameters 73.2 nm and 80.9 nm, correspondingly. In [25], there is no explanation of how the particles probes for the SEM analysis were prepared in order to represent total size distribution and why only about 100 particles were measured. An example of particle size distribution for 50 nm sample is shown in Figure 7. It can be recognized that there are particles, which size is approximately twice larger than the medium one. Consequently, when comparing the results corresponding to different samples, it is necessary to take account of the presence of the highest size particles. The wide distribution of particle sizes might be a reason for weak dependency of burning time on the nanoparticles' size. Another reason could be different origin and properties of particles from different sources: mass content of Al and oxide-coating thickness.

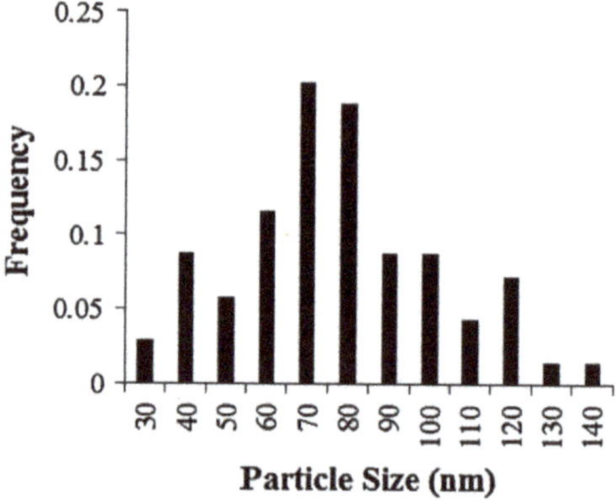

Figure 7. Particle size distribution (50 nm nominal size, SkySpring Nanomaterials). Reproduced from [25], with permission from Elsevier, 2014.

One more reason for uncertainty in determination of the nanoparticles' ignition and burning times can be aggregation and agglomeration of original particles leading to formation of bigger size particles with longer operation time. In [25], the authors discuss this issue and state that in conditions of shock waves the strong shear forces could be effective at breaking weak agglomeration in nanoparticles. Unfortunately, the evidences of effective disaggregation of particles are not presented. At the same time, there are known cases of studying the reactive characteristics of metal nanoparticles in laminar flames with the injection of particles in a high temperature zone, where the particles' agglomeration may have significant effect. As an example, Figure 8 presents a setup used in [26] for studying isolated boron particles ignition and combustion in fluorine-containing environments. The boron particles were injected in a spatially uniform post-flame region from a fluidized-bed particle feeder. The particles mixture in a bed consisted of boron (1–3 µm) and large silica particles (70–150 µm in diameter) at 1:20 average mass ratio. Coarse silica particles are assumed to break up existing boron particle agglomerates. No special analysis has been made to determine real particles size in the flame. A similar approach for feeding the boron particles of 2.3 µm average size was used in [27].

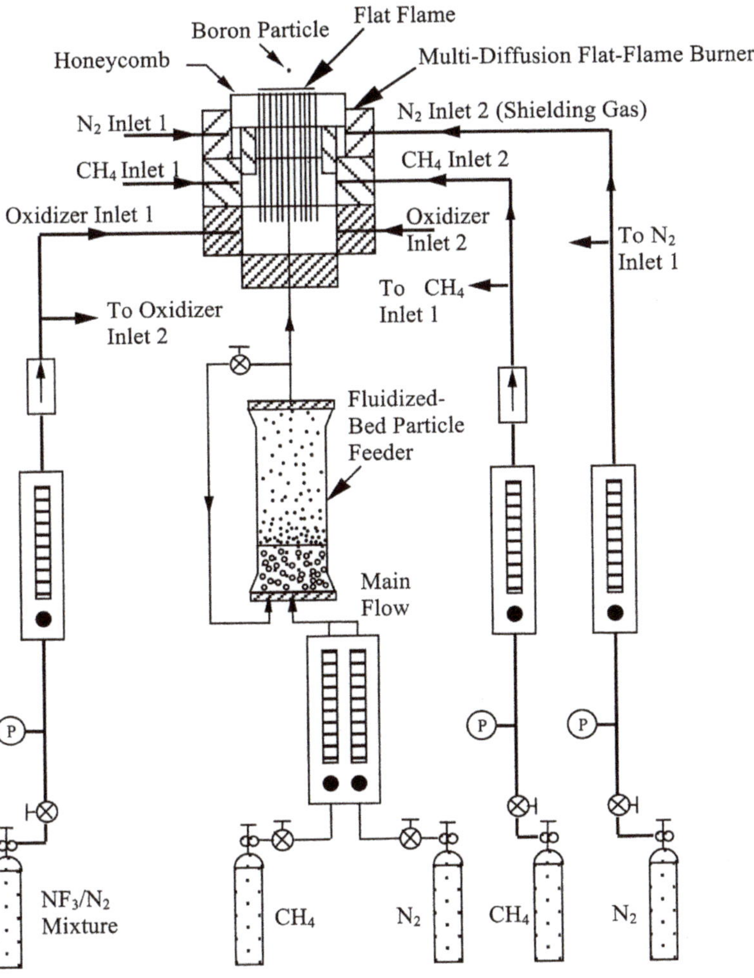

Figure 8. Schematic of used the burner setup and gas supply system. Reproduced from [26], with permission from Elsevier, 2001.

A detailed examination of metallized particles ignition and combustion behavior was conducted in the beginning stage of such research in [28]. The particles were characterized by sizes provided by manufacturers; they included nanoAl in the range of sizes from 24 to 450 nm and micron-sized Al and Ti: Al, TiH$_2$, Ta, Hf, and B particles. Two types of particles introduction into the flame were employed: with use of commercial Meinhard™ nebulizer and a homemade pneumatic system. In the last case, the fuel sample powder was placed between two aluminum foil burst disks and injected shortly into the hot zone of steam generator after the disks bursting. In the first case, the nebulizer generated liquid droplets of average size of 20 microns that provided an average weight of aluminum in a drop equal to that in an approximately 2 micron diameter particle. The sizes of individual (non-aggregated) particles remained unknown. In the case of the pneumatic feeder, the dispersibility of ignited and burning particles was also unknown. All data on reactive characteristics were discussed in terms of effective sizes of original particles. The authors of [28] reported about possible errors in the results of research due to wide particle size distributions of aluminum samples and the agglomeration effect and mentioned that it is nearly impossible to entirely disperse nanosized particles.

During past two decades, there were undertaken several attempts by different groups to elaborate effective methods of obtaining narrow size particle distribution samples of Me nanoparticles and methods of in situ measuring size of individual particles. Some positive and promising results were obtained in this direction by the group of Prof. M.R. Zachariah at the University of Maryland.

In the paper devoted to kinetic measurements of size-resolved aluminum nanoparticles' oxidation, it was formulated [29] that the intrinsic reactivity of nanoparticles would be better to explore in the absence of other rate-limiting kinetic effects associated with the heat and mass transfer. For this end, it is necessary to produce isolated particles and expose them to reactants in conditions of controlled environment with simultaneous measurement of size and residence time. Individual aluminum nanoparticles were generated by means of DC arc discharge or laser ablation (Figure 9) that allowed the authors to obtain free-of-oxygen particles.

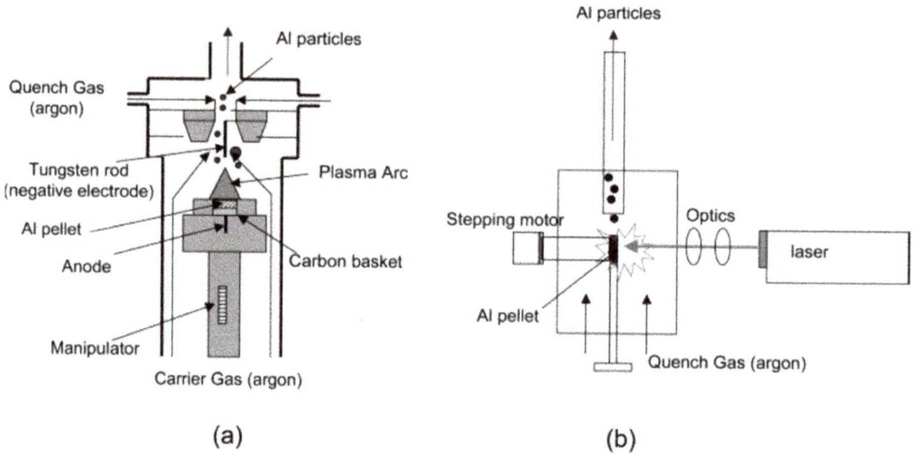

Figure 9. Methods used for generating nanoparticles of Al: (**a**) through DC arc discharge, (**b**) via laser ablation. Reprinted from [29] with permission from J. Phys. Chem. B. Copyright 2005 American Chemical Society.

The particles were oxidized at different temperatures (up to 1100 °C) in an aerosol flow reactor. To measure the particles' elemental composition and size as a function of the furnace temperature, the aerosols after reactor were sampled into the single particle mass spectrometer (SPMS). It is composed of an aerodynamic lens inlet, three-stage differential vacuum system, free-firing pulsed laser for ionization of a particle, and linear time-of-flight mass spectrometer. This provides an efficient simultaneous measurement of individual particles composition and size in terms of metal conversion into oxide extent. In front of the SPMS, it was installed a differential mobility analyzer (DMA) for sending monodisperse aerosols directly in the SPMS in order to derive the particle size and ion signal intensity relationship. Additionally, to measure particle size distribution, the DMA was coupled with a condensation particle counter. Finally, to explore the morphology of aluminum particles, a transmission electron microscopy (TEM) was employed. The size of individual particles selected by DMA (particle mobility size) was also measured using a time-of-flight mass spectrometer and compared with real size of sampled particles measured by TEM. It was revealed that aerosol particles (mobility size) consist of about 30 primary particles, and all kinetic data were referred to mobile size, which was determined at varying flow reactor conditions. Nevertheless, it can be concluded that in this case researchers had deal with the narrowest particle size distributions and obtained some very important information about the reactive characteristics of metal nanoparticles.

In particular, it was established in [29] that when primary and mobility particle sizes decrease, the reactivity of aluminum nanoparticles increases. It was also found that activation energy of the aluminum oxidation decreases when particle size decreases. Another important finding was that there exists significant difference of SPMS results with those obtained with classical non-isothermal gravimetric measurements method (TGA-ThermoGravimetric Analysis). Opposite to the onset temperature for aluminum nanoparticle oxidation below the melting point (510–530 °C, TGA), the SPMS indicates that oxidation starts just above the aluminum melting point. Besides, measured with the SPMS oxidation reaction rates became significantly different from those that were obtained from the conventional TGA procedures. All these differences are attributed to inherent features of bulk thermal methods application that resulted from ensemble effects associated with the relatively large mass of bulk samples.

Interesting results regarding combustion mechanism of metal nanoparticles were obtained with SPMS application in [30,31] when having deal with study of Ti and Zr particles. Due to high boiling points (Ti: 3560 K, Zr: 4650 K), which exceed the volatilization temperatures of corresponding oxides, the processes of Ti and Zr combustion are mainly dominated by surface reactions (Glassman's criterion). Laser ablation was used to generate nanoparticles of Ti and Zr in an inert environment and then the particles were introduced into the post-flame zone of a laminar $CH_4/O_2/N_2$ diffusion flame with the temperature from 1700 K to 2500 K. A modified DMA device employed as a size selection tool was combined with a high-speed camera for registration of combustion characteristics. In addition, the metal particles were sampled by a nanometer aerosol sampler at different heights above the burner and characterized by TEM. Analysis of sampled "fresh" Ti particles revealed that they consist of aggregates composed of primary particles with size of 10.3 ± 0.4 nm. After passing the high temperature burner zone, the resulting particle morphology changed from aggregate to isolated sphere. Similar behavior demonstrated particles of Zr. The measured burning times were plotted at diagrams for particles size measured by DMA (mobile size) or estimated after sintering (TEM size). An example of such diagrams for Ti particles is presented in Figure 10.

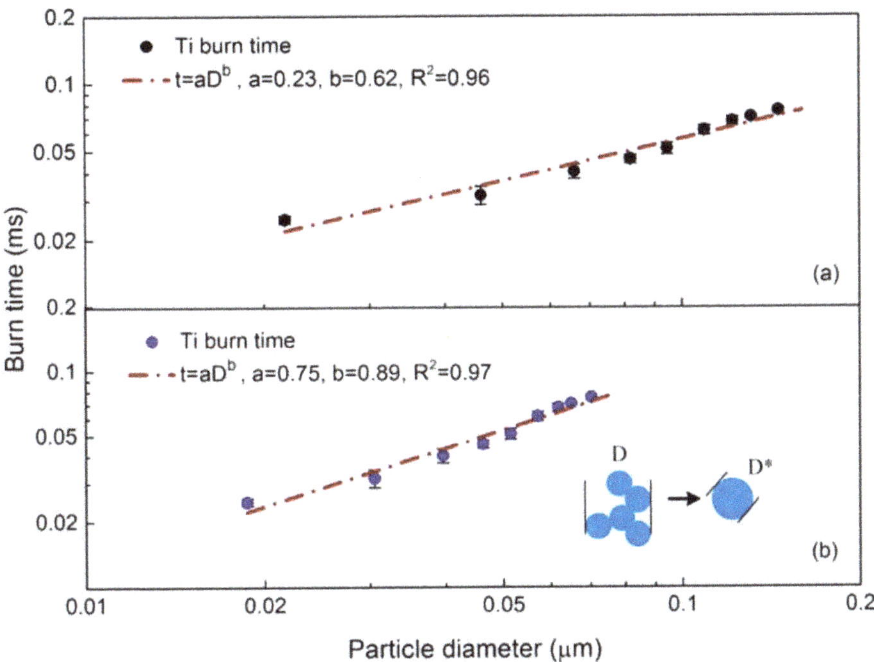

Figure 10. Ti particles burning time vs. particle size diagrams: (**a**) based on the peak DMA selected particle size, (**b**) based on the TEM measured diameter after sintering. Reprinted from [30] with permission from J. Phys. Chem. A. Copyright 2015 American Chemical Society.

It is seen that the power law exponent for burning time vs. particle size dependency equals 0.62 for DMA size and 0.89 for sintering size. These values are higher than typical values (ca. 0.3–0.5) for metal nanoparticles and for micron-size Ti and Zr particles [32]. Corresponding values of exponent for Zr particles comprise 0.53 and 0.77, correspondingly.

The visual observations of Ti and Zr nano particles combustion demonstrated very short streaks of the particle aggregates. This allows us to assume that their combustion proceeds in a practically isothermal environment. Using the measured oxidation zone temperature profiles, the temperature difference on an average streak length was estimated to be about 20 K. Therefore, isoconversion techniques were used to obtain apparent Arrhenius parameters for the nanoparticles under study oxidation reaction. The estimations were performed for chosen mobile particle size (145.9 nm) via analysis of the streaks recorded at various flame conditions. They were determined for Ti pre-exponential factor of 7.5×10^5 s^{-1} and activation energy of 56 kJ/mol and, for Zr, 3.4×10^5 s^{-1} and 43 kJ/mol, respectively. Actually, these results became available due to essential advancements in developing techniques, which allowed us to obtain "monodispersal" aerosols of metal nanoparticles.

4. Correct Determination of Energy Accommodation Coefficient

As is mentioned in Section 2, the conduction heat transfer for nanoparticles obeys the free-molecular regime. It was underlined in [25] and other works that the poor efficiency of this type of heat transfer is caused by the low value of energy accommodation coefficient α. This coefficient characterizes the behavior of gas molecules in their collisions with a solid or liquid body surface. It is calculated as the ratio of the gas molecule energy transferred during a collision to the theoretical maximum value under complete energy accommodation:

$$a = (E_0 - E_1)/(E_0 - E_s),$$

where E_0 and E_1 are the average energies of incident and scattered gas molecules, respectively, and E_s the average energy of gas molecules in thermal equilibrium with the surface. The accommodation coefficient value depends on the nature of particle surface and its state as well as on the gas mixture composition and pressure.

The correct knowledge of accommodation coefficient is important for objective description of nanoparticles heat transfer that may help to substantiate possible mechanism of particles' reacting. The particle energy conservation equation can be written as

$$m_p C_p \frac{dT}{dt} = \dot{Q}_{gen} - \dot{Q}_{rad} - \dot{Q}_{cond},$$

where m_p is the mass of particle.

The chemical heat generation rate is described by the formula

$$\dot{Q}_{gen} = A_p \phi N_{ox} c q / 4,$$

where A_p is the particle surface area, ϕ the sticking probability, N_{ox} the oxidizer molecules number concentration, c the molecular velocity, and q the reaction heat. The sticking probability characterizes the fraction of collisions that results in chemical reactions.

Radiation heat transfer is described by the formula

$$\dot{Q}_{rad} = \varepsilon_p \sigma A_p \left(T_p^4 - T_a^4 \right),$$

where ε_p is the emissivity of the particle.

The free-molecular heat conduction regime is described by the formula

$$\dot{Q}_{cond} = \alpha \pi D^2 \frac{p_a \sqrt{8k_B T_a / \pi m_a}}{8}\left(\frac{\gamma+1}{\gamma-1}\right)\left(\frac{T_p}{T_a}-1\right),$$

where α is the energy accommodation coefficient, k_B the Boltzmann constant, m_a the average mass of the gas molecule, and γ the ratio of specific heats. The subscripts a and p stand for the ambient gas and particle, correspondingly.

The attempts to describe theoretically the interaction of gas and surface atoms were undertaken in the past in several works [33–35] and later the estimate for the upper bound for the energy accommodation coefficient was presented in [36]. The principle of detailed balance has been employed to derive the following expression for calculating energy accommodation coefficient:

$$a < \Theta^2(\gamma-1)/(\gamma+1)T_a T_p,$$

where Θ is the Debye temperature. The calculations [36] showed that for Θ = 300 K and for the temperatures in gas T_g = 3000 K and in condensed phase T_s = 300 K, the accommodation coefficient α = 1/400 (monatomic gas) and α = 1/600 (diatomic gas). Thus, for typical values of metal particles parameters in flame the estimated value of α is of the order of 0.001, which is substantially (anomalously) lower of known in literature values [37]. Such an extremely low accommodation coefficient value was used in [25] to fit properly the available experimental ignition delay data with calculation results of Al nanoparticles heat transfer in conditions of shock wave. This approach was also cited in reviews [15,17] justifying the statement [36] about nanoparticles' "thermal isolation" in a free-molecular heat-transfer regime.

However, recent experimental and theoretical data do not support the statement about significant thermal isolation of nanoparticles in a gas environment. In detailed experiments [30,31] on size resolved high-temperature oxidation of Ti and Zr nanoparticles (20–150 nm size), it was revealed that use of a very low value of energy accommodation coefficient (0.005) does not provide good correlation with several listed models: shrinking core-kinetic (with a kinetic controlled reaction), shrinking core-diffusion through ash layer (with fast kinetics and diffusion rate limiting), and Avrami–Erofeev nucleation model (with oxygen dissolution in the unreacted core). When treating energy accommodation coefficient as a best-fit free parameter the good result was achieved with the value α = 0.3. This is shown in Figure 11 for 40 nm Ti particle. Similar results were obtained for different sizes of Ti and Zr particles.

A sort of comprehensive theoretical analysis of determination of non-equilibrium energy accommodation coefficients for aluminum-inert gas systems is presented in [38]. Calculations of the coefficients for aluminum surface temperature of 300 K and gas (helium, argon, and xenon) temperatures within the range 1000–3000 K were performed in the framework of molecular dynamics approach. Additionally, based on density functional theory, the simulations of the gas–surface interaction potentials were accurately conducted, and the effects were determined by gas temperature and molecular weight on the accommodation coefficient. The calculated coefficients were found to be of the order of 0.1 (0.2–0.3 at 1000–3000 K argon temperature) and became essentially larger of the values predicted for similar conditions by Altman's model [36]. Opposite of Altman's model, the analysis [38] predicted weaker gas temperature dependency of the accommodation coefficient.

These estimates correlate well with classical representations [37,39] regarding dependencies of accommodation coefficient value on properties of gas and solid surface, and they do not depend (opposite of [25]) on assumptions made in the frameworks of ignition and combustion models. It can be mentioned that in the series of studies performed by a group with Prof. E.L. Dreizin the relatively large values of accommodation coefficient have been used to calculate the Al particle temperature history (α = 0.87 [40], α = 1 [10,41]). In [42], experimental dependency of ignition temperature on Al particle diameter [28] has been matched with theoretical calculations when using α = 0.5 value.

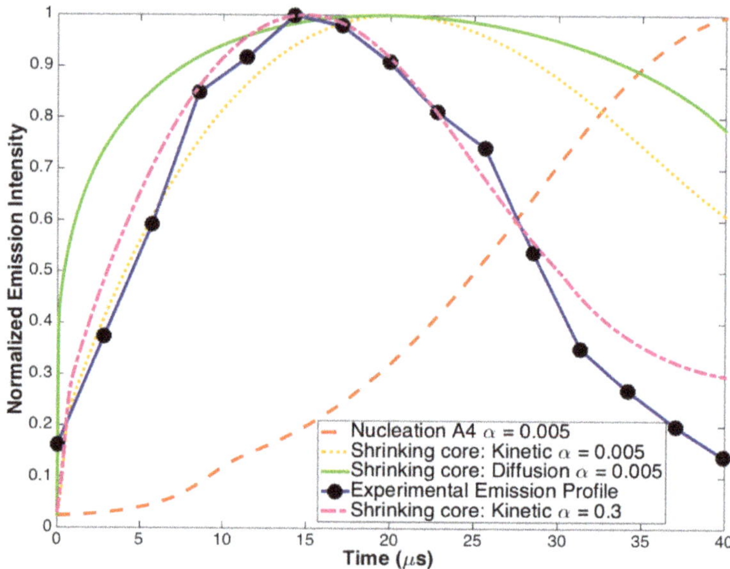

Figure 11. Model simulations for 40 nm Ti particle with α = 0.005 and α = 0.3. Reprinted from [30] with permission from J. Phys. Chem. A. Copyright 2015 American Chemical Society.

It was stressed in [10] that using a smaller-than-unity value of accommodation coefficient—not taking account of the finite rate of surface reactions—in calculations would lead to predicting an overestimated temperature of particles. To avoid this, it would be necessary to essentially reduce both the thermal accommodation coefficient and sticking coefficient (fraction of collisions of oxidizer molecules with the particle surface resulting in a chemical reaction). Namely, that was done in [25], where anomalously low values of α = 0.0035 and ϕ = 0.0009 were chosen to fit experimental data to calculations when using the chosen kinetic model. Obviously, use of such coefficient values could not justify the kinetic models employed in calculations, and future work is needed for quantifying the thermal accommodation and sticking coefficients. This may become important also for micron-sized particles, which are ignited and burned in a transition regime.

Despite the existence of numerous data on energy accommodation in different gas–solid systems, these could not be used in conditions that are of interest in combustion processes. Direct measurements of the accommodation coefficient are still very scarce. It should be noted that the data available in the literature about α-values mostly relate to pure substances. There exist discrepancies between the experimental data obtained under "surface science" conditions (ultrahigh vacuum, low coverage, low temperatures) and conditions of nanoparticles oxidation in practically important cases. The factors, such as admixtures on the evaporation surface or the chemical reactions between the vapor and components of gas environment, which proceed simultaneously with the evaporation, are believed to significantly affect the accommodation coefficient value. In addition, in the case of polyatomic gas molecules, it is necessary to make account of changing both translational and internal modes of energy of gas molecules transfer [37]. It is assumed in many cases that there is equilibrium between translational and internal freedom degrees of a polyatomic gas and thus a single accommodation coefficient can be employed. However, such an assumption is not well substantiated because the relaxation usually is faster for the translational degree of freedom.

To take account of that, it can be suggested using separate partial accommodation coefficients for the translational (α_{tr}) and for internal (α_{rot}) energy [37]. Their determination is a rather difficult technical task that takes special arrangement of the experiment. Such measurements were conducted in [43,44] with gilded tungsten wires of 8.3 μm diameter in a low-density wind tunnel. The initial pressure was about 2.7 Pa, stagnation pressure $1000 \div 2100$ Pa, and stagnation temperature 300 K. The experiments were performed with monoatomic gases Ar and He and polyatomic ones H_2, N_2, CH_4, and CO_2. In the treatment of experiments with polyatomic gases, it was taken into account that, at the room stagnation temperature and wire temperature below 400 °C, only rotational freedom degrees are active. The values of accommodation coefficients were obtained by accurate numerical solving of the heat balance equation for wire of which the temperature was precisely measured with the use of the resistance method. There were obtained the values of α_{tr} comprising 0.29 (He) and 0.8 (Ar) as well as 0.67 (N_2), 0.78 (CH_4), and 0.9 (CO_2). In addition, the values of α_{rot} were measured as 0.56 (N_2), 0.73 (CH_4), and CO_2 (0.79). The data on α_{tr} are in a qualitative agreement with the data known from the literature. However, there are no similar data available in the literature regarding α_{rot}. It is seen that the values of α_{tr} slightly exceed the values of α_{rot} and knowledge of both allows describing more precisely the heat exchange of gas with a solid surface. Most interesting experimental result of [43,44] is that for H_2, the value of α_{rot} was found to be negative (-0.15). This contradicts to traditional definition of internal energy accommodation coefficients and can be explained by partial conversion of rotational energy into kinetic energy of scattered gas molecules. Note that in the literature one can find the discussion on the limits of accommodation coefficient magnitude [45,46]. It is stated there that these limits are $0 \leq \alpha \leq 2$, and correct description of gas–surface heat exchange has to include an account of translational, rotational, and vibrational regimes for gas molecules.

5. Determination of the Ignition and Extinction Time Instants

The data on metal particles ignition delay and burning time are necessary for both practical applications and establishing and substantiating proper ignition and combustion mechanisms. In the case of nanoparticles, there arise specific problems due to their diminishing sizes, which do not allow for making direct photography of individual particles. Therefore, temporal characteristics of nanoparticles motion and reaction are recorded mainly by measuring the length of tracks or the luminosity duration of reacting material.

One of the first detailed investigations [18] of Al nanoparticles ignition under shock wave conditions discussed different methods of determining the burning time. These methods include the constant intensity threshold, the full-width at half-maximum (FWHM), and the percent integrated light intensity approaches. As the light-intensity signals often have several peaks, especially at low oxidizer concentrations (or weak oxidizers use), it is believed that the method of percent-area may provide the most unambiguous determination of burn time. An example of application of this method is presented in Figure 12 for Al nanoparticles ignited in a shock wave in the mixture of CO_2/N_2 (50/50).

The reaction (burning) time is denoted in Figure 12 as the time corresponding period between points 10 and 90% of the total integrated intensity signal. The same procedure was also used to determine burning time in experiments with the mixtures of O_2 with N_2, where a single peak of light-intensity signal has been recorded. Note that the recorded light signals contained a relatively constant background emission signal, which must be deleted from the total record. This fitting procedure together with the procedure of choosing the characteristic points on the light intensity trace lead to a random uncertainty of approximately 20% in the value of burn time. The experiments were performed at pressures of 8 and 32 atm, with the temperatures behind the reflected shock being varied in the range of 1200–2100 K. As a result of the Arrhenius fitting for 80 nm Al particles ($O_2/N_2 = 1{:}1$), the activation energies were obtained, which comprised 71.6 ± 24.6 kcal/mol at 8 atm and 50.6 ± 15.2 kcal/mol at 32 atm. Additionally, it was revealed that when the method of FWHM was used for recording the burning time, the obtained energies of activation became about 50% higher than in the case of measuring with the 10–90% area method.

In later research, when working with nano- and micron-sized Al particles [20], it was suggested to use for detecting the burning time the light emission from the burning particles at wavelength 486 nm corresponding to AlO, which is intermediate of aluminum combustion. Its appearance characterizes the ignition, while disappearance stands for extinction, correspondingly. The burning time was measured with use of 486 nm records employing the 10–90% percent-area method. It has established very little AlO emission for nano-particle combustion that indicates occurring reactions mainly in the condensed phase. Therefore, in subsequent shock tube research with nanoAl, only the visible light intensity has been recorded [25,47].

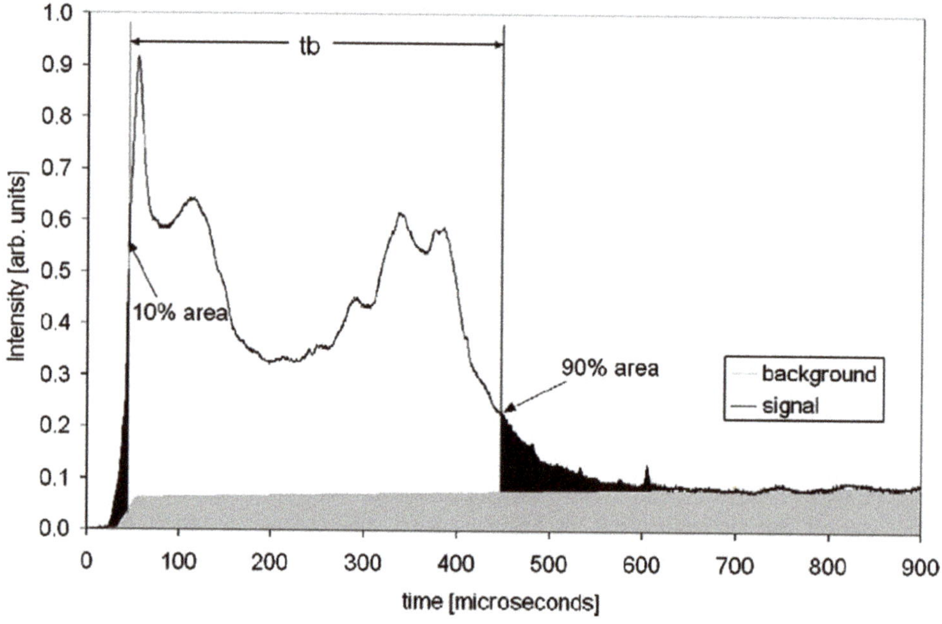

Figure 12. Visible light intensity vs. time traces for nanoaluminum in 50% CO_2 and 50% N_2 at 1760 K and 32.4 atm. Reproduced from [18], with permission from Elsevier, 2006.

Interesting information about burning time and behavior of nanoAl particles was recently reported in [48]. In this work the particles were injected at atmospheric pressure along the centerline of the burner directly into the products of the flat diffusion flame ($CH_4/O_2/N_2$). The experiments were conducted with commercial aluminum nanoparticles (ALEX) of 50 nm primary size and with a 1.6 μm mean size electrospray generated mesoparticles composed of original ALEX particles and nitrocellulose. For tracking the burning particles, a high-speed video camera was used focused directly at the burner centerline. Consequently, the burning time was determined by recording and counting the number of individual particles frames. It was established in detailed experiments that the burning time of mesoparticles is practically one order of magnitude shorter than the burning time of ALEX particles. According to scheme presented in Figure 13, this is caused by the fact that the primary ALEX particles already exist in the form of hard agglomerates with intraparticle necking, which may sinter immediately under heating that results in the formation of much larger particles participating in the real combustion process. In contrast, assembled with nitrocellulose mesoparticles burn as fast as the original smallest hard aggregates in the nanopowder. These results correlate with experimental findings [19], which revealed that for Al nanoparticles burning time is about 500 μs in a shock tube at 8 atm pressure, 1400 K, and a 50% O_2 environment. It is believed that the shock-induced breakup of the large agglomerates and thus the shock tube results correspond to the combustion of the smallest nanoAl aggregates in the aerosol.

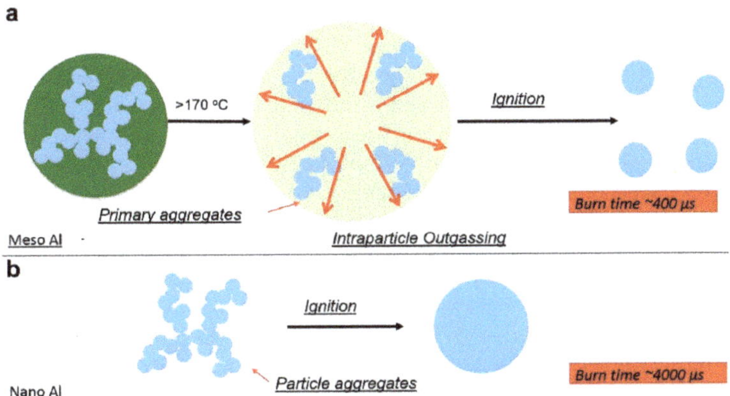

Figure 13. Overall pictures of the events occurring in the combustion of: (**a**) Al/NC mesoparticles, (**b**) Al nanoparticles ALEX. Reproduced from [48], with permission from Elsevier, 2016.

The pioneering study of the ignition and combustion of single crystalline boron particles was conducted more than 50 years ago [49]. In that work, the particles of boron with average diameters of 34.5 and 44.2 μm were introduced at atmospheric pressure in the post-flame (CO or propane with O_2) zone of a flat-flame burner. The experimental procedure of particle combustion studies has been described previously in earlier work [50]. To produce introduction of particles, a 250 micron bore hypodermic needle is mounted axially terminating at the upper surface of the burner disc. Initial particle velocity comprises few cm/s and the burnt-gas velocity is of the order of 1 m/s. The particle's travel distances on photographs are converted to residence times on the basis of known gas flow velocity, and a certain correction has to be made to account for initial acceleratory period. The estimation of the correction coefficient [50] for typical parameters of gas flow showed substantial correction value (about 1.5–2.0). Like in several previous studies, it was established [49] that the combustion of boron particles exhibits two successive stages. The first combustion stage is associated with the burning of boron particles initially covered with an oxide layer, while second combustion stage is associated with fully fledged combustion of the bare boron particle. Such representation of boron particles' combustion mechanism was confirmed and examined in detail later in numerous research dealing with micron- and nano-sized boron particles [26,51]. In [26], the burning time of 1 μm amorphous boron particle was determined, similar to [49], by dividing the length of the burning streak by the average particle velocity.

Another approach to measure the ignition and burning time of crystalline boron particles at pressures 30–150 atm was presented in [52]. The particles of 24 μm mean size were injected into a constant volume combustion bomb filled with the combustion products of N_2 diluted hydrogen/oxygen mixtures (O_2 excess concentrations 5–20% and temperatures 2440–2830 K). Note that in these conditions, only single-stage combustion has been observed. Experimental diagnostics consisted of dynamic pressure measurement, particle injection timing, and optical emission detection. The latter was done at BO_2 molecular band (interference filter at 546.1 ± 4 nm), corresponding to intermediary gas-phase species in B combustion. Figure 14 shows graphical procedure for determining the particle combustion characteristic time.

Following a methodology similar to that described earlier in [53], the ignition delay is defined as the time period from the instant of particle injection to the point of ignition indicated as the half of emission maximum value. The burning time is defined as the time from ignition to the moment when the filtered photodiode output signal diminishes to half maximum value. For comparison, a signal of pure B_2O_3 is shown indicating that the oxide emission is significantly lower even though $BO_2(g)$ molecules are produced by the B_2O_3 particles injected in a hot gas in separate experiments. Taking into account previously determined size distribution of particles, it was suggested to choose the half height of the net particle signal as the ignition point because this moment corresponds to mean-sized particles emittance, while the time of first light corresponds to the signal from smallest size particles comprising a small percent of the total mass. The same considerations were taken into account when defining the burning time. Further, determined in this way, the characteristic times were compared to predictions from two ignition models. In particular, excellent agreement of experimental burning times with predictions by a detailed-chemistry Princeton/Aerodyne model in 20% O_2 mixtures and discrepancy (over prediction) for lower X_{O2} mixtures was established. That discrepancy is believed to arise from the experimental interpretation of the instant when particle combustion is completed.

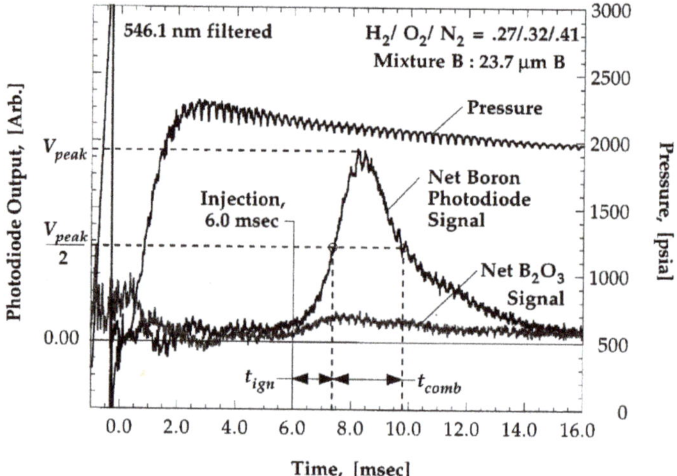

Figure 14. Filtered at 546.1 nm signal of B particle burning in gas mixture ($H_2/O_2/N_2$ = 27/32/41). Reproduced from [52], with permission from Elsevier, 1999.

When dealing with nano-sized B particles, the problem arises of how to distinguish correctly the first and second combustion stages. To solve this problem, it was suggested [9] to analyze high-speed video images of glowing particles made with and without interference filter, centered at 546 nm. Earlier, the emission on this band was not detected during the first stage of combustion [54]. Therefore, the filtered images are assumed to provide the spatial location of second stage in B combustion, whereas unfiltered images indicate starting location for the first combustion stage. An example of unfiltered image of SB99 (72.2% pure B) nanopowder combustion is shown in Figure 15. This powder has a primary particles size about 62 nm and aggregates size 200–320 nm.

Note that in [9], the particles were injected perpendicular to the burner flow and traversed the diameter of the burner. To obtain the velocity profiles of the injected particles in the post-flame region, the detailed PIV particle image velocimetry experiments were conducted, which allowed making reasonable estimates of ignition and burning times. It is seen in Figure 15 that there are two distinct parts of image luminosity where the first one (yellow) corresponds to ignition and the second (white glow) to fully fledged combustion. The qualitative treatment of this two-dimensional picture is rather difficult task, and it was suggested to convert it into one-dimensional by summing the columns (or the *y*-dimensions) of the images with subsequent constructing the profile in the *x*-direction. Then an area-based method was employed in order to provide the reliable determination of the burning time [55]. As is shown in Figure 16, the burning time was detected as occupying 95% of the area of the original intensity profile, with the blue line representing the original profile of filtrated image and the red line representing 95% of the total area. Subsequently, the burning time t_2 defined in this way was used to compare with the measurements of different authors (see Figure 2, present paper) and to derive the apparent energy activation from Arrhenius burning law correlation in the form of $1/t_2 \sim \text{const} \, \exp(-E/RT)$.

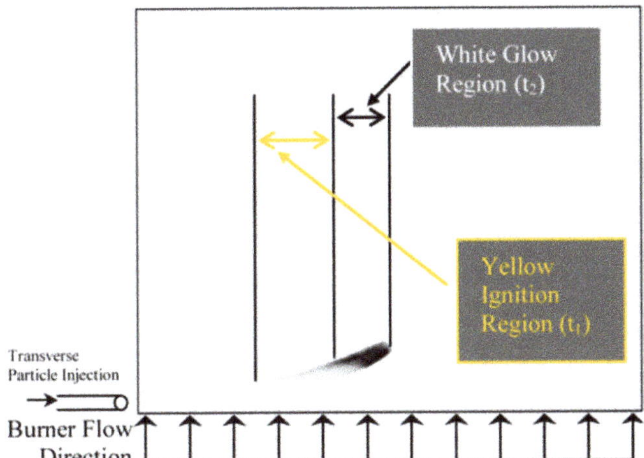

Figure 15. Unfiltered image of SB99 nanopowder combustion. Reproduced from [9], with permission from Elsevier, 2009.

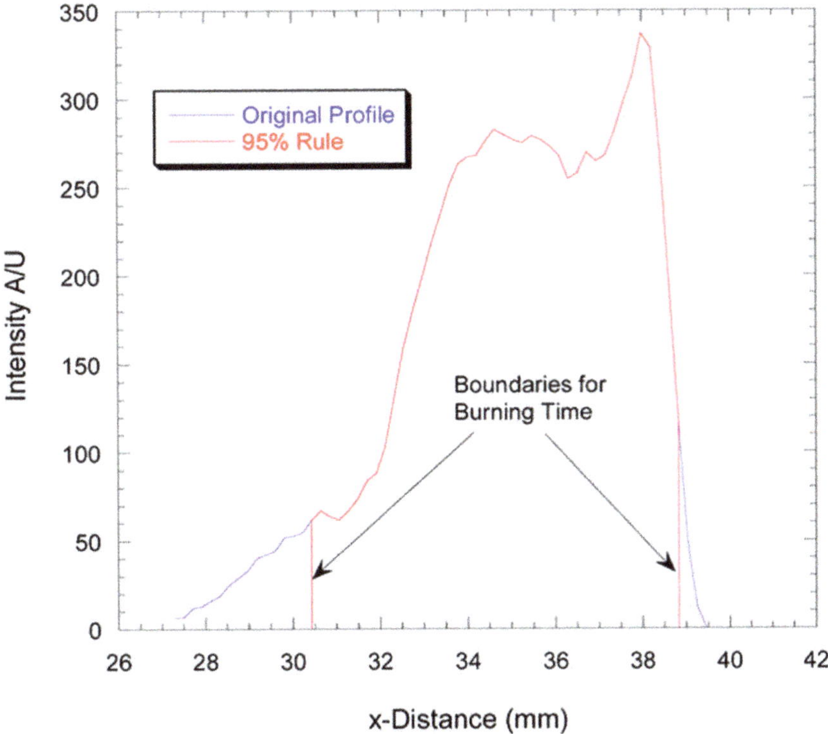

Figure 16. Illustration of t_2 burning time determination (X_{O_2} = 0.2, T = 1808 K). Reproduced from [9], with permission from Elsevier, 2009.

A similar approach was used in [30,31] for determining characteristic times of high-temperature reactions of nano-sized (20–150 nm) titanium and zirconium particles. Nanoparticles were produced by a metal target pulsed laser ablation and then the size was selected with use of a differential mobility analyzer. To evaluate the total burning time, a high-speed camera recorded the particles luminosity streaks with an exposure greatly exceeding the particle burning time. In this case, the whole streak length was used to determine the burning time on the basis of known gas velocity. The experiments were conducted with highly diluted aerosol of the size-selected metal particles, which were injected into a high temperature (1700–2500 K) oxidizing zone produced by a laminar $CH_4/O_2/N_2$ diffusion flame. It might be noted that for the size-selected burning times, the data scatter was noticeably small (2~9%) indicating narrow range of burning times for the particles of a given size. This serves an advantage of developed approach and strengthens fidelity of experimental data.

To illustrate a variety of existing methods to determine metal particles' burning time, one can consider fresh information presented in [56]. Fuel-rich boron B composites containing bismuth fluoride (B·BiF$_3$) and bismuth (B·Bi) were prepared by arrested reactive milling and then characterized using field emission scanning electron microscope to obtain the particles' size distributions. The aerosolized particles were directed into the focal zone of CO_2 laser producing particles ignition at room temperature in air at atmospheric pressure. A photomultiplier tube (PMT) was used to record a light emission of ignited particles. The examples of the PMT records for composites with different content of boron (90–69%) are presented in Figure 17. The instant of ignition was determined from high-speed videography and the burning time was defined, as it was done previously in [57], as the period of time over which the PMT signal exceeds 10% of the peak maximum.

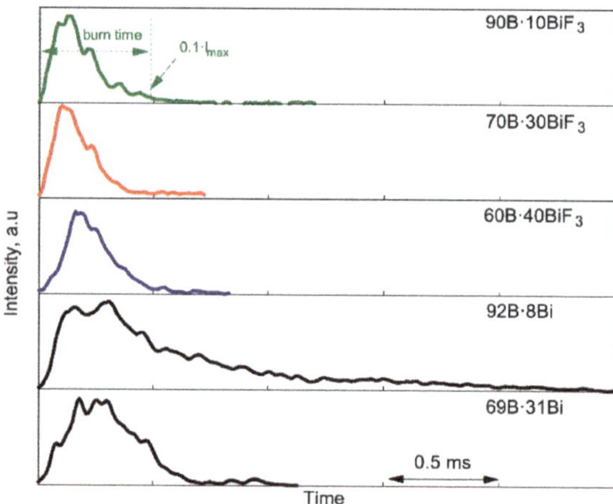

Figure 17. Characteristic emission records of the burning in air particles of B·BiF$_3$ and B·Bi composites. Reproduced from [56], with permission from Elsevier, 2020.

For obtaining the effect of particle size on burning time, the correlations between statistical distributions of the particle sizes and measured burning times were examined. Figure 18 presents the results obtained in the form of burning time–particle size diagrams for B·BiF$_3$ and B·Bi composite powders and for pure boron [57] and 50B·50BiF$_3$ samples [58], respectively.

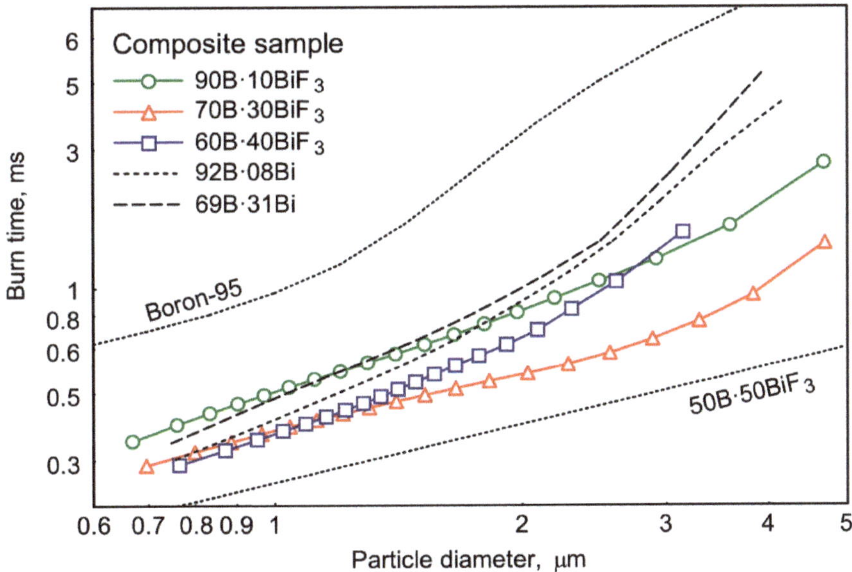

Figure 18. The burning time–particle size correlations for B·BiF$_3$ and B·Bi composite powders burning in air (Boron-95 [57], 50B·50BiF$_3$ [58]). Reproduced from [56], with permission from Elsevier, 2020.

6. Laser Heating of Nanoparticles

When describing the temperature history of metal particles under irradiation, the issue of radiation absorption has to be carefully discussed. The matter is that the radiation absorption efficiency does depend on the particle size, and this should be taken into account in the case of fine metal particles. Following [40], the laser energy delivered to the particle can be calculated as $q_{Laser} = 0.25\pi D^2 \eta\,(\lambda, D, n_i)\,I_{rad}$, where η is the particle laser absorption efficiency, which depends on the laser wavelength λ, D the particle diameter, n_i the material complex refractive index; and I_{rad} is the laser power density.

The energy absorption efficiency η is described in detail for spherical metal particles in [59]. Based on this approach, the absorption efficiency was calculated in [40] for the temperatures below the Al melting point using Mie's scattering theory. The calculations revealed that for Al the absorption efficiency peak is realized at 3.37 µm particle diameter.

Similar calculations performed for other metals revealed that for a given laser wavelength (10.6 µm) the absorption efficiency estimated on the basis of Mie's scattering theory has a maximum at approximately the same particle size ($D = 3.37$ µm). It can be supposed that due to the particle size selective heating, the particles corresponding to the peak absorption efficiency will be ignited mostly under laser irradiation. Later the model was modified to take into account the effect of melting [60] on the absorption efficiency. It was found that the particle absorption efficiency experiences a jump on melting, which is caused by abrupt density change.

The results of the calculations clearly demonstrate that the absorption efficiency depends on the particle size and its temperature. More information can be found in specialized literature sources. In common, the particle radiation absorptivity is characterized [59] by the absorption cross section, C_a, which is calculated as the total amount of absorbed energy divided by the incident radiation intensity. Further, the efficiency to absorb radiation can be characterized by absorption efficiency of the particle based on its cross-sectional area, $\eta = C_a/(\pi D^2/4)$, and the volumetric absorption efficiency based on its volume, $\eta_v = 6C_a/\pi D^3$.

Detailed calculations for different metals showed that the absorption efficiency essentially depends on the temperature. It increases by about four, three, and two times for Ni, Al, and Cr particles when the temperature increases from 300 K to 900 K. Thus, the particles' preheating can effectively enhance absorption of radiation in metallic particles.

In accordance with calculations, the micron- and submicron-size particles can absorb radiation more efficiently than larger particles because of strong diffraction effect at the particle surface. For nanometer particles, radiation absorption becomes less efficient. For vanishingly small particles (e.g., d < 10 nm), η_v approaches a constant, which is the Rayleigh limit of light scattering by small particles. In case of metallic particles, the maximum η_v is by about two orders of magnitude larger than the Rayleigh-limiting value.

From the point of view of practical applications, the knowledge of fine particles' radiation absorption efficiency is important for estimation of ignition delays of energetic materials containing inclusions of highly absorbing substances. In particular, it may have sense in the case of laser heating of explosives seeded by nanoparticles of metal. Based on the Mie theory, the calculations were conducted [61] for inclusions of in total 12 different metals in the matrices of silver azide, lead azide, and pentaerythrite (PETN). The wavelength of radiation was 1064 nm and particle sizes varied in the range 0–600 nm. Figure 19 presents calculated dependencies of the radiation absorption efficiency on the radius r of inclusions of silver ($n_i = 0.15 - 6.0i$) in a silver azide matrix ($n = 2$), lead ($n_i = 1.416 - 5.742i$) in a lead azide matrix ($n = 1.85$), and lead and aluminum ($n_i = 0.978 - 8.030i$) in a PETN matrix ($n = 1.55$). Here, n_i is the complex refractive index of the metal.

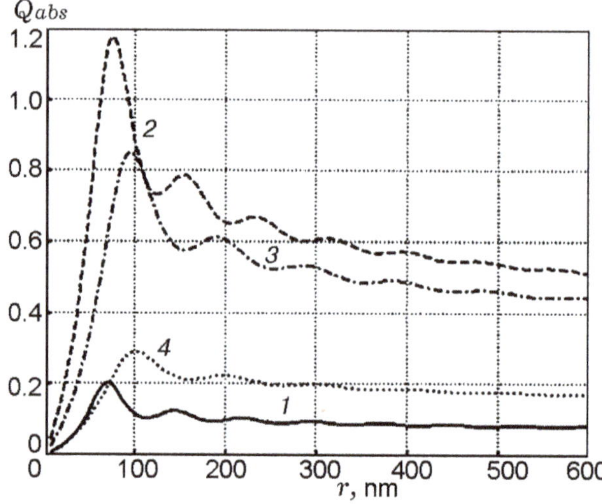

Figure 19. Efficiency of radiation absorption ($\lambda = 1064$ nm) by metallic inclusions in a matrix of energetic materials vs. radius of inclusions: Ag in AgN$_3$ matrix (*1*), Pb in PbN$_6$ matrix (*2*), and Pb and Al in PETN matrix (*3* and *4*). Reproduced from [61], with permission from Combust., Explos., Shock Waves, 2012.

It is seen from Figure 19 that silver inclusions in a silver azide matrix absorb radiation ineffectively. The maximum absorption efficiency η equals ca. 0.204, whereas for lead in a lead azide matrix, it comprises ca. 1.18. Small η value in the case of silver inclusions is related to the pronounced metal properties of silver and corresponds to low value of the real part of refractive index and high value of its imaginary part. Correspondingly, the larger η value in the case of lead inclusion is due to its less-pronounced metallic property. Aluminum demonstrates intermediate η values. Thus, when calculating the heating history of the inclusion, it is necessary to take into account the nature of the metal and use corresponding absorption efficiency.

An example of such calculations is presented in Figure 20. The calculations were conducted for a very short duration of laser pulse (30 ns) when the heat exchange with matrix and the effect of its chemical reaction can be ignored. The comparison is shown between the time history of heating with the account of real dependency radiation absorption on inclusion radius and in the case when it is assuming $\eta = 1$. It is evident that in the case of a "weak metal" (Pb) the maximum heating temperature is essentially lower when taking into account the real value of its radiation absorptivity.

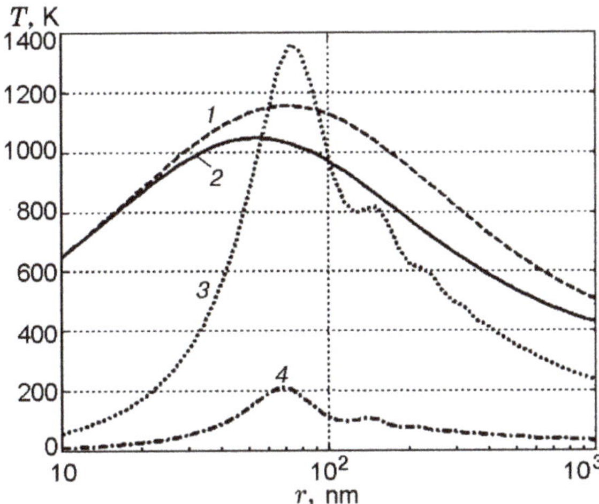

Figure 20. Maximum heating temperature of metal nanoparticles in an inert matrix having thermophysical properties of PbN$_6$ (laser pulse length 30 ns and energy density of 50 mJ/cm^2): curve *1* stands for Ag ($\eta = 1$) and curve *2* to Pb ($\eta = 1$); curves *3* and *4* refer to Ag and Pb with account of the dependency $\eta(r)$. Reproduced from [61], with permission from Combust., Explos., Shock Waves, 2012.

7. Concluding Remarks

As is mentioned in the Introduction, the metal nanoparticles have a wide list of applications in various explosive and propulsion systems, including development of effective modifiers of the burning rate of solid propellants, advanced components of pyrotechnic compositions, etc. The research in that direction is conducted intensively in USA, Russia, China, and other countries. Different aspects of such applications are discussed in the books published in Elsevier in 2016 [62] and 2019 [63]. In particular, use of metal nanoparticles has allowed obtaining great burning rates in MoO$_3$–Al nanothermites, which reach 790 m/s and provide perspectives for developing microscale rocket propellants [64]. Nanostructured energetic materials such as Al/Ni or Al/CuO can be used for fabrication of igniting bridges to facilitate ignition process with augmented output energy and igniting ability that allows initiation of high explosives [65].

An important advantage of studying the nanoparticles' performance is that it allows exploring the intrinsic mechanism of heterogeneous reactions. For example, recently, some detailed studies [66] revealed that, in nano-Al-based thermite reactions, the major part of heat release is contributed by a condensed phase mechanism. Obviously, these studies take development of the proper approaches and techniques.

The present paper focuses attention on the analysis of specific features of studying the metal particles ignition and combustion, because the literature data contain sometimes ambiguous information on the characteristics of these processes that do not allow making substantiated statements about their mechanism and predicting reliably its temporal behavior. First of all, there is no possibility to study the reaction behavior of a single nanoparticle due to technical problems of direct recording (photography) parameters of very fine objects and due to extremely fast [13] sintering of small-size particles leading to formation of the hard aggregates. Even in the case of advanced techniques of laser ablation or arc discharge [29], the data obtained correspond to relatively small-size aggregates of original nanoparticles. Nevertheless, the advantage of such techniques is that they allow us to indicate the real size of an aggregate in contrast to the earlier techniques with injection of finite size distribution powders into the shock tube or post-flame zone of the burner.

Another problem is the correct determination of ignition and burning times. It is evident that various research groups use different methods to detect ignition and extinction instances (10% of maximum luminosity signal; 90 or 95% of integrated signal). Moreover, in most of the literature sources, it is assumed that particles immediately enter into a hot gas zone with ignorance of the effects of carrying gas flow on the temperature and the oxygen concentration in environmental gas. Obviously, the account of these effects may give more substantiate interpretation of observed experimental facts. For instance, it was recently explained [16] that the enhancement of the burning rate of boron nanoparticles in a wet hot gas zone (with water vapors) can occur due to changing the heat-exchange intensity (increase in the Nusselt number due to increase in the velocity of gas exhaust from the burner). Note that no increase in the compact boron samples burning rate in wet gas environment was experimentally observed in [67]. It was also underlined in [16] that there exist intrinsic difficulties in detecting the final stage of large particles combustion because the amount of energy released at the particle trace where the particle size is below 3 μm becomes smaller at least by a factor of 1000 and an equipment tuned to detect signals of large particles may fail to capture such low-intensity signal.

It is necessary to mention here the issue of correct interpretation of experimental burning rate vs. time dependencies related to correct using the values of the energy accommodation coefficient. It was proposed recently [68] to employ for Al nanoparticles logarithmic burning law $t_b{\sim}lnD$ instead of commonly used power law $t_b{\sim}D^{0.3}$. The last law simply represents an experimental fit to data, and the power value 0.3 does not have physical meaning. The logarithmic law is derived on the basis of a simplified model based on the energy balance on a burning metal nanoparticle surface. This model utilizes a small value of energy accommodation coefficient previously used to explain shock tube experimental data [25]. The original idea on two different regimes of metal nanoparticles' combustion, which may exist depending on the value of energy accommodation coefficient, is argued in a fresh paper [69]. The illustration of a new idea is presented in Figure 21. Red and green triangles represent burning times of 4 μm Al particles in two combustion regimes. Solid lines correspond to available literature data, while dashed line corresponds to a hypothetical regime.

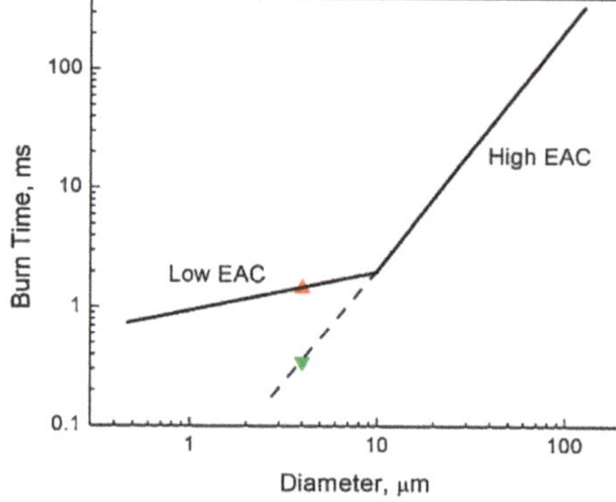

Figure 21. Burning time vs. Al particle diameter dependency. Reproduced from [69], with permission from Elsevier, 2020.

It is assumed that if a small size Al particle were burning with a high energy accommodation coefficient, the burning time-particle diameter dependency would be much stronger than the experimentally observed one. The background for such an assumption can be found in experiments with pre-stressed via fast cooling the heated Al particles [70]. It was established that the shell-core delamination associated with the fast cooling the Al particles facilitates diffusion reactions. This results in reducing a burning time. Thus, the altering of the shell-core particles' mechanical properties leads to altering of their reaction mechanism. The considerations related to electron energy transfer on the interface between a gas and metal and its contribution to the energy accommodation coefficient are discussed in [71]. Numerical calculations show that the coefficient magnitude decrease by a couple of orders of magnitude is possible. However, the approach used has a semi-empirical nature, and as is mentioned in [68], it does not pretend to be a part of comprehensive theory of the nanoparticle combustion but can be employed for the development of its objective background.

Finally, global problems exist regarding how to compare correctly and generalize the data on the ignition and burning time vs. particle size dependencies, which are obtained with use of different techniques of particles' injection into a hot gas zone and use of different methods of reaction time measurement. Besides, reported data are obtained for particles of different metal purity and different states of oxide layer. Obviously, at present, such a comparison can be made mainly on a qualitative level, and it is necessary to characterize in detail all relevant parameters of a particle's material quality and measurement techniques. In addition, it is necessary to continue developing the advanced approaches for obtaining and characterization of narrow fractions of nanoparticles and for detailed description of the particles' dynamic behavior in hot gas environment.

Author Contributions: Conceptualization, V.Z.; resources, V.Z. and A.G.; writing—review and editing, V.Z. and A.G.; funding acquisition, V.Z. and A.G. All authors have read and agreed to the published version of the manuscript.

Funding: This research was partially funded by Ministry of Science and Higher Education RF, grant number 2020-1902-01-7141 and by Tomsk State University competitiveness improvement programme, grant No. 8.2.12.2018.

Acknowledgments: The authors thank Djalal Trache for valuable technical assistance in preparing the manuscript.

Conflicts of Interest: The authors declare no conflict of interest.

References

1. Faraday, M. Experimental relations of gold (and other metals) to light. *Philos. Trans. R. Soc.* **1847**, *147*, 145.
2. Gen, M.Y.; Ziskin, M.S.; Petrov, Y.I. Research of dispersion of aerosols of aluminum depending on conditions of their formation. *USSR Acad. Sci. Proc.* **1959**, *127*, 366–368.
3. Ivanov, G.V.; Tepper, F. Activated aluminium as a stored energy source for propellants. In Proceedings of the 4th International Symposium on Special Topics in Chemical Propulsion, Stockholm, Sweden, 27–31 May 1996; pp. 636–644.
4. Zachariah, M. Nanoenergetics: Hype, reality and future. *Propellant Explos. Pyrotech.* **2013**, *38*, 7.
5. Sundaram, D.; Yang, V.; Yetter, R.A. Metal-based nanoenergetic materials: Synthesis, properties, and applications. *Prog. Energy Combust. Sci.* **2017**, *61*, 293–365. [CrossRef]
6. Maček, A. Combustion of boron particles: Experiment and theory. *Symp. (Int.) Combust.* **1973**, *14*, 1401–1411. [CrossRef]
7. Yeh, C.L.; Kuo, K.K. Ignition and combustion of boron particles. *Prog. Energy Combust. Sci.* **1996**, *22*, 511–541. [CrossRef]
8. Li, S.C.; Williams, F.A. Ignition and combustion of boron particles. In *Combustion of Boron Based Solid Propellants and Fuels*; Kuo, K.K., Pein, R., Eds.; CRC Press: Boca Raton, FL, USA, 1993; pp. 248–271.
9. Young, G.; Sullivan, K.; Zachariah, M.R.; Yu, K. Combustion characteristics of boron nanoparticles. *Combust. Flame* **2009**, *156*, 322–333. [CrossRef]
10. Badiola, C.; Dreizin, E.L. Combustion of micron-sized particles of titanium and zirconium. *Proc. Combust. Inst.* **2013**, *34*, 2237–2243. [CrossRef]
11. Molodetsky, I.E.; Dreizin, E.L.; Law, C.K. Evolution of particle temperature and internal composition for zirconium burning in air. *Symp. (Int.) Combust.* **1996**, *26*, 1919–1927. [CrossRef]

12. Ermoline, A.; Yildiz, D.; Dreizin, E.L. Model of heterogeneous combustion of small particles. *Combust. Flame* **2013**, *160*, 2982–2989. [CrossRef]

13. Chakraborty, P.; Zachariah, M.R. Do nanoenergetic particles remain nano-sized during combustion? *Combust. Flame* **2014**, *161*, 1408–1416. [CrossRef]

14. Sundaram, D.S.; Yang, V.; Zarko, V.E. Combustion of nano aluminum particles (review). *Combust. Explos. Shock Waves* **2015**, *51*, 173–196. [CrossRef]

15. Sundaram, D.S.; Puri, P.; Yang, V. A general theory of ignition and combustion of nano- and micron-sized aluminum particles. *Combust. Flame* **2016**, *169*, 94–109. [CrossRef]

16. Ermolaev, G.V.; Zaitsev, A.V. Diffusion model of combustion of large boron particles. *Combust. Explos. Shock Waves* **2018**, *54*, 442–449. [CrossRef]

17. Chang, C.H.H.; Dahm, W.J.A.; Tryggvason, G. Lagrangian model simulations of molecular mixing, including finite rate chemical reactions, in a temporally developing shear layer. *Phys. Fluids A* **1991**, *3*, 1300–1311. [CrossRef]

18. Bazyn, T.; Krier, H.; Glumac, N. Combustion of nanoaluminum at elevated pressure and temperature behind reflected shock waves. *Combust. Flame* **2006**, *145*, 703–713. [CrossRef]

19. Bazyn, T.; Krier, H.; Glumac, N. Evidence for the transition from the diffusion-limit in aluminum particle combustion. *Proc. Combust. Inst.* **2007**, *31*, 2021–2028. [CrossRef]

20. Glassman, I. *Combustion*, 3rd ed.; Academic Press: Orlando, FL, USA, 1996.

21. Glassman, I. Combustion of metals: Physical considerations. In *Solid Propellant Rocket Research, ARS Progress in Astronautics and Rocketry*; Academic Press: New York, NY, USA, 1960; Volume 1, pp. 253–258.

22. Dreizin, E.L. Metal-based reactive nanomaterials. *Prog. Energy Combust. Sci.* **2009**, *35*, 141–167. [CrossRef]

23. Yetter, R.A.; Risha, G.A.; Son, S.F. Metal particle combustion and nanotechnology. *Proc. Combust. Inst.* **2009**, *32*, 1819–1838. [CrossRef]

24. Shpara, A.P.; Yagodnikov, D.A.; Sukhov, A.V. Effect of particle size on boron combustion in air. *Combust. Explos. Shock Waves* **2020**, *56*, 471–478. [CrossRef]

25. Allen, D.; Krier, H.; Glumac, N. Heat transfer effects in nano-aluminum combustion at high temperatures. *Combust. Flame* **2014**, *161*, 295–302. [CrossRef]

26. Ulas, A.; Kuo, K.K.; Gotzmer, C. Ignition and combustion of boron particles in fluorine-containing environments. *Combust. Flame* **2001**, *127*, 1935–1957. [CrossRef]

27. Ao, W.; Yang, W.; Wang, Y.; Zhou, J.; Liu, J.; Cen, K. Ignition and combustion of boron particles at one to ten standard atmosphere. *J. Propuls. Power* **2014**, *30*, 760–764. [CrossRef]

28. Parr, T.P.; Johnson, C.; Hanson-Parr, D.; Higa, K.; Wilson, K. Evaluation of Advanced Fuels for Underwater Propulsion. In Proceedings of the 39th JANNAF Combustion Subcommittee Meeting, Colorado Springs, CO, USA, 1–5 December 2003.

29. Park, K.; Lee, D.; Rai, A.; Mukherjee, D.; Zachariah, M.R. Size-resolved kinetic measurements of aluminum nanoparticle oxidation with single particle mass spectrometry. *J. Phys. Chem. B* **2005**, *109*, 7290–7299. [CrossRef]

30. Zong, Y.C.; Jacob, R.J.; Li, S.; Zachariah, M.R. Size resolved high temperature oxidation kinetics of nano-sized titanium and zirconium particles. *J. Phys. Chem. A* **2015**, *119*, 6171–6178. [CrossRef]

31. Jacob, R.J.; Zong, Y.; Yang, Y.; Li, S.; Zachariah, M.R. Measurement of size resolved burning of metal nanoparticles for evaluation of combustion mechanisms. In Proceedings of the 54th AIAA Aerospace Sciences Meeting, San Diego, CA, USA, 4–8 January 2016; pp. 4–8.

32. Huang, Y.; Risha, G.A.; Yang, V.; Yetter, R.A. Combustion of bimodal nano/micron-sized aluminum particle dust in air. *Proc. Combust. Inst.* **2007**, *31*, 2001–2009. [CrossRef]

33. Baule, B. Theoretische behandlung der erscheinungen in verdünnten gasen. *Ann. Phys.* **1914**, *349*, 145–176. [CrossRef]

34. Goodman, F.O. Thermal accommodation. *Prog. Surf. Sci.* **1974**, *5*, 261–375. [CrossRef]

35. Goodman, F.O.; Wachman, H.Y. *Dynamics of Gas-Surface Scattering*; Academic Press: New York, NY, USA, 1976.

36. Altman, I. On heat transfer between nanoparticles and gas at high temperatures. *J. Aerosol Sci.* **1999**, *30*, S423–S424. [CrossRef]

37. Saxena, S.C.; Joshi, R.K. *Thermal Accommodation and Absorption Coefficient of Gases*; Hemisphere Pub. Corp: New York, NY, USA, 1989.

38. Mane, T.; Bhat, P.; Yang, V.; Sundaram, D.S. Energy accommodation under non-equilibrium conditions for aluminum inert gas systems. *Surf. Sci.* **2018**, *677*, 135–148. [CrossRef]

39. Mohan, S.; Trunov, M.A.; Dreizin, E.L. On possibility of vapor-phase combustion for fine aluminum particles. *Combust. Flame* **2009**, *156*, 2213–2216. [CrossRef]

40. Mohan, S.; Trunov, M.A.; Dreizin, E. Heating and ignition of metallic particles by a CO_2 laser. *J. Propuls. Power* **2008**, *24*, 199–205. [CrossRef]

41. Mohan, S.; Trunov, M.A.; Dreizin, E.L. Heating and ignition of metal particles in the transition heat transfer regime. *J. Heat Transf.* **2008**, *130*, 104505. [CrossRef]

42. Ermoline, A. Thermal theory of aluminum particle ignition in continuum, free-molecular, and transition heat transfer regimes. *J. Appl. Phys.* **2018**, *124*, 054301. [CrossRef]

43. Rebrov, A.K.; Morozov, A.A.; Plotnikov, M.Y.; Timoshenko, N.I.; Shishkin, A.V. The accommodation of the translational and rotational energy of a gas in a Knudsen flow past a thin wire. *J. Exp. Theor. Phys.* **2003**, *97*, 738–744. [CrossRef]

44. Rebrov, A.K.; Morozov, A.A.; Plotnikov, M.Y.; Timoshenko, N.I.; Shishkin, A.V. Using a thin wire in a free-molecular flow for determination of accommodation coefficients of ranslational and internal energy. In *CP663, Rarefied Gas Dynamics: 23rd International Symposium*; American Institute of Physics 0-7354-0124-1/03/$20.00; Ketsdener, A.D., Muntz, E.P., Eds.; AIP Publishing: University Park, MD, USA, 2003.

45. Kolenchits, O.A. *Thermal Accommodation of Gas-Solid Systems*; Nauka i Tekhnika: Minsk, Belarus, 1977.

46. Goodman, F.O.; Wachman, H.Y. Restrictions on the values of energy accommodation coefficients. *Surf. Sci* **1974**, *43*, 306–308. [CrossRef]

47. Allen, D.J. Dynamics of Nanoparticle Combustion. Ph.D. Thesis, University of Illinois at Urbana-Champaign, Champaign, IL, USA, 2015.

48. Jacob, R.J.; Wei, B.; Zachariah, M.R. Quantifying the enhanced combustion characteristics of electrospray assembled aluminum mesoparticles. *Combust. Flame* **2016**, *167*, 472–480. [CrossRef]

49. Mačeic, A.; Semple, J.M. Combustion of boron particles at atmospheric pressure. *Combust. Sci. Technol.* **1969**, *1*, 181–191. [CrossRef]

50. Friedman, R.; Macek, A. Ignition and combustion of aluminum particles in hot ambient gases. *Combust. Flame* **1962**, *6*, 9. [CrossRef]

51. Kuo, K.K.; Pein, R. *Combustion of Boron Based Solid Propellants and Fuels*; CRC Press: Boca Raton, FL, USA, 1993.

52. Foelsche, R.O.; Burton, R.L.; Krier, H. Boron particle ignition and combustion at 30–150 atm. *Combust. Flame* **1999**, *117*, 32–58. [CrossRef]

53. Roberts, T.A.; Burton, R.L.; Krier, H. Ignition and combustion of aluminum magnesium alloy particles in O2 at high pressures. *Combust. Flame* **1993**, *92*, 125–143. [CrossRef]

54. Li, S.C.; Williams, F.A.; Takahashi, F. An investigation of combustion of boron suspensions. *Symp. (Int.) Combust.* **1988**, *22*, 1951–1960. [CrossRef]

55. Bazyn, T.; Krier, H.; Glumac, N. Oxidizer and pressure effects on the combustion of 10-micron aluminum particles. *J. Propuls. Power* **2005**, *21*, 577–582. [CrossRef]

56. Valluri, S.K.; Ravi, K.K.; Schoenitz, M.; Dreizin, E.L. Effect of boron content in $B \cdot BiF_3$ and $B \cdot Bi$ composites on their ignition and combustion. *Combust. Flame* **2020**, *215*, 78–85. [CrossRef]

57. Chintersingh, K.L.; Schoenitz, M.; Dreizin, E.L. Boron doped with iron: Preparation and combustion in air. *Combust. Flame* **2019**, 286–295. [CrossRef]

58. Valluri, S.K.; Schoenitz, M.; Dreizin, E.L. Combustion of composites of boron with bismuth fluoride and cobalt fluoride in different environments. *Combust. Sci. Technol.* **2019**, *1*, 16. [CrossRef]

59. Qiu, T.Q.; Longtin, J.P.; Tien, C.L. Characteristics of radiation absorption in metallic particles. *J. Heat Transf.* **1995**, *117*, 340–345. [CrossRef]

60. Graham, S.A. Absorptivity of several metals at 10.6 μm: Empirical expressions for the temperature dependence computed from Drude theory. *Appl. Opt.* **1984**, *23*, 1434–1436.

61. Kriger, V.G.; Kalenskii, A.V.; Zvekov, A.A.; Zykov, I.Y.; Aduev, B.P. Effect of laser radiation absorption efficiency on the heating temperature of inclusions in transparent media. *Combust. Explos. Shock Waves* **2012**, *48*, 705–708. [CrossRef]

62. Zarko, V.E.; Gromov, A.A. *Energetic Nanomaterials: Synthesis, Characterization, and Application*; Elsevier: Amsterdam, The Netherlands, 2016; pp. 1–374. [CrossRef]

63. Yan, Q.L.; He, G.Q.; Liu, P.J.; Gozin, M. *Nanomaterials in Rocket Propulsion Systems*; Elsevier: Amsterdam, The Netherlands, 2019; pp. 1–567. [CrossRef]
64. Klapotke, T.M. *Chemistry of High-Energy Materials*; Walter de Gruyter: Berlin, Germany, 2011.
65. Zhu, P.; Shen, R.; Ye, Y.; Fu, S.; Li, D. Characterization of Al/CuO nanoenergetic multilayer films integrated with semiconductor bridge for initiator applications. *J. Appl. Phys.* **2013**, *113*, 184505. [CrossRef]
66. Jacob, R.J.; Jian, G.Q.; Guerieri, P.M.; Zachariah, M.R. Energy release pathways in nanothermites follow through the condensed state. *Combust. Flame* **2015**, *162*, 258–264. [CrossRef]
67. Yoshida, T.; Yuasa, S. Effect of water vapor on ignition and combustion of boron lumps in an oxygen stream. *Proc. Combust. Inst.* **2000**, *28*, 2735–2741. [CrossRef]
68. Altman, I. Burn time of metal nanoparticles. *Materials* **2019**, *12*, 1368. [CrossRef] [PubMed]
69. Altman, I.; Demko, A.; Hill, K.; Pantoya, M. On the possible coexistence of two different regimes of metal particle combustion. *Combust. Flame* **2020**, *221*, 416–419. [CrossRef]
70. Hill, K.J.; Pantoya, M.L.; Washburn, E.; Kalman, J. Single particle combustion of pre-stressed aluminum. *Materials* **2019**, *12*, 1737. [CrossRef]
71. Altman, I. On energy accommodation coefficient of gas molecules on metal surface at high temperatures. *Surf. Sci.* **2020**, *698*, 121609. [CrossRef]

MDPI
St. Alban-Anlage 66
4052 Basel
Switzerland
Tel. +41 61 683 77 34
Fax +41 61 302 89 18
www.mdpi.com

Nanomaterials Editorial Office
E-mail: nanomaterials@mdpi.com
www.mdpi.com/journal/nanomaterials